Introduction to Critical Phenomena in Fluids

Topics in Chemical Engineering: A Series of Textbooks and Monographs

Series Editor
Keith E. Gubbins, North Carolina State University

Associate Editors
Mark A. Barteau, University of Delaware
Douglas A. Lauffenburger, MIT
Manfred Morari, ETH
W. Harmon Ray, University of Wisconsin
William B. Russel, Princeton University

Chimowitz, *Introduction to Critical Phenomena in Fluids*

Deen, *Analysis of Transport Phenomena*

Doraiswamy, *Organic Synthesis Engineering*

Floudas, *Nonlinear and Mixed Integer Optimization: Fundamentals and Applications*

Friedlander, *Smoke, Dust, and Haze, Second Edition*

Fuller, *Optical Rheometry of Complex Fluids*

Harrison, Todd, Rudge, and Petrides, *Bioseparations Science and Engineering*

Larson, *The Structure and Rhelogy of Complex Fluids*

Lauffenburger and Linderman, *Receptors: Models for Binding, Trafficking, and Signalling*

Morrison, *Understanding Rheology*

Ogunnaike and Ray, *Process Dynamics, Modeling, and Control*

Pearson, *Discrete-Time Dynamic Models*

Phan-Thein and Kim, *Microstructures in Elastic Media*

Pozrikidis, *An Introduction to Theoretical and Computational Fluid Dynamics*

Pozrikidis, *Numerical Computation in Science and Engineering*

Schmidt, *The Engineering of Chemical Reactions*

Varma and Morbidelli, *Mathematical Methods in Chemical Engineering*

Introduction to CRITICAL PHENOMENA IN FLUIDS

Eldred H. Chimowitz

University of Rochester
New York

OXFORD
UNIVERSITY PRESS

2005

OXFORD
UNIVERSITY PRESS

Oxford University Press, Inc., publishes works that further
Oxford University's objective of excellence
in research, scholarship, and education.

Oxford New York
Auckland Cape Town Dar es Salaam Hong Kong Karachi
Kuala Lumpur Madrid Melbourne Mexico City Nairobi
New Delhi Shanghai Taipei Toronto

With offices in
Argentina Austria Brazil Chile Czech Republic France Greece
Guatemala Hungary Italy Japan Poland Portugal Singapore
South Korea Switzerland Thailand Turkey Ukraine Vietnam

Published by Oxford University Press, Inc.
198 Madison Avenue, New York, New York 10016

www.oup.com

Library of Congress Cataloging-in-Publication Data
Chimowitz, Eldred H.
Introduction to critical phenomena in fluids / Eldred H. Chimowitz.
p. cm.
Includes bibliographical references and index.
ISBN-13 978-0-19-511930-5
ISBN 0-19-511930-4
1. Critical phenomena (Physics) I. Title.

QC173.4 C74 C46 2004
530.4′24—dc22 2003024214

9 8 7 6 5 4 3 2 1
Printed in the United States of America
on acid-free paper

Preface

My interest in writing this book grew out of research in the supercritical fluids area that I have been pursuing for some time. The term supercritical fluid is used quite loosely in the literature, often simply to describe a high-pressure fluid above its critical temperature. However, many of the most interesting phenomena in these systems arise only with close proximity to the critical point itself. As a consequence, my own knowledge of the subject, such as it is, has required me to steadily familiarize myself with concepts from much farther afield than is traditionally considered in engineering books dealing with molecular thermodynamics. This text describes some of what I have learned during this process. My objective has been to describe the critical behavior of supercritical fluid systems within a statistical mechanical framework.

The treatment is introductory and meant to provide the relatively uninitiated reader with as transparent an analysis as I could possibly come up with of some of the major fundamental ideas underpinning the physics of critical phenomena. The systems of interest fall within the Ising universality class, which covers, as far as we understand, the simple fluids that are usually of interest in supercritical fluid chemical process technology. The focus on basic ideas has been motivated by my own belief that the better one's mastery of fundamentals, the more likely one is to both enjoy reading the wider literature on the subject and engage in creative teaching and research. A good example of what I mean by this is provided by the Boltzmann energy distribution in systems exhibiting fluctuations. This important result, though often described in the first few pages of texts that deal with statistical mechanics, to a large extent also provides the intellectual foundations for many newer areas of scientific investigation. A good example of this is the growing importance of a field like molecular simulation, where Monte Carlo methods, which rely heavily on Boltzmann sampling ideas, are now used in myriad different ways.

This text is divided into two main sections. Chapters 1–6 deal mainly with a macroscopic description of thermodynamic stability theory, and its consequences for describing phase transitions and critical behaviors in fluid systems. Chapter 1 is an attempt to express thermodynamic stability theory using a few fundamental theorems taken from linear algebra. The purpose is to provide a formulation of the topic with expanded "reach," which will hopefully then allow many of the more unfamiliar results in stability theory to be relatively easily derived. For this part of the book, in particular, I am indebted to the late Francisco Munoz, whose untimely passing was a severe personal and professional loss. Chapter 3 introduces many of the fundamental concepts concerned with critical scaling behavior in pure systems, with the extension of these ideas to multicomponent mixtures described in chapter 4. Chapters 5 and 6 are concerned with practical topics relevant to engineering applications that use supercritical fluid solvents. Chapter 5 deals with solvation behaviors and describes some unique solution behaviors in the critical region. These include retrograde phenomena, and the ancillary crossover effect, which is potentially useful in designing chemical separation processes.

Chapters 7–12 deal with more theoretical topics developed in the context of fluid systems falling within the Ising universality class. Chapter 7 deals with mean-field theory, while chapter 8 discusses the important role of fluctuations in the critical region. Chapter 9 introduces the reader to finite-size effects in critical systems and focuses upon the important role that finite-size scaling theories have come to play in computer simulations in the critical region. Chapter 10 is an introduction to the renormalization-group method, which is not only intellectually exciting, even in its simplest formulations, but also now an integral part of modern theoretical physics. My own forays into this area have relied upon several prior texts and, when these treatments have proven particularly illuminating, I have tried to weave some of this into the narrative given here. Chapter 11 deals with the role of pore-confinement on the critical behavior of fluids. Much of the analysis in this chapter has come from our own group's research over the past few years, especially the developments leading to the pore-confined fluid equation of state given in the chapter. Chapter 12 deals with transport phenomena in the critical region. It introduces the reader to the interesting phenomenon of critical slowing down and a significant part of this chapter again describes recent work from our research group. In particular, we describe a relaxation-dynamics simulation method that enables the efficient, accurate simulation of both dynamical and static proprieties in the critical regions of fluids. Exercises are interspersed throughout the chapters, with additional ones provided at the end of each one. The bibliographies provided are not meant to be exhaustive, but rather represent work that I thought could most closely be tied to the flow of ideas presented here.

In carrying out this project I am deeply indebted to many people with whom I have been fortunate to interact over the years. From an educational perspective, my graduate students have been extraordinary partners who have helped to sharpen ideas and maintain the forward momentum of our group's research efforts. These former students include Ken Pennisi, Doug Kelley, Ta Wei Li, Francisco Munoz, George Afrane, Frank van Puyvelde, Carlos Tapia, Vikram Kumaran, Subhranil De, and An Chen, who have, at one time or another, each made significant contributions to material described in this text. These efforts have largely been supported by the U.S. National Science Foundation, for which I am grateful. I also want to acknowledge support from faculty colleagues and administrative staff in the chemical engineering department at the University of

Rochester; in particular, Jacob Jorne and Yonathan Shapir for their indispensable humor and counsel. I fondly remember the late Professor Leroy Stutzman, teacher and a founder of the Control Data Corporation, who epitomized the generosity of this country to a new graduate student, and Mr. Wyngaard and Stanley Schur, who demonstrated to me, as a young man, the twin virtues of analysis and persistence. The editorial staff at Oxford University Press and Keyword Publishing Services have also provided invaluable input while helping to steer this project to a successful conclusion. This group has, at various times, included Bob Rogers, Peter Gordon, Cliff Mills, Lisa Stallings, Sue Nicholls, and the copyeditor, John Bentin. Without their encouragement all of this would have been far less fun.

Finally, this work has been accomplished within a much larger context and, for their support and encouragement, I am indebted to my wife Maria, sons André and Michael, and, of course, Shadow. As a small measure of my thanks I dedicate this book to them.

Rochester, New York

Contents

Introduction to Critical Phenomena in Fluids

1 Fundamentals of Thermodynamic Stability

The second law of thermodynamics states that the entropy change in any spontaneous adiabatic process is greater than or equal to zero. It is a disarmingly simple statement but one that is a cornerstone of scientific theories. It is instrumental in describing the extent and direction of all physical and chemical transformations and contains within it the essential ideas for developing thermodynamic stability theory. Stability theory concerns itself with answering questions such as (1) What is a stable thermodynamic state? (2) Which conditions define the limit to this state beyond which the system becomes unstable? (3) How does the instability manifest itself ? In a real sense, stability theory provides the underlying framework for a macroscopic understanding of phase transitions and critical phenomena, the subject of this text.

Many of the results of stability theory related to phase equilibria are well known; an example is the condition that, for a pure fluid in a stable state, the quantity $-(\partial P/\partial V)_{T,N}$ must be greater than or equal to zero, with the equality condition holding at the *limit of stability*. Many other facets of thermodynamic stability theory, however, are relatively unfamiliar. For example, any of the well-known thermodynamic potentials E, H, A, and G can be used to develop stability criteria for a given system. Are these criteria always equivalent, or do some take precedence over others? If so, what are the implications of this for understanding phase transformations in physicochemical systems? It is questions of this sort that we take up in this chapter, where we lay the macroscopic foundations for the material developed throughout the rest of the text. In this analysis, we rely heavily upon results taken from *linear algebra*, a branch of mathematics that provides an ideal tool for developing a comprehensive description of thermodynamic stability concepts.

1.1 Mathematical and Thermodynamic Preliminaries

Energy as a Thermodynamic Potential

The combination of the first and second law of thermodynamics for a closed system leads to the well-known equation:

$$dE = TdS - PdV \tag{1.1}$$

where $E(S, V)$ represents the system *energy* as a function of the independent variables S and V. This function of the entropy and volume is called a *thermodynamic potential* and is a function from which all other thermodynamic properties can be obtained. For example, we have that the pressure and temperature are given by the equations

$$T = \left(\frac{\partial E}{\partial S}\right)_V \quad \text{and} \quad P = -\left(\frac{\partial E}{\partial V}\right)_S \tag{1.2}$$

In another example, we have that the heat capacity C_V can be derived from $E(S, V)$ as follows:

$$C_V = \left(\frac{\partial E}{\partial T}\right)_V = T\left(\frac{\partial S}{\partial T}\right)_V = T\left(\frac{\partial T}{\partial S}\right)_V^{-1} = T\left(\frac{\partial^2 E}{\partial S^2}\right)_V^{-1} \tag{1.3}$$

However useful a thermodynamic potential like $E(S, V)$ may actually be, this particular set of independent variables may not be the most convenient to use. For example, the entropy S is not a variable we can directly measure in the laboratory and in an experimental setting the independent variables T and V (or T and P) might be preferable to use. While the transformation between $E(S, V)$ and $E(T, V)$ is straightforward mathematically, it is easy to verify that the function $E(T, V)$ contains less information than $E(S, V)$. By this we mean that the function $E(T, V)$ does not provide as complete a description of the system as does $E(S, V)$—see exercise 1.1.

EXERCISE 1.1

Given the following form for $E(T, V, N)$:

$$E = \tfrac{3}{2}NkT \tag{1.4}$$

try and determine an expression for the absolute pressure relying only on information derived from equation (1.4) and thermodynamic identities.

Solution

From equations (1.9) and (1.4), we have

$$\left(\frac{\partial P}{\partial T}\right)_V = \frac{P}{T} + 0 \tag{1.5}$$

thus

$$\frac{dP}{P} = \frac{dT}{T} \tag{1.6}$$

or

$$\frac{P_2}{P_1} = \frac{T_2}{T_1} \tag{1.7}$$

Note: only pressure change is available from this analysis.

To illustrate this point further, if we were to try to obtain the pressure from $E(T, V)$, we would arrive at the following equation:

$$dE = C_V\,dT + \left[T\left(\frac{\partial P}{\partial T}\right)_V - P\right]dV \tag{1.8}$$

from which we have that

$$P = T\left(\frac{\partial P}{\partial T}\right)_V - \left(\frac{\partial E}{\partial V}\right)_T \tag{1.9}$$

We see from equation (1.9) that the pressure must be obtained by integration and thus can only be found to within an arbitrary constant. Clearly equation (1.4) does not contain the complete thermodynamic information for this particular system (an ideal gas), and it is consequently impossible to extract the ideal-gas equation of state from the function $E(T, V, N)$ only. Thus, the question that arises is whether or not there are other thermodynamic potentials that contain all the necessary thermodynamic information but which depend on independent variables other than T and P. If so, is there a systematic way to generate these thermodynamic potentials? Such an approach is provided by the *Legendre transformation* method which we now describe.

Legendre Transforms

To begin, consider a function of only one variable: $y = f(x)$. A Legendre transformation produces a new function $w(v)$ whose independent variable v is the derivative dy/dx. The Legendre transformation $w(v)$ is given by the mapping,

$$L[y(x)] = w(v) \equiv y - vx \tag{1.10}$$

with $v = dy/dx$. The variable v is referred to as the *conjugate* of the variable x. The Legendre transformation of $y(x)$, namely $w(v)$, is completely equivalent to $y(x)$ but expressed as a function of the slope of $y(x)$. Suppose that we were given only the function $w(v)$, can we determine the function $y(x)$? We can readily do this inversion by calculating the Legendre transformation of $w(v)$ as follows:

$$L[w(v)] \equiv w(v) - \frac{dw(v)}{dv} v \tag{1.11}$$

From equation (1.10) we have that

$$\frac{dw(v)}{dv} = \frac{d(y - vx)}{dv} = \frac{dy}{dv} - x - v\frac{dx}{dv}$$

Since $dy = v\,dx$, this implies that

$$\frac{dw(v)}{dv} \equiv -x \tag{1.12}$$

Using equations (1.10)–(1.12), it follows that

$$L[w(v)] \equiv y(x) \tag{1.13}$$

This illustrates that $y(x)$ and $w(v)$ contain completely equivalent information, with $w(v)$ representing the original function using a different independent variable. Legendre transformations thus provide us with a procedure to generate "equivalent" functions from $E(S, V)$ depending on different independent variables.

Using Legendre transformation, it is a straightforward exercise to generate all the common thermodynamic potentials. To replace *entropy* by temperature, for example, we take the Legendre transform as follows:

$$L[E(S)] \equiv E - \left(\frac{\partial E}{\partial S}\right)_V S = E - TS = A(T, V) \tag{1.14}$$

where A is the familiar *Helmholtz free energy* function whose differential follows:

$$dA = -S\,dT - P\,dV \tag{1.15}$$

Another straightforward example would be to replace S by T, and P by V, in the function $E(S, V)$ to yield

$$L[E(S, V)] \equiv E - \left(\frac{\partial E}{\partial S}\right) S - \left(\frac{\partial E}{\partial V}\right) V = E - TS + PV = G(T, P) \tag{1.16}$$

where G is the *Gibbs free energy* given by

$$G(T, P) = E - TS + PV \tag{1.17}$$

with a differential form

$$dG = -S\,dT + V\,dP \tag{1.18}$$

As a final example, we replace V by P in $E(S, V)$ leading to

$$L[E(V)] = E - \left(\frac{\partial E}{\partial V}\right) V = E + PV = H(S, P) \tag{1.19}$$

where $H(S, P)$ is the *enthalpy* with the differential form:

$$dH = T\,dS + V\,dP \tag{1.20}$$

Multicomponent Open Systems

The above discussion for a single-component system can be readily extended to multicomponent *open* systems. The extension of equation (1.1) to a multicomponent open system is given as

$$dE = T\,dS - P\,dV + \sum_{i=1}^{c} \left(\frac{\partial E}{\partial N_i}\right)_{S,V,N_j \neq i} dN_i \tag{1.21}$$

If we define

$$\mu_i = \left(\frac{\partial E}{\partial N_i}\right)_{S,V,N_j \neq i} \tag{1.22}$$

it follows that

$$dE = T\,dS - P\,dV + \sum_{i=1}^{c} \mu_i\,dN_i \tag{1.23}$$

where μ_i is the *chemical potential* of component i. Equation (1.23) expresses the energy as a function of S, V, N_i, or $E(S, V, N_1, \ldots, N_c)$. The set S, V, $N_1, \ldots, N_c$ is just an extended set of independent variables, and the Legendre transformations of E discussed in the previous section apply in exactly the same way here. Thus equations (1.14), (1.17), and (1.20) apply for multicomponent open systems as well. The differential forms of these functions are given by the equations

$$dA = -S\,dT - P\,dV + \sum_{i=1}^{c} \mu_i\,dN_i \tag{1.24}$$

$$dG = -S\,dT + V\,dP + \sum_{i=1}^{c} \mu_i\,dN_i \tag{1.25}$$

$$dH = T\,dS + V\,dP + \sum_{i=1}^{c} \mu_i\,dN_i \tag{1.26}$$

From these above equations we get a series of equivalent definitions for the chemical potential μ_i as follows:

$$\mu_i = \left(\frac{\partial E}{\partial N_i}\right)_{S,V,N_{j\neq i}} = \left(\frac{\partial A}{\partial N_i}\right)_{T,V,N_{j\neq i}} = \left(\frac{\partial G}{\partial N_i}\right)_{T,P,N_{j\neq i}} = \left(\frac{\partial H}{\partial N_i}\right)_{S,P,N_{j\neq i}} \tag{1.27}$$

Euler's Theorem

This theorem states that, if $f(x_1, \ldots, x_n)$ is a *first-order homogeneous function* of $x_1, \ldots, x_n$ (i.e., $f(\lambda x_1, \ldots, \lambda x_n) = \lambda f(x_1, \ldots, x_n)$), then $f(x_1, \ldots, x_n) = \sum_{i=1}^{n} (\partial f/\partial x_i)_{x_j} x_i$.

In order to introduce other useful Legendre transformations, it is helpful to first apply Euler's theorem to $E(S, V, N_i)$:

$$E(S, V, N_1, \ldots, N_c) = S\left(\frac{\partial E}{\partial S}\right)_{V,N_k} + V\left(\frac{\partial E}{\partial V}\right)_{S,N_k} + \sum_{i=1}^{c} N\left(\frac{\partial E}{\partial N_i}\right)_{S,V,N_{k\neq i}} \tag{1.28}$$

or

$$E(S, V, N_1, \ldots, N_c) \equiv TS - PV + \sum_{i=1}^{c} N_i \mu_i \tag{1.29}$$

which implies that

$$A(T, V, N_1, \ldots, N_c) \equiv -PV + \sum_{i=1}^{c} N_i \mu_i \tag{1.30}$$

For completeness, we also have the expression for G given by

$$G(T, P, N_1, \ldots, N_c) \equiv \sum_{i=1}^{c} N_i \mu_i \tag{1.31}$$

Now applying Legendre transformations to $A(N_1, \ldots, N_c)$ in order to replace all the independent variables $N_1, \ldots, N_c$, we get:

$$L[A(N_1, \ldots, N_c)] \equiv A - \sum_{i=1}^{c} \mu_i N_i \equiv -PV \tag{1.32}$$

Equation (1.32) is another very useful Legendre transformation which we will use throughout this text. We denote this transformation by

$$L[A(N_1, \ldots, N_c)] = \Phi(T, V, \mu_1, \ldots, \mu_c) = -PV \tag{1.33}$$

EXERCISE 1.2

If we define $\beta = 1/kT$, where k is the *Boltzmann* constant, we can write dimensionless forms of the thermodynamic potentials. What is interesting about these dimensionless forms is that their differentials provide direct relationships between the respective thermodynamic potentials and *mechanical variables*. For example, we have

$$\left(\frac{\partial(\beta A)}{\partial\beta}\right)_{V,N_i} = \left(\frac{\partial(A/T)}{\partial(1/T)}\right)_{V,N_i} = E \tag{1.34}$$

$$\left(\frac{\partial(\beta G)}{\partial\beta}\right)_{P,N_i} = \left(\frac{\partial(G/T)}{\partial(1/T)}\right)_{P,N_i} = E + PV = H \tag{1.35}$$

$$\left(\frac{\partial(\beta PV)}{\partial\beta}\right)_{V,\beta\mu_i} = \left(\frac{\partial(PV/T)}{\partial(1/T)}\right)_{V,\mu_i/T} = -E \tag{1.36}$$

and so on. The common theme of the derivatives in these equations is that each yields a mechanical variable. Mechanical variables are well-defined physical quantities in many-particle systems, whose evaluation does not necessarily require the use of thermodynamics. In statistical-mechanical terms, these variables can be calculated as time-average properties. The above equations provide the necessary connections for developing changes in the thermodynamic potentials such as G by integration of a mechanical variable with respect to temperature.

1.2 Thermodynamic Stability Theory

In this section, we focus upon the thermodynamic stability of systems. The starting point for our analysis is the energy equation characterized by the entropy S, the volume V, and the number of molecules of each species N_i. These are all extensive variables that we arrange in a vector $\underline{X}$. As just described, we see that the internal energy, and all of the other well-known thermodynamic potentials such as the Helmholtz energy A, the enthalpy H, and the Gibbs energy G, can be related to E through Legendre transforms.

Let Φ stand for the internal energy, or any Legendre transform of it, which is a natural function of the extensive variables $X_1, X_2, \ldots, X_r$ and the intensive variables $I_{r+1}, \ldots, I_n$. Then it is easily shown [1] that

$$\Phi = \Phi(X_1, \ldots, X_r, I_{r+1}, \ldots, I_n) \tag{1.37}$$

$$d\Phi = \sum_{i=1}^{r} I_i\, dX_i - \sum_{j=r+1}^{n} X_j\, dI_j \tag{1.38}$$

where

$$I_i \equiv \left(\frac{\partial\Phi}{\partial X_i}\right)_{X_{j\neq i}, I_k} \tag{1.39}$$

More recognizable forms of equation (1.38) can be seen upon substitution of the corresponding potentials. For example,

$$dE = T\,dS - p\,dV + \sum_i \mu_i\,dN_i, \quad \underline{X} = (S, V, \underline{N}) \tag{1.40}$$

$$dA = -S\,dT - p\,dV + \sum_i \mu_i\,dN_i, \quad \underline{X} = (V, \underline{N}) \tag{1.41}$$

$$dH = T\,dS + V\,dp + \sum_i \mu_i\,dN_i, \quad \underline{X} = (S, \underline{N}) \tag{1.42}$$

$$dG = -S\,dT + V\,dp + \sum_i \mu_i\,dN_i, \quad \underline{X} = \underline{N} \tag{1.43}$$

The variational statements of the second law [1] establish that, at a stable equilibrium point,

$$\Delta E_{S,V,\underline{N}} > 0 \tag{1.44}$$

$$\Delta H_{S,P,\underline{N}} > 0 \tag{1.45}$$

$$\Delta A_{T,V,\underline{N}} > 0 \tag{1.46}$$

$$\Delta G_{T,P,\underline{N}} > 0 \tag{1.47}$$

At a stable equilibrium point in a homogeneous phase, the intensive variables are considered uniform throughout the system, and any of these foregoing statements can be restated as satisfying the condition

$$\Delta\Phi(X_1, X_2, \ldots, X_r) > 0 \tag{1.48}$$

where the respective *extensive variables* are the ones natural to the function Φ. For example, if $\Phi = G$, then $\underline{X} = N_1, N_2, \ldots, N_n$. From a mathematical standpoint, equation (1.48) implies that, at a stable point, Φ must be a *convex function* of the extensive variables $X_1, \ldots, X_r$. This means that each thermodynamic potential associated with a set of $X_i (i = 1, \ldots, r)$ can be associated with an $r \times r$ *Hessian matrix* $\boldsymbol{A}$ defined as

$$\boldsymbol{A} \equiv \left\{\left(\frac{\partial^2\Phi}{\partial X_i\,\partial X_j}\right)\right\} \tag{1.49}$$

which, at a stable equilibrium point, must be *finite* and *positive definite*.

A limit of stability by definition occurs when the matrix $\boldsymbol{A}$ becomes semidefinite. (The matrix $\boldsymbol{A}$ is said to be *positive semidefinite* if $\mathbf{y}^{\mathrm{T}}\boldsymbol{A}\mathbf{y}$ is nonnegative, where there is at least one nonzero vector y for which the equality holds.) We note that, in the cases we consider, the Hessian matrices arise from the second-order derivatives of thermodynamic potentials with respect to extensive variables and, as such, are *finite real symmetric matrices*.

We now introduce a number of definitions and theorems from linear algebra that greatly facilitates thermodynamic stability analysis.

DEFINITION 1.1: Positive Definiteness

A matrix $\boldsymbol{A}$ is *positive definite* if and only if the *inner product* $\mathbf{y}^{\mathrm{T}}\boldsymbol{A}\mathbf{y}$ is positive for all nonzero vectors $\mathbf{y}$. In a geometric sense, positive definiteness of a Hessian matrix implies a bowl-shaped (convex) function. Thus, at a stable equilibrium point, the Hessian matrices of various potentials E, H, A, G defined in terms of their natural extensive variables are convex.

DEFINITION 1.2: Nonsingular Matrices

A square matrix $\boldsymbol{A}$ that possesses an inverse is said to be *nonsingular*. Conversely, if it does not possess an inverse, it is termed *singular*.

DEFINITION 1.3: Cofactors of Matrices

The determinant of the $(n-1)\times(n-1)$ matrix formed from an $n \times n$ matrix $\boldsymbol{A}$ by eliminating its ith row and jth column is called its *minor* M_{ij}. The number A_{ij} formed by the equation

$$A_{ij} = (-1)^{i+j} M_{ij} \tag{1.50}$$

is called the (i, j) cofactor, and these factors play an important role in linear algebra.

The relationship between a singular matrix and its determinant is stated in the following theorem.

THEOREM 1

(Singularity and the determinant of a matrix $\boldsymbol{A}$) An $n \times n$ matrix $\boldsymbol{A}$ is singular if and only if its determinant, designated det $\boldsymbol{A}$ or $|\boldsymbol{A}|$, is equal to zero.

Although the concept of positive definiteness appears, on the face of it, to be a simple one, it permeates the analysis that follows in this chapter. It is also important to be clear about the relationship between positive definiteness and the singularity of a matrix (see exercise 1.3).

EXERCISE 1.3

Verify the following statements:

- If a matrix $\boldsymbol{A}$ is positive definite, then det $\boldsymbol{A} \geq 0$.
- If det $A > 0$, this does not necessarily imply that $\boldsymbol{A}$ is positive definite.
- If det $\boldsymbol{A} < 0$, then $\boldsymbol{A}$ cannot be positive definite.
- If det $\boldsymbol{A} = 0$, then $\boldsymbol{A}$ is singular but it can be either positive or negative semidefinite.

We now present two useful theorems from linear algebra concerned with the properties of $\boldsymbol{A}$ that make it positive definite [2].

THEOREM 2

(Positive-definiteness conditions for $\boldsymbol{A}$) If $\boldsymbol{A}$ is positive definite, then all of the diagonal elements a_{kk} of $\boldsymbol{A}$ must be positive.

THEOREM 3

(Positive definiteness conditions for $\boldsymbol{A}$) A necessary and sufficient condition for an $n \times n$ matrix $\boldsymbol{A}$ to be positive definite is that the *principal minors*, consisting of the determinants of the $k \times k$ matrices in the top left-hand corner of $\boldsymbol{A}$ $(k = 1, \ldots, n)$, are all positive; that is,

$$\mathrm{A} \equiv \begin{bmatrix} a_{11} & a_{12} & \cdots & a_{1n} \\ a_{21} & a_{22} & \cdots & a_{2n} \\ \vdots & \vdots & \ddots & \vdots \\ a_{n1} & a_{n2} & \cdots & a_{nn} \end{bmatrix} \text{ is positive definite} \tag{1.51}$$

if and only if

$$a_{11} > 0, \quad \begin{vmatrix} a_{11} & a_{12} \\ a_{21} & a_{22} \end{vmatrix} > 0, \quad \begin{vmatrix} a_{11} & a_{12} & a_{13} \\ a_{21} & a_{22} & a_{23} \\ a_{31} & a_{32} & a_{33} \end{vmatrix} > 0, \quad \ldots, \quad \det \boldsymbol{A} > 0 \tag{1.52}$$

The results of theorem 3 may simply be used; however, details provided in the proof given in the appendix to this chapter are useful for an understanding of many results in this chapter, making it a worthwhile analysis for the reader to follow.

Positive Definiteness and Local Stability

We comment at this point about the relationship between positive definiteness (convexity of the potential) and stability. While a stable phase requires the property of positive definiteness in the Hessian of the relevant thermodynamic potential, the converse is not necessarily true. Since convexity of the potential function is a *local* (*point*) *property* it does not exclude the existence of other locally convex points at some other state point. If the system has access to these other states, a phase transition may occur. Each coexisting phase has the properties defined at a locally convex point of the free-energy function. Such locally stable states are called *metastable.* Hence theorems 2 and 3 provide necessary conditions for both stable and metastable states which are considered "equivalent" in terms of a limit of stability analysis *per se.* We refer to both types as stable states for the sake of conciseness. In figure 1.1, the metastable region (shaded) is shown in the schematic showing the phase diagram for a typical fluid. Its boundary is limited by a curve referred to as the *spinodal curve* which extends all the way to the critical point.

Some Well-Known Thermodynamic Stability Conditions

Theorem 2 can immediately be used to formulate certain well-known inequality relationships representing necessary conditions for stability. By theorem 2, at a stable

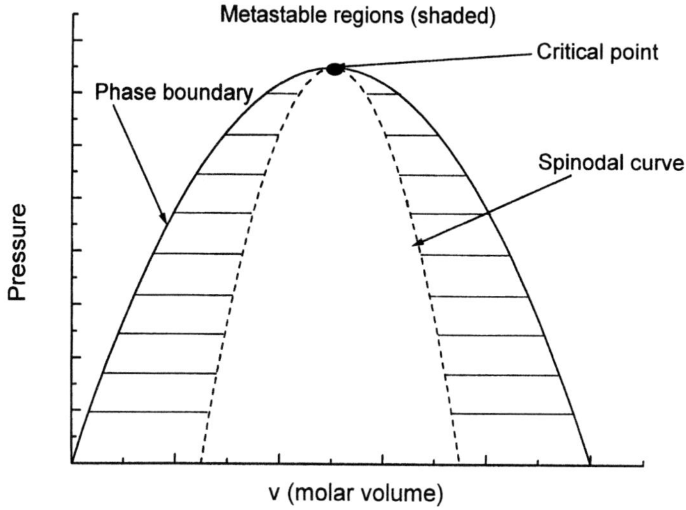

Figure 1.1 Co-existence curve schematic for a fluid system

equilibrium point, with the Hessian matrix of any of the thermodynamic potentials Φ positive definite, we must have that

$$\left(\frac{\partial^2 \Phi}{\partial X_i^2}\right)_{X_{j \neq i}} > 0 \quad \text{for any } i \tag{1.53}$$

Since $(\partial \Phi / \partial X_i)_{X_{j \neq i}} \equiv I_i$, this means that

$$\left(\frac{\partial I_i}{\partial X_i}\right)_{X_{j \neq i}} > 0 \tag{1.54}$$

The use of inequality (1.54) applied to equations (1.40)–(1.43) leads to many of the most familiar necessary conditions for a thermodynamically stable point, for example,

$$\left(\frac{\partial \mu_i}{\partial N_i}\right)_{T,V,N_{j \neq i}} > 0 \quad (i = 1, \ldots, n) \tag{1.55}$$

$$-\left(\frac{\partial P}{\partial V}\right)_{T,N_i} > 0 \quad \text{(etc.)} \tag{1.56}$$

$$\left(\frac{\partial T}{\partial S}\right)_{P,N_i} \equiv \frac{T}{C_P} > 0 \tag{1.57}$$

$$\left(\frac{\partial \mu_i}{\partial N_i}\right)_{T,P,N_{j \neq i}} > 0 \quad (i = 1, \ldots, n) \tag{1.58}$$

EXERCISE 1.4

If the inequalities in (1.55) and (1.56) are developed for a pure fluid, which of them, if any, is satisfied at a so-called limit of stability?

1.3 Thermodynamic Conditions at the Limit of Stability

By a *limit of stability* we mean a point at which the associated $r \times r$ Hessian matrix $\boldsymbol{A}$ of the thermodynamic potential of interest is on the verge of losing its positive-definiteness property.

To discuss this further, we first define $\boldsymbol{A}_i$ to be a principal minor submatrix of $\boldsymbol{A}$. This is the $i \times i$ submatrix consisting of the initial i elements in each of the first i rows and i columns of $\boldsymbol{A}$. From theorem 3, it is evident that a limit of stability will occur when, for one or more i, the following condition occurs:

$$\det \boldsymbol{A}_i = 0 \tag{1.59}$$

where $1 \leq i \leq r$. We now pose examples of questions of the sort we would like to answer:

Is it possible to determine which of these determinants vanishes first?
Can a number of them vanish simultaneously?

Questions like these are important because their answers provide an unambiguous statement of the precise thermodynamic conditions that can prevail at a limit of stability. We now pursue this issue using a series of theorems which we mostly state; for the interested reader, the more intricate proofs of these results are given in the appendix to the chapter.

THEOREM 4

(Euler's theorem and the $r \times r$ stability matrix $\boldsymbol{A}$) The $r \times r$ matrix $\boldsymbol{A}$ defined as $\{(\partial^2\Phi/\partial X_i \partial X_j)\}$ $(i,j = 1, \ldots, r)$ *for any thermodynamic potential* Φ *is singular.*

Proof

Since Φ is a homogeneous function of the first degree of the extensive variables $X_1, \ldots, X_r$, Euler's theorem yields

$$\Phi = \sum_{i=1}^{r} X_i \left(\frac{\partial \Phi}{\partial X_i}\right)_{X_{i \neq j}, I_k} = \sum_{i=1}^{r} X_i I_i \tag{1.60}$$

Taking the partial derivative in equation (1.60) with respect to one of the extensive variables, we have

$$\left(\frac{\partial \Phi}{\partial X_l}\right)_{X_{j \neq l}, I_k} = \left(\frac{\partial \Phi}{\partial X_l}\right)_{X_{j \neq l}, I_k} + \sum_{i=1}^{r} X_i \left(\frac{\partial^2 \Phi}{\partial X_i \partial X_l}\right)_{X_{j \neq i,l}, I_k} \tag{1.61}$$

Thus,

$$\sum_{i=1}^{r} X_i \left(\frac{\partial^2 \Phi}{\partial X_i \partial X_l}\right)_{X_{j \neq i,l}, I_k} = 0 \qquad (l = 1, \ldots, r) \tag{1.62}$$

This set of equations implies that the $r \times r$ matrix $\boldsymbol{A}$ defined as

$$\boldsymbol{A} = \left\{\left(\frac{\partial^2 \Phi}{\partial X_i \partial X_j}\right)\right\} = \{a_{ij}\} \tag{1.63}$$

is *singular* with $\det \boldsymbol{A} = 0$.

Note that the results of theorem 4 do not preclude the fact that the matrix $\boldsymbol{A}$ can be positive definite since it can be shown that, for a stable system,

$$\delta^2 \Phi = \sum_{i=2}^{r} \frac{\det \boldsymbol{A}_i}{\det \boldsymbol{A}_{i-1}} y_i^2 > 0 \text{ for all } y_2, \ldots, y_r \neq 0 \tag{1.64}$$

Thus, provided that the principal minor determinants from 1 to $r - 1$ are positive, $\boldsymbol{A}$ can be positive definite even though $\det \boldsymbol{A} = 0$. In other words, a matrix can be positive definite and also singular.

THEOREM 5

(Order of the matrix defining the stability limit when *only a single determinant of a principal minor of* $\boldsymbol{A}_r$ *vanishes*) In a thermodynamic system represented by a potential $\Phi(X_1, \ldots, X_r)$, with associated Hessian matrix $\boldsymbol{A}_r$ assuming that, as the limit of stability is approached, no two (or more) principal minor determinants vanish simultaneously, then the principal minor determinant that vanishes first is $\det \boldsymbol{A}_{r-1}$, namely that of order $r - 1$.

Proof outline

Use standard relationships between determinant expansions for the matrices $\boldsymbol{A}_{r-1}$ and $\boldsymbol{A}_r$, respectively.

THEOREM 6

(Structure of the matrix at the limit of stability) Given a real symmetric $r \times r$ matrix $\boldsymbol{A}_r$ that represents the Hessian matrix of a thermodynamic potential $\Phi(X_1, \ldots, X_r)$, the determinants of any of the $(r - 1) \times (r - 1)$ submatrices found by eliminating one *row and one column* of $\boldsymbol{A}$ will vanish simultaneously at a limit of stability.

Proof

The proof is given in the appendix to this chapter.

The Size (Order) and Structure of the Matrix Defining the Thermodynamic Stability Limit

Theorems 4 and 5 establish that one cannot use the $r \times r$ stability matrix $\boldsymbol{A}_r$ to determine whether or not the system is stable. To do this, we need to verify whether or not the principal minor matrix $\boldsymbol{A}_{r-1}$ is positive definite (why?). If it is, then the system is guaranteed to be stable. However, the question remains as to which of the principal minor matrices of $\boldsymbol{A}_r$ is to be used to determine the limit of system stability.

The results of theorem 5 are not conclusive about this issue, since there are many possible choices for the matrices $\mathbf{A}_{r-1}$ depending upon which of the extensive variables is deleted from the system representation. For example, in an n component system using the Helmholtz energy representation for Φ, denoted by $A(T, V, N_1, \ldots, N_n)$, there are $n+1$ possible choices for the stability matrix $\mathbf{A}_{r-1}$. For example, two of these choices in a ternary system are

$$\mathbf{A}_{r-1} = \begin{bmatrix} \left(\dfrac{\partial^2 A}{\partial V^2}\right) & \left(\dfrac{\partial^2 A}{\partial V \partial N_1}\right) & \left(\dfrac{\partial^2 A}{\partial V \partial N_2}\right) \\ \left(\dfrac{\partial^2 A}{\partial N_1 \partial V}\right) & \left(\dfrac{\partial^2 A}{\partial N_1^2}\right) & \left(\dfrac{\partial^2 A}{\partial N_1 \partial N_2}\right) \\ \left(\dfrac{\partial^2 A}{\partial N_2 \partial V}\right) & \left(\dfrac{\partial^2 A}{\partial N_2 \partial N_1}\right) & \left(\dfrac{\partial^2 A}{\partial N_2^2}\right) \end{bmatrix}$$

$$\mathbf{A}_{r-1} = \begin{bmatrix} \left(\dfrac{\partial^2 A}{\partial N_1^2}\right) & \left(\dfrac{\partial^2 A}{\partial N_1 \partial N_2}\right) & \left(\dfrac{\partial^2 A}{\partial N_1 \partial N_3}\right) \\ \left(\dfrac{\partial^2 A}{\partial N_2 \partial N_1}\right) & \left(\dfrac{\partial^2 A}{\partial N_2^2}\right) & \left(\dfrac{\partial^2 A}{\partial N_2 \partial N_3}\right) \\ \left(\dfrac{\partial^2 A}{\partial N_3 \partial N_1}\right) & \left(\dfrac{\partial^2 A}{\partial N_3 \partial N_2}\right) & \left(\dfrac{\partial^2 A}{\partial N_3^2}\right) \end{bmatrix} \tag{1.65}$$

However, the results of theorem 6 show that we are at liberty to eliminate *any row and column from the matrix* $\mathbf{A}_r$ *to arrive at the requisite* $\mathbf{A}_{r-1}$ *matrix that defines the limit of stability*. We note that the previous results seemed to suggest that $\mathbf{A}_{r-1}$ was found from $\mathbf{A}_r$ by eliminating the rth row and column from $\mathbf{A}_r$. This in turn is equivalent to the removal of the extensive variable X_r from the system description. Theorem 6 provides results that show this to be unnecessarily restrictive.

Thermodynamic Stability Hierarchy

An interesting question that extends the reach of theorem 4 concerns the issue of whether, as the limit of stability is approached, more than one of the principal minors can vanish simultaneously. Theorem 7 stated below addresses this issue and shows that, in situations where this occurs, a strict hierarchy exists between the various conditions that are satisfied at the stability limit.

THEOREM 7

(Stability-condition hierarchy) When more than one principal minor vanishes simultaneously at a stability limit, we can show that the following hierarchy establishes an ordering among stability conditions that defines the limit of thermodynamic stability.

$$\det \mathbf{A}_1 \geq \alpha \det \mathbf{A}_2 \geq \alpha^2 \det \mathbf{A}_3 \geq \cdots \geq \alpha^{r-2} \det \mathbf{A}_{r-1} \geq 0 \tag{1.66}$$

where $1/\alpha \equiv \text{trace } \mathbf{A}$.

This hierarchy establishes that, if at the limit of stability,

$$\det \boldsymbol{A}_i = 0, \quad \text{for } i < r - 1 \tag{1.67}$$

then all the determinants $\det \boldsymbol{A}_j$ for $j = i, \ldots, r-1$ also vanish simultaneously.

Summary of Results

The important results from the theorems just stated can be succinctly summarized:

> Given a real symmetric $r \times r$ matrix $\boldsymbol{A}_r$ that represents the Hessian matrix of a thermodynamic potential $\Phi(X_1, \ldots, X_r)$, the limit of stability can be found from the vanishing of the determinant of any of the $(r-1) \times (r-1)$ submatrices found by eliminating one row and one column from $\boldsymbol{A}_r$.

Theorems 4–7 address stability criteria *within* a specific representation using any thermodynamic potential. The next question we address concerns the equivalence of stability criteria *between* various potentials using the common ones to make the analysis more familiar and useful.

1.4 Equivalence of Stability Criteria between Different Thermodynamic Potentials

The central result here is based upon the following theorem 8. We include the proof for this theorem since it uses Legendre transformations to show new, useful relationships between thermodynamic potentials and important derivative quantities.

THEOREM 8

(Equivalence between stability criteria with the commonly used thermodynamic potentials) The limit of stability in a thermodynamic system is *the equivalent state point using any of the thermodynamic potentials* ***E***, ***A***, ***G***, *and* ***H***.

Proof

Define $\boldsymbol{E}$, $\boldsymbol{A}$, $\boldsymbol{G}$, and $\boldsymbol{H}$, as the Hessian matrices of the internal energy, the Helmholtz energy, the Gibbs energy, and the enthalpy, respectively, so that

$$\begin{aligned} \boldsymbol{E} &\equiv \left\{ \left(\frac{\partial^2 E}{\partial X_i \partial X_j} \right) \right\}, \quad \boldsymbol{A} \equiv \left\{ \left(\frac{\partial^2 A}{\partial X_i \partial X_j} \right) \right\} \\ \boldsymbol{G} &\equiv \left\{ \left(\frac{\partial^2 G}{\partial X_i \partial X_j} \right) \right\}, \quad \boldsymbol{H} \equiv \left\{ \left(\frac{\partial^2 H}{\partial X_i \partial X_j} \right) \right\} \end{aligned} \tag{1.68}$$

Note that, for a system of n components, $\boldsymbol{E}$ is an $(n+2) \times (n+2)$ matrix, $\boldsymbol{A}$ is an $(n+1) \times (n+1)$ matrix, $\boldsymbol{G}$ is an $n \times n$ matrix, and $\boldsymbol{H}$ is an $(n+1) \times (n+1)$ matrix. We now show that the following relationships hold between the *principal minors* of the Hessian matrices in equation (1.68):

$$\det \boldsymbol{E}_{i+1} = \left(\frac{\partial^2 E}{\partial S^2}\right)_{V,N_k} \det \boldsymbol{A}_i \tag{1.69}$$

$$\det \boldsymbol{A}_{i+1} = \left(\frac{\partial^2 A}{\partial V^2}\right)_{T,N_k} \det \boldsymbol{G}_i \tag{1.70}$$

$$\det \boldsymbol{E}_{i+1} = \left(\frac{\partial^2 E}{\partial V^2}\right)_{S,N_k} \det \boldsymbol{H}_i \tag{1.71}$$

To prove equation (1.69), we relate the elements of matrices $\boldsymbol{E}$ and $\boldsymbol{A}$. Since

$$\left(\frac{\partial E}{\partial X_i}\right)_{S,X_{l\neq i}} = \left(\frac{\partial A}{\partial X_i}\right)_{T,X_{l\neq i}} \tag{1.72}$$

then, from the calculus of multivariable functions,

$$\left(\frac{\partial^2 A}{\partial X_i \partial X_j}\right)_{T,X_l} = \left(\frac{\partial^2 E}{\partial X_i \partial X_j}\right)_{S,X_l} + \left(\frac{\partial^2 E}{\partial S \partial X_i}\right)_{X_l} \left(\frac{\partial S}{\partial X_j}\right)_{T,X_l} \tag{1.73}$$

with X_i and X_j representing any of the extensive variables on which the Helmholtz energy depends (i.e., $V, N_1, \ldots, N_n$). Using the cyclical identity

$$\left(\frac{\partial x}{\partial y}\right)_z \left(\frac{\partial y}{\partial z}\right)_x \left(\frac{\partial z}{\partial x}\right)_y = -1 \tag{1.74}$$

we have

$$\begin{aligned}\left(\frac{\partial S}{\partial X_j}\right)_{T,X_l} &= -\left(\frac{\partial T}{\partial X_j}\right)_{S,X_l} \Big/ \left(\frac{\partial T}{\partial S}\right)_{X_l} \\ &= -\left(\frac{\partial^2 E}{\partial S \partial X_j}\right)_{X_l} \Big/ \left(\frac{\partial^2 E}{\partial S^2}\right)_{X_l}\end{aligned} \tag{1.75}$$

Using (1.75) in (1.73) we have

$$\begin{aligned}\left(\frac{\partial^2 A}{\partial X_i X_j}\right)_{T,X_l} &= \left(\frac{\partial^2 E}{\partial X_i \partial X_j}\right)_{S,X_l} \\ &- \left(\frac{\partial^2 E}{\partial S \partial X_i}\right)_{X_l} \left(\frac{\partial^2 E}{\partial S \partial X_j}\right)_{X_l} \Big/ \left(\frac{\partial^2 E}{\partial S^2}\right)_{X_l} \quad (i, j = 2, \ldots, n)\end{aligned} \tag{1.76}$$

Note that if we carry out a *Gaussian LU* decomposition of the Hessian matrix of the internal energy—namely matrix **E**—then equation (1.76) is the *Gaussian elimination prescription* for carrying out one iteration in the LU decomposition [2]. Thus, we can represent all the possible relationships shown in (1.76) by the following matrix product:

$$\mathbf{E} = \begin{bmatrix} \left(\frac{\partial^2 E}{\partial S^2}\right) & \left(\frac{\partial^2 E}{\partial S \partial V}\right) & \left(\frac{\partial^2 E}{\partial S \partial N_1}\right) & \cdots \\ \left(\frac{\partial^2 E}{\partial V \partial S}\right) & \left(\frac{\partial^2 E}{\partial V^2}\right) & \left(\frac{\partial^2 E}{\partial V \partial N_1}\right) & \cdots \\ \left(\frac{\partial^2 E}{\partial S \partial N_1}\right) & \left(\frac{\partial^2 E}{\partial V \partial N_1}\right) & \left(\frac{\partial^2 E}{\partial N_1^2}\right) & \cdots \\ \vdots & \vdots & \vdots & \ddots \end{bmatrix}$$

$$= \begin{bmatrix} 1 & 0 & 0 & \cdots \\ \left(\frac{\partial^2 E}{\partial S \partial V}\right)\left(\frac{\partial^2 E}{\partial S^2}\right)^{-1} & 1 & 0 & \cdots \\ \left(\frac{\partial^2 E}{\partial S \partial N_1}\right)\left(\frac{\partial^2 E}{\partial S^2}\right)^{-1} & 0 & 1 & \cdots \\ \vdots & \vdots & \vdots & \ddots \end{bmatrix} \tag{1.77}$$

$$\times \begin{bmatrix} \left(\frac{\partial^2 E}{\partial S^2}\right) & \left(\frac{\partial^2 E}{\partial S \partial V}\right) & \left(\frac{\partial^2 E}{\partial S \partial N_1}\right) & \cdots \\ 0 & \left(\frac{\partial^2 A}{\partial V^2}\right) & \left(\frac{\partial^2 A}{\partial V \partial N_1}\right) & \cdots \\ 0 & \left(\frac{\partial^2 A}{\partial V \partial N_1}\right) & \left(\frac{\partial^2 A}{\partial N_1^2}\right) & \cdots \\ \vdots & \vdots & \vdots & \ddots \end{bmatrix}$$

The result in equation (1.77) may be generalized to any number of components, and a statement of this generalization is in fact equation (1.69), the objective of the proof. Similar arguments beginning with the relationship between **A** and **G**, and between **E** and **H**, lead to equations (1.70) and (1.71) and are left as an exercise for the reader.

The results of theorem 8 taken together establish the equivalence of limiting stability conditions using any of the common thermodynamic potentials.

EXERCISE 1.5

Show that the (2, 2) element of the matrix product given in equation (1.77) is consistent with the relationship shown in (1.76).

Illustration of the Use of Theorem 8 in a Binary System

To further illustrate the use of this theorem, let us examine equation (1.69) for a binary system where $n = 2$. Here $\boldsymbol{E}$ and $\boldsymbol{A}$ are the full Hessian matrices of the energy and Helmholtz energy potentials, respectively, with $\boldsymbol{E}$ a 4×4 matrix and $\boldsymbol{A}$ a 3×3 matrix. In the interest of uncluttered nomenclature, we omit the subscripted variables held constant in the following derivatives. These should be evident from the context of the potentials used.

In this case,

$$\boldsymbol{E} = \begin{bmatrix} \left(\dfrac{\partial^2 E}{\partial S^2}\right) & \left(\dfrac{\partial^2 E}{\partial S \partial V}\right) & \left(\dfrac{\partial^2 E}{\partial S \partial N_1}\right) & \left(\dfrac{\partial^2 E}{\partial S \partial N_2}\right) \\ \left(\dfrac{\partial^2 E}{\partial S \partial V}\right) & \left(\dfrac{\partial^2 E}{\partial V^2}\right) & \left(\dfrac{\partial^2 E}{\partial V \partial N_1}\right) & \left(\dfrac{\partial^2 E}{\partial V \partial N_2}\right) \\ \left(\dfrac{\partial^2}{\partial S \partial N_1}\right) & \left(\dfrac{\partial^2 E}{\partial V \partial N_1}\right) & \left(\dfrac{\partial^2 E}{\partial N_1^2}\right) & \left(\dfrac{\partial^2 E}{\partial N_1 \partial N_2}\right) \\ \left(\dfrac{\partial^2 E}{\partial S \partial N_2}\right) & \left(\dfrac{\partial^2 E}{\partial V \partial N_2}\right) & \left(\dfrac{\partial^2 E}{\partial N_1 \partial N_2}\right) & \left(\dfrac{\partial^2 E}{\partial N_2^2}\right) \end{bmatrix}$$

$$= \begin{bmatrix} 1 & 0 & 0 & 0 \\ \left(\dfrac{\partial^2 E}{\partial S \partial V}\right)\left(\dfrac{\partial^2 E}{\partial S^2}\right)^{-1} & 1 & 0 & 0 \\ \left(\dfrac{\partial^2 E}{\partial S \partial V}\right)\left(\dfrac{\partial^2 E}{\partial S^2}\right)^{-1} & 0 & 1 & 0 \\ \left(\dfrac{\partial^2 E}{\partial S \partial V}\right)\left(\dfrac{\partial^2 E}{\partial S^2}\right)^{-1} & 0 & 0 & 1 \end{bmatrix} \tag{1.78}$$

$$\times \begin{bmatrix} \left(\dfrac{\partial^2 E}{\partial S^2}\right) & \left(\dfrac{\partial^2 E}{\partial S \partial V}\right) & \left(\dfrac{\partial^2 E}{\partial S \partial N_1}\right) & \left(\dfrac{\partial^2 E}{\partial S \partial N_2}\right) \\ 0 & \left(\dfrac{\partial^2 A}{\partial V^2}\right) & \left(\dfrac{\partial^2 A}{\partial V \partial N_1}\right) & \left(\dfrac{\partial^2 A}{\partial V \partial N_2}\right) \\ 0 & \left(\dfrac{\partial^2 A}{\partial V \partial N_1}\right) & \left(\dfrac{\partial^2 A}{\partial N_1^2}\right) & \left(\dfrac{\partial^2 A}{\partial N_1 \partial N_2}\right) \\ 0 & \left(\dfrac{\partial^2 A}{\partial V \partial N_2}\right) & \left(\dfrac{\partial^2 A}{\partial N_1 \partial N_2}\right) & \left(\dfrac{\partial^2 A}{\partial N_2^2}\right) \end{bmatrix}$$

The determinant of $\boldsymbol{E}$ is equal to the determinant of the product of the two matrices on the right-hand side of equation (1.78). This follows from a standard result in linear algebra that states that, if $\boldsymbol{P}$ and $\boldsymbol{Q}$ are two square matrices, of the same size, then

$$\det(\boldsymbol{PQ}) = \det \boldsymbol{P} \det \boldsymbol{Q} \tag{1.79}$$

Furthermore, since the determinant of both upper and lower triangular matrices is equal to the product of their respective diagonal elements, we have that

$$\det \boldsymbol{E} = \left(\frac{\partial^2 E}{\partial S^2}\right) \det \boldsymbol{A} \tag{1.80}$$

or, more generally,

$$\det \boldsymbol{E}_4 = \left(\frac{\partial^2 E}{\partial S^2}\right) \det \boldsymbol{A}_3 \tag{1.81}$$

$$\det \boldsymbol{E}_3 = \left(\frac{\partial^2 E}{\partial S^2}\right) \det \boldsymbol{A}_2 \tag{1.82}$$

$$\det \boldsymbol{E}_2 = \left(\frac{\partial^2 E}{\partial S^2}\right) \det \boldsymbol{A}_1 \tag{1.83}$$

where $\boldsymbol{E}_4$ is the 4×4 matrix on the left-hand side of equation (1.78), $\boldsymbol{A}_3$ is the 3×3 submatrix containing the derivatives of the Helmholtz energy on the right hand side of equation (1.78), and $\boldsymbol{E}_3$, $\boldsymbol{E}_2$, $\boldsymbol{A}_2$, and $\boldsymbol{A}_1$ are principal submatrices of the Hessian matrices $\boldsymbol{E}$ and $\boldsymbol{A}$, respectively.

From theorem 4, applied to a two-component system, $\det \boldsymbol{E}_4$ and $\det \boldsymbol{A}_3$ are always zero; however, this does not apply to the determinants of the principal submatrices $\boldsymbol{E}_3$, $\boldsymbol{E}_2$, $\boldsymbol{A}_2$, $\boldsymbol{A}_1$. In this case, the limit of thermodynamic stability is reached in the $\boldsymbol{E}$ representation when $\det \boldsymbol{E}_3 = 0$. Thus, from equation (1.82), if $(\partial^2 E/\partial S^2)_{V,N_k} \neq 0$, then $\det \boldsymbol{A}_2 = 0$, showing that the Helmholtz energy condition for limiting stability is reached simultaneously with the energy representation. The assumption that $(\partial^2 E/\partial S^2)_{V,N_k} \neq 0$ is ensured by both theorem 1 and the stability hierarchy presented earlier in theorem 7, which shows that the determinants $\det \boldsymbol{E}_i$ will vanish prior to the second derivative $(\partial^2 E/\partial S^2)_{V,N_k}$.

EXERCISE 1.6

Show that the necessary and sufficient stability criteria for a binary system based on the Helmholtz energy and enthalpy potentials are interrelated by

$$\begin{aligned} &\left(\frac{\partial^2 E}{\partial S^2}\right)_{V,N_k} \begin{bmatrix} \left(\frac{\partial^2 A}{\partial V^2}\right) & \left(\frac{\partial^2 A}{\partial V \partial N_1}\right) \\ \left(\frac{\partial^2 A}{\partial V \partial N_1}\right) & \left(\frac{\partial^2 A}{\partial N_1^2}\right) \end{bmatrix} = \left(\frac{\partial^2 E}{\partial V^2}\right)_{S,N_k} \\ &\qquad \times \begin{bmatrix} \left(\frac{\partial^2 H}{\partial S^2}\right) & \left(\frac{\partial^2 H}{\partial S \partial N_1}\right) \\ \left(\frac{\partial^2 H}{\partial S \partial N_1}\right) & \left(\frac{\partial^2 H}{\partial N_1^2}\right) \end{bmatrix} \geq 0 \end{aligned} \tag{1.84}$$

1.5 Further Use of the Combined Theorems

We continue to illustrate the uses of these theorems to derive relationships of interest in stability theory that go beyond some of the most commonly recognized results.

We look at *binary mixtures* using the energy, enthalpy, Helmholtz energy, and Gibbs energy representations given in terms of their respective extensive variables:

$$
\begin{aligned}
E &= E(S, V, N_1, N_2); && \text{four extensive variables} \\
H &= H(S, P, N_1, N_2); && \text{three extensive variables} \\
A &= A(T, V, N_1, N_2); && \text{three extensive variables} \\
G &= G(T, P, N_1, N_2); && \text{two extensive variables}
\end{aligned}
$$

Energy Representation

In the case of the energy a common form of the stability condition is given by

$$
D = \begin{bmatrix}
\left(\dfrac{\partial^2 E}{\partial S^2}\right) & \left(\dfrac{\partial^2 E}{\partial S \partial V}\right) & \left(\dfrac{\partial^2 E}{\partial S \partial N_1}\right) \\
\left(\dfrac{\partial^2 E}{\partial S \partial V}\right) & \left(\dfrac{\partial^2 E}{\partial V^2}\right) & \left(\dfrac{\partial^2 E}{\partial V \partial N_1}\right) \\
\left(\dfrac{\partial^2 E}{\partial S \partial N_1}\right) & \left(\dfrac{\partial^2 E}{\partial V \partial N_1}\right) & \left(\dfrac{\partial^2 E}{\partial N_1^2}\right)
\end{bmatrix} > 0 \tag{1.85}
$$

Note from the use of theorem 6 that an equally valid criterion in the energy representation is

$$
F = \begin{bmatrix}
\left(\dfrac{\partial^2 E}{\partial V^2}\right) & \left(\dfrac{\partial^2 E}{\partial V \partial N_1}\right) & \left(\dfrac{\partial^2 E}{\partial V \partial N_2}\right) \\
\left(\dfrac{\partial^2 E}{\partial V \partial N_1}\right) & \left(\dfrac{\partial^2 E}{\partial N_1^2}\right) & \left(\dfrac{\partial^2 E}{\partial N_1 \partial N_2}\right) \\
\left(\dfrac{\partial^2 E}{\partial V \partial N_2}\right) & \left(\dfrac{\partial^2 E}{\partial N_1 \partial N_2}\right) & \left(\dfrac{\partial^2 E}{\partial N_2^2}\right)
\end{bmatrix} > 0 \tag{1.86}
$$

EXERCISE 1.7

Prove that, given any real symmetric $r \times r$ matrix $\boldsymbol{A}_r$ (with extensive variables designated by $X_l, \ldots, X_r$) with det $\boldsymbol{A}_r = 0$, its various cofactors C_{ij} are related as follows:

$$
\frac{C_{kl}}{X_k X_l} = \frac{C_{ij}}{X_i X_j} \quad (i, j, k, l = 1, \ldots, r) \tag{1.87}
$$

A solution of this exercise is presented in the appendix to this chapter: see identity 2.

Enthalpy Representation

In the case of the enthalpy, we have only three extensive variables, instead of four as is the case with the energy. We can readily write

$$\begin{bmatrix} \left(\dfrac{\partial^2 H}{\partial S^2}\right) & \left(\dfrac{\partial^2 H}{\partial S \partial N_1}\right) \\ \left(\dfrac{\partial^2 H}{\partial S \partial N_1}\right) & \left(\dfrac{\partial^2 H}{\partial N_1^2}\right) \end{bmatrix} = \left(\frac{N_2}{N_1}\right)^2 \begin{bmatrix} \left(\dfrac{\partial^2 H}{\partial S^2}\right) & \left(\dfrac{\partial^2 H}{\partial S \partial N_2}\right) \\ \left(\dfrac{\partial^2 H}{\partial S \partial N_2}\right) & \left(\dfrac{\partial^2 H}{\partial N_2^2}\right) \end{bmatrix}$$

$$= \left(\frac{N_2}{S}\right)^2 \begin{bmatrix} \left(\dfrac{\partial^2 H}{\partial N_1^2}\right) & \left(\dfrac{\partial^2 H}{\partial N_1 \partial N_2}\right) \\ \left(\dfrac{\partial^2 H}{\partial N_1 \partial N_2}\right) & \left(\dfrac{\partial^2 H}{\partial N_2^2}\right) \end{bmatrix} > 0 \tag{1.88}$$

where (1.88) follows from the results in exercise 1.7:

$$C_{33} = \left(\frac{X_3}{X_2}\right)^2 C_{22} = \left(\frac{X_3}{X_1}\right)^2 C_{11} \tag{1.89}$$

in which, $X_1 \equiv S$, $X_2 \equiv N_1$, $X_3 \equiv N_2$, and the $C_{ij}s$ are cofactors of the corresponding stability matrices.

In addition to these symmetric determinant relationships, there are three more off-diagonal ones, which can be used to test the stability of the system. As stated earlier in theorem 6, all of these determinants vanish simultaneously when the limit of stability is reached.

Helmholtz Energy Representation

In the case of the Helmholtz energy, we also have three extensive variables as with the enthalpy. In the following example, we relate two stability criteria commonly used in the literature. In the statistical-mechanics literature, a stability condition often used is

$$\det \boldsymbol{B} = \begin{bmatrix} \left(\dfrac{\partial \mu_1}{\partial N_1}\right)_{T,V,N_2} & \left(\dfrac{\partial \mu_1}{\partial N_2}\right)_{T,V,N_1} \\ \left(\dfrac{\partial \mu_1}{\partial N_2}\right)_{T,V,N_1} & \left(\dfrac{\partial \mu_2}{\partial N_2}\right)_{T,V,N_1} \end{bmatrix} > 0 \tag{1.90}$$

whereas other texts [3] use an alternative form:

$$\det \boldsymbol{B}' = \begin{bmatrix} -\left(\dfrac{\partial P}{\partial V}\right)_{T,N_1 N_2} & -\left(\dfrac{\partial P}{\partial N_1}\right)_{T,V,N_2} \\ -\left(\dfrac{\partial P}{\partial N_1}\right)_{T,V,N_2} & \left(\dfrac{\partial \mu_1}{\partial N_1}\right)_{T,V,N_2} \end{bmatrix} > 0 \tag{1.91}$$

One way to show the equivalence of these two criteria is to express the elements of $\boldsymbol{B}'$ in terms of the elements of $\boldsymbol{B}$ using multivariable calculus transformations. This is a very laborious procedure, which leads ultimately to the following equation:

$$\det \boldsymbol{B}' = \left(\frac{N_2}{V}\right)^2 \det \boldsymbol{B} \tag{1.92}$$

Instead, we remark that this result follows immediately from the use of the results given in theorem 6. The condition given by equation (1.90) is found by eliminating $X_1 \equiv V$ from the Hessian matrix of the Helmholtz energy, while equation (1.92) follows from the elimination of $X_3 \equiv N_2$ for this same matrix. This example shows how a good understanding of the theorems presented earlier can be used to quickly establish useful results in this area.

Gibbs Energy Representation

In this case we have only two extensive variables. Note that one may be tempted to use the following determinant as a stability criterion:

$$D = \begin{vmatrix} \left(\frac{\partial \mu_1}{\partial N_1}\right)_{T,P,N_2} & \left(\frac{\partial \mu_1}{\partial N_2}\right)_{T,P,N_1} \\ \left(\frac{\partial \mu_1}{\partial N_2}\right)_{T,P,N_1} & \left(\frac{\partial \mu_2}{\partial N_2}\right)_{T,P,N_1} \end{vmatrix} > 0 \tag{1.93}$$

However, such a determinant cannot be used since

$$D = \det \boldsymbol{A}_r = 0 \tag{1.94}$$

as shown earlier in theorem 3. Now, using the identity given in exercise 1.7, we can write

$$C_{22} = \left(\frac{X_2}{X_1}\right)^2 C_{11} = \frac{X_2}{X_1} C_{12} \tag{1.95}$$

where $X_1 \equiv N_1$ and $X_2 \equiv N_2$. Thus, in a stable system,

$$\left(\frac{\partial \mu_1}{\partial N_1}\right)_{T,P,N_2} = \left(\frac{N_2}{N_1}\right)^2 \left(\frac{\partial \mu_2}{\partial N_2}\right)_{T,P,N_1} = -\left(\frac{N_2}{N_1}\right)\left(\frac{\partial \mu_1}{\partial N_2}\right)_{T,P,N_1} > 0 \tag{1.96}$$

which proves that, in these systems,

$$\left(\frac{\partial \mu_1}{\partial N_2}\right)_{T,P,N_1} < 0 \tag{1.97}$$

Note that we could have used the *Gibbs–Duhem* equation to derive the equality in equation (1.96). However, the use of stability theory to obtain equation (1.96) is more general, since it can be applied to all the other well-known thermodynamic potentials, while the Gibbs–Duhem equation can only be used in the case of the Gibbs energy.

The Relationship between the Mechanical and Material Stability Coefficients in Mixtures

The relationship between mechanical and material stability coefficients in mixtures is an interesting one that can be explored with the aid of these theorems. For a pure fluid, both coefficients vanish simultaneously, but this is generally not the case in mixtures.

In their book, Rowlinson and Swinton [4] state that a Helmholtz-energy-based stability criterion in a binary mixture is given by

$$D = \begin{vmatrix} \left(\dfrac{\partial^2 a}{\partial v^2}\right)_{T,x} & \left(\dfrac{\partial^2 a}{\partial v \partial x}\right)_T \\ \left(\dfrac{\partial^2 a}{\partial v \partial x}\right)_T & \left(\dfrac{\partial^2 a}{\partial x^2}\right)_{T,v} \end{vmatrix} > 0, \quad a \equiv \frac{A}{N} \tag{1.98}$$

which was further represented as

$$D = -\left(\frac{\partial p}{\partial v}\right)_{T,x} \left(\frac{\partial^2 g}{\partial x^2}\right)_{T,P} > 0, \quad g \equiv \frac{G}{N} \tag{1.99}$$

where x is the mole fraction of component 1, and v is the specific volume V/N.

The term $-(\partial P/\partial v)_T$ is referred to as *the mechanical stability* coefficient, while the quantity $(\partial \mu/\partial x)_{T,P}$ is referred to as the *material stability* coefficient, Clearly, at a stability limit, $D = 0$ (from equation (1.99)), but this does not tell us anything about which of the terms in equation (1.99) goes to zero first or, in fact, whether both go to zero simultaneously. The answer to these questions establishes a precedence order between the mechanical and material thermodynamic stability criteria in binary mixtures. It can be resolved with the aid of the theorems previously presented.

We first prove the results presented in equation (1.98) and (1.99) using the theorems. A representation of the system energy in terms of mole fractions may be obtained from the following description for E:

$$E = E(S, V, N_1, N_2, \ldots, N_{n-2}, N_{n-1}, N) \tag{1.100}$$

with $N = \sum_{i=1}^{n} N_i$, where n is the number of components. With the choice of extensive variables in equation (1.100) the pairs of conjugate variables are

$$(S, T), \ (-P, V), \ (\eta_i, N_i), \ (\eta_n, N) \tag{1.101}$$

with

$$\eta_i \equiv \mu_i - \mu_n \quad (i = 1, \ldots, n-1) \tag{1.102}$$

$$\eta_n = \mu_n \tag{1.103}$$

The results of theorem 6 show that stability conditions are independent of the choice of extensive variables. If we choose the Helmholtz energy representation, then choosing

V, N_1, and N as our independent variables in a binary system, we get the following equivalence from theorem 8 and equation (1.70), in particular:

$$\det \mathbf{A}_2 = \begin{vmatrix} \left(\frac{\partial^2 A}{\partial V^2}\right)_{T,N_1,N} & \left(\frac{\partial^2 A}{\partial V \partial N_1}\right)_{T,N} \\ \left(\frac{\partial^2 A}{\partial V \partial N_1}\right)_{T,N} & \left(\frac{\partial^2 A}{\partial N_1^2}\right)_{T,V,N} \end{vmatrix}$$

$$= \left(\frac{\partial^2 A}{\partial V^2}\right)_{T,N_1,N} \left(\frac{\partial^2 G}{\partial N_1^2}\right)_{T,P,N} > 0 \tag{1.104}$$

Multiplying both sides of equation (1.104) by N^4 yields the result given in (1.99):

$$D = \begin{bmatrix} \left(\frac{\partial^2 a}{\partial v^2}\right)_{T,x} & \left(\frac{\partial^2 a}{\partial v \partial x}\right)_T \\ \left(\frac{\partial^2 a}{\partial v \partial x}\right)_T & \left(\frac{\partial^2 a}{\partial x^2}\right)_{T,v} \end{bmatrix} = -\left(\frac{\partial p}{\partial v}\right)_T \left(\frac{\partial^2 g}{\partial x^2}\right)_{T,P} > 0 \tag{1.105}$$

Precedence of the Material Stability Coefficient

Next, we address the issue of precedence in reaching the thermodynamic stability limit. Let us represent by $\mathbf{A}$ the Hessian matrix of the Helmholtz energy. Then, for a binary system, we have from theorem 1 that

$$\mathbf{A}_1 = \left(\frac{\partial^2 A}{\partial V^2}\right)_{T,N_1,N_2} > 0 \tag{1.106}$$

Also

$$\det \mathbf{A}_2 = \begin{vmatrix} \left(\frac{\partial^2 A}{\partial V^2}\right)_{T,N_1,N_2} & \left(\frac{\partial^2 A}{\partial V \partial N_1}\right)_{T,N_2} \\ \left(\frac{\partial^2 A}{\partial V \partial N_1}\right)_{T,N_2} & \left(\frac{\partial^2 A}{\partial N_1^2}\right)_{T,V,N_2} \end{vmatrix} > 0 \tag{1.107}$$

and we showed above, using theorem 8, that

$$\det \mathbf{A}_2 = \left(\frac{\partial^2 A}{\partial V^2}\right)_{T,N_1,N_2} \left(\frac{\partial^2 G}{\partial N_1^2}\right)_{T,P,N_2} \tag{1.108}$$

The usual mechanical stability condition is

$$\left(\frac{\partial^2 A}{\partial V^2}\right)_{T,N_1,N_2} > 0 \tag{1.109}$$

and, if the system is materially stable, then

$$\left(\frac{\partial^2 G}{\partial N_1^2}\right)_{T,P,N_2} > 0 \tag{1.110}$$

The question we pose is: Which of these limits of stability is reached first? The stability hierarchy relation presented in theorem 7 establishes that

$$\det \boldsymbol{A}_1 \geq \alpha \det \boldsymbol{A}_2 \geq 0 \tag{1.111}$$

with $\alpha = 1/\mathrm{trace}\boldsymbol{A} > 0$. From equation (1.111) there are only two possibilities:

1. det $\boldsymbol{A}_2$ vanishes prior to det $\boldsymbol{A}_1$, or
2. det $\boldsymbol{A}_1$ and det $\boldsymbol{A}_2$ vanish simultaneously.

For case (1), we have from equations (1.106) and (1.108) that $(\partial^2 G/\partial N_1^2)_{T,P,N_2}$ must vanish prior to $(\partial^2 A/\partial V^2)_{T,N_1,N_2}$; *in other words, the material stability condition takes precedence over the mechanical stability condition.*

For case (2), we have that both derivatives $(\partial^2 G/\partial N_1^2)_{T,P,N_2}$ and $(\partial^2 A/\partial V^2)_{T,N_1,N_2}$ vanish simultaneously, which occurs at a *critical azeotrope*, discussed in chapter 2.

1.6 Chapter Review

In this chapter, we developed the basic mathematical framework for describing macroscopic thermodynamic stability theory using a few key theorems from linear algebra. Legendre transforms were used to link the fundamental energy equation to all other thermodynamic potentials. Convexity properties of these functions, based upon second-law considerations, led to a set of necessary and sufficient conditions for thermodynamic stability defined with respect to the Hessian matrix (stability matrix) of any given potential.

The theorems presented concern mainly the properties of this stability matrix including, among other things, its order (rank) and structure in terms of the natural extensive thermodynamic variables for any given potential. Equations defining a limit of stability were developed, and a hierarchy established, showing which conditions assume precedence as this limit is approached. Furthermore, results were presented showing the equivalence of stability criteria between various common thermodynamic potentials.

We did not consider whether the stability limit thus reached is a stable one or not. If we require it to be a stable point, then an additional set of requirements arises. These stable limits of stability are known as *critical points*, and the conditions prevailing at such points are discussed in the next chapter.

1.7 Additional Exercises

General Questions

1. How do we decide upon the specification of thermodynamic variables for a chemical system? Is there any fundamental law that prescribes this?
2. Polymer models are often thought of as consisting of coiled segments. If there are M_i coiled units each of length l_i and the polymer can be stretched by applying a force τ, write an equation for the thermodynamic energy in the system. Develop a

set of stability conditions for the polymer and discuss what type of transition might occur if the system becomes unstable.

3. Consider a gaseous system consisting of molecules with dipoles each of which is represented as two point charges of magnitude $q(+)$ and $q(-)$ separated by a distance l. If we apply an external field D that orients the dipoles in its direction, then we can write the system energy as follows:

$$dE = T\,dS - p\,dV - M\,dD + \mu\,dN$$

where M is the electrical polarization of the system defined as the sum of the permanent and induced dipole moments in the fluid:

$$M = \sum_{i=1}^{N} [|\mu| + |\mu^*|]_i$$

where $|\mu| = ql$ and $|\mu^*| = \alpha D$, with α the polarizability. Write a set of thermodynamic conditions at the limit of stability of the system. What happens to M at a stability limit?

4. The usual specification for thermodynamic energy in a gaseous system involves the variables $S, V, N_1, \ldots, N_n$. In what situation would it be more useful to use additionally the (dependent) variable N, and write the set $E(S, V, N_1, \ldots, N_{n-1}, N_i)$, where $N = \sum_{i=1}^{n} N_i$?

Stability in Pure Fluids

5. Two specific heats in pure fluids are defined as follows:

$$C_V \equiv T\left(\frac{\partial S}{\partial T}\right)_{N,V}$$

$$C_P \equiv T\left(\frac{\partial S}{\partial T}\right)_{P,N}$$

What does thermodynamic stability theory tell us about the signs and magnitudes of these properties in a stable fluid? Which Legendre functions are relevant to C_V and C_P, respectively?

6. A well-known mechanical stability condition in a pure fluid is given by the expression

$$-\left(\frac{\partial P}{\partial V}\right)_{T,N} \geq 0$$

Is the inequality $-\left(\frac{\partial P}{\partial V}\right)_{S,N} \geq 0$ a stability coefficient and, if so, are both of these inequality conditions satisfied simultaneously at a stability limit?

7. Prove the thermodynamic identity

$$C_P - C_V = -T\left(\frac{\partial P}{\partial V}\right)_{T,N}\left[\left(\frac{\partial V}{\partial T}\right)_{P,N}\right]^2$$

In a stable phase, it follows that $C_P > C_V$. What can be concluded about the relative magnitudes of C_P and C_V at a limit of stability?

8. Given the definitions

$$\alpha_P \equiv \frac{1}{V}\left(\frac{\partial V}{\partial T}\right)_{P,N}, \quad K_T \equiv -\frac{1}{V}\left(\frac{\partial V}{\partial P}\right)_{T,N}$$

is the expansion coefficient α_P a stability coefficient—and if not, why not? Prove that

$$\frac{\alpha_P}{K_T} = \left(\frac{\partial P}{\partial T}\right)_{N,V}$$

9. Prove the identity

$$C_v = \alpha_p v T\left[\left(\frac{\partial P}{\partial T}\right)_s - \left(\frac{\partial P}{\partial T}\right)_v\right]$$

10. In a pure component, is C_V *necessarily finite and positive* at a first order (unstable) limit of stability? How about C_P?

Thermodynamic Stability in Mixtures

11. Can a so-called *ideal binary solution*, defined as being one where the chemical potential for component i is given by the equation $\mu_i = \mu_i^0 + RT\ln x_i$, exhibit phase equilibria between two or more phases? Here the mole fraction x_i is given by $x_i \equiv N_i/\sum_{i=1}^{n} N_i$ and μ_i^0 is the ideal-gas chemical potential.

12. In a ternary system, prove that the following derivatives are "stability" coefficients:

$$-\left(\frac{\partial P}{\partial V}\right)_{T,\mu_2,N_3} > 0, \quad -\left(\frac{\partial \mu_3}{\partial N_3}\right)_{T,\mu_2,V} > 0$$

$$\left(\frac{\partial T}{\partial S}\right)_{P,\mu_2,N_3} > 0, \quad -\left(\frac{\partial P}{\partial V}\right)_{\mu_2,\mu_3,S} > 0$$

Are any two of these violated simultaneously at a stability limit? In each case, write out the appropriate equations defining the limit of stability.

13. In a binary system, at constant composition and temperature, prove that at constant pressure the *spinodal curve*, which represents the locus of limiting stability points, is encountered before the mechanical limit of stability is reached.

$$\left[\text{Hint: Show that} \begin{vmatrix} A_{VV} & A_{VN_1} \\ A_{N_1V} & A_{N_1N_1} \end{vmatrix} \text{reaches zero before } A_{VV}.\right]$$

14. Prove that the stability criteria for a ternary mixture can be written as

$$\left(\frac{\partial \mu_1}{\partial N_1}\right)_{T,P,N_2} \left(\frac{\partial \mu_2}{\partial N_2}\right)_{T,P,N_1} = \left(\frac{\partial \mu_1}{\partial N_2}\right)^2_{T,P,N_1}$$

15. The equilibrium *K* value for species *i* in a *vapor–liquid equilibrium* (VLE) system is defined as follows:

$$K_i \equiv \frac{y_i}{x_i}$$

where x_i and y_i are the liquid-phase and vapor-phase mole fractions of component i, respectively. In a binary vapor–liquid system, the Gibbs phase rule specifies that there are two thermodynamic intensive degrees of freedom; given this, find expressions for the following derivatives (σ designates a property along the phase coexistence boundary.)

$$\left(\frac{\partial K_i}{\partial T}\right)_{P,\sigma}, \quad \left(\frac{\partial K_i}{\partial P}\right)_{T,\sigma}$$

If an azeotrope (i.e., equal compositions in both phases) exists along the phase envelope, what are the values of these derivatives there?

16. Consider a thermodynamic potential Φ with extensive variables $X_1, \ldots, X_r$. Its successive Legendre transformation $\Phi^{(1)}$ is obtained by swapping X_1 for the intensive variable I_1, where $I_1 \equiv (\partial \Phi / \partial X_1)_{X_l}$. Thus we have that

$$\Phi^{(1)}(I_1, \ldots, X_r) = \Phi(X_1, \ldots, X_r) - I_1 X_1$$

Show that, in general,

$$\left(\frac{\partial \Phi^{(1)}}{\partial X_i}\right)_{I_1, X_{j \neq i}} = \left(\frac{\partial \Phi}{\partial X_i}\right)_{X_{j \neq i}}$$

Also, prove that

$$\left(\frac{\partial^2 \Phi^{(1)}}{\partial I_1^2}\right)_{X_K} = -\frac{1}{(\frac{\partial^2 \Phi}{\partial X_1^2})_{X_{K \neq 1}}}$$

Use this approach to develop linkages between ***E***, ***A***, ***H***, and ***G*** that follow these identities.

17. Use the identities of the previous question to prove that

$$G_{TP}/G_{TT} = -G_{SP}$$

where the subscripts indicate partial differentiation.

Basic Linear Algebra

18. The inner product of a vector $\boldsymbol{x}$ is defined as

$$(\boldsymbol{x}, \boldsymbol{x}) = \sum_{i=1}^{n} x_i^2 = |\boldsymbol{x}|^2$$

where $|\boldsymbol{x}|$ is the vector length. Prove, for any matrix $\boldsymbol{A}$, that $(\boldsymbol{A}, \boldsymbol{y}) = (\boldsymbol{x}, \boldsymbol{A}^{\mathsf{T}}\mathbf{y})$, where $\boldsymbol{A}^{\mathsf{T}}$ means the transpose of $\boldsymbol{A}$.

19. Prove that, if $\boldsymbol{A}^{\mathsf{T}} = \boldsymbol{A}$ ($\boldsymbol{A}$ is said to be self-adjoint), and if $\boldsymbol{P}$ is the eigenvector matrix of $\boldsymbol{A}$ normalized so that $|\boldsymbol{P}|^2 = 1$ for each column $\boldsymbol{P}$ of $\boldsymbol{P}$, then

$$\boldsymbol{P}^{\mathsf{T}} = \boldsymbol{P}^{-1}$$

What does this result imply about the eigenvectors of $\boldsymbol{A}$?

20. Prove that, at a stable thermodynamic state, the eigenvalues of the Hessian matrix of a thermodynamic potential must all be greater than or equal to zero. What about at a metastable state?

21. The eigenvalues of a matrix $\boldsymbol{A}$ are the scalar λ that satisfy the vector equation

$$\boldsymbol{A}\boldsymbol{x} = \lambda \boldsymbol{x}$$

or equivalently are the solutions in λ to the determinant equation

$$\det(\boldsymbol{A} - \lambda \boldsymbol{I}) = 0$$

Using this latter equation, find an expression for the eigenvalues of the matrix

$$\begin{bmatrix} a_{11} & a_{12} \\ a_{21} & a_{22} \end{bmatrix}$$

Use this result with the energy potential for a pure fluid to generate thermodynamic inequalities that must be obeyed in a stable pure phase.

22. Prove or disprove the following statements:
    (a) If det $\boldsymbol{A}$ is positive, then $\boldsymbol{A}$ is necessarily positive definite.
    (b) If $\boldsymbol{A}$ is positive definite, then it follows that det $\boldsymbol{A}$ is positive.

23. Show that, at a stable state, the variation of any thermodynamic potential $F(\mathbf{y})$ in response to a fluctuation in its natural extensive variables, denoted by $\Delta\mathbf{y}$, is given by the expression:

$$F = \lambda_1 \, \Delta y_1^2 + \cdots + \lambda_n \, \Delta y_n^2$$

for any vector $\Delta\mathbf{y}$. The eigenvalues of the Hessian of the potential $F(\mathbf{y})$ are given by the λ_i. Does the Hessian matrix of the *Legendre transform* of $F(\mathbf{y})$, which we denote by $H(I_1, \ldots, y_N)$ share a common eigenvalue with those of the Hessian matrix of $F(\mathbf{y})$? [Hint: look at the convexity of the functions $F(y_1, \ldots, y_N)$ and $H(I_1, \ldots, y_N)$, where $I_1 = (\partial F/\partial Y_1)_{y_k \neq 1}$].

24. Given the Legendre transform of a function $L[f(x, y)] = g(z, y)$, show that, if f is positive definite at a point, the same is not true for g there. Why does this result not contradict the stability-theory results of this chapter?

25. Prove that the cofactors $C_{i,v+1}$ of an $(v+1) \times (v+1)$ matrix $\boldsymbol{A}$ is related to the cofactors $G_{i,l}$ of the $v \times v$ matrix $\boldsymbol{B}$ found by deleting the ith row and $(v+1)$th column of $\boldsymbol{A}$ by the equation

$$C_{i,v+1} = -1 \sum_{l=1}^{v} a_{v+1,l} G_{i,l} \tag{1.112}$$

Appendix: Linear Algebra

Some Basic Definitions and Concepts in Linear Algebra

Determinants, minors, cofactors, and adjoints

Given the general nonsingular matrix $\boldsymbol{A}$ defined as

$$\boldsymbol{A} \equiv \begin{bmatrix} a_{11} & a_{12} & \cdots & a_{1n} \\ a_{21} & a_{22} & \cdots & a_{2n} \\ \vdots & \vdots & \ddots & \vdots \\ a_{n1} & a_{n2} & \cdots & a_{nn} \end{bmatrix}$$

The determinant of the $(n-1) \times (n-1)$ matrix formed by omitting the ith row and the jth column of the nonsingular $(n \times n)$ matrix $\boldsymbol{A}$ is called the *minor* of the term a_{ij}, denoted as M_{ij}. The cofactor of a_{ij} is denoted as A_{ij} and defined as

$$A_{ij} \equiv (-1)^{i+1} M_{ij}$$

The matrix of cofactors of $\boldsymbol{A}$ is related to the matrix inverse of $\boldsymbol{A}$ as follows:

$$\boldsymbol{A}^{-1} = \frac{\text{adj } \boldsymbol{A}}{\det \boldsymbol{A}}$$

where adj $\boldsymbol{A}$ is a matrix called the *adjoint* of $\boldsymbol{A}$, which is defined by

$$\text{adj } \boldsymbol{A} \equiv [A_{ij}]^{\mathsf{T}}$$

where T denotes the transpose.

Eigenvalues, eigenvectors, and similarity transformations

An important theorem in linear algebra states that a *similarity transformation* results in a diagonal matrix $\boldsymbol{\Lambda}$ for any nonsingular matrix $\boldsymbol{A}$, where $\boldsymbol{A}$ and $\boldsymbol{\Lambda}$ are connected as follows:

$$\boldsymbol{P}^{-1} = \boldsymbol{A}\boldsymbol{P} = \boldsymbol{\Lambda} \tag{A.1}$$

The diagonal elements of $\boldsymbol{\Lambda}$ are the eigenvalues of the matrix $\boldsymbol{A}$. If we further stipulate that the original matrix $\boldsymbol{A}$ is symmetric, it can also be shown that the columns of $\boldsymbol{P}$ are corresponding eigenvectors of $\boldsymbol{A}$ and further that $\boldsymbol{P}$ can be chosen such that its columns are *orthonormal*, which means that

$$\boldsymbol{P}^{-1}\boldsymbol{P}\boldsymbol{P}^{\mathrm{T}}\boldsymbol{P} = \boldsymbol{I} \tag{A.2}$$

or equivalently

$$\boldsymbol{P}^{-1}\boldsymbol{P}^{\mathrm{T}} \tag{A.3}$$

Further, $\boldsymbol{P}$ satisfies

$$|\det \boldsymbol{P}| = |\det \boldsymbol{P}^{-1}| = 1 \tag{A.4}$$

Matrices of the type $\boldsymbol{P}$ satisfying (A.3) are called *orthogonal.*

Proofs of Some Relevant Identities

Identity 1

Here we prove that

$$\sum_{i=1}^{v} a_{(v+1)i} G_{ki} = -C_{k(v+1)} \tag{A.5}$$

where C_{ij} are the cofactors of matrix $\boldsymbol{A}_{v+1}$ and G_{ij} are the cofactors of matrix $\boldsymbol{A}_v$. (The condition $|\boldsymbol{A}_v| = 0$ is not required.)

Proof

Let M_{ij} be minors of matrix $\boldsymbol{A}_{v+1}$; then

$$C_{i(v+1)} = (-1)^{i+v+1} M_{i(v+1)} \tag{A.6}$$

And we can write the minor in equation (A.6) in terms of the cofactors G_{ij} of matrix $\boldsymbol{A}_v$:

$$M_{i(v+1)} = (-1)^{(v-i)} \sum_{l=1}^{v} a_{(v+1)l} G_{il} \quad (i = 1, \ldots, v) \tag{A.7}$$

The factor $(-1)^{(v-i)}$ takes care of the change in sign of the determinant upon reordering of the rows necessary to obtain the correct determinant for $M_{i(v+1)}$. Using (A.7) in (A.6), we obtain equation (A.5).

Identity 2

We show that given the real symmetric matrix $\boldsymbol{A}_r$, its various cofactors C_{ij} are related by the identity

$$\frac{C_{kl}}{X_k X_l} = \frac{C_{ij}}{X_i X_j} \quad \text{for all } k, l, i, j = 1, \ldots, r \tag{A.8}$$

Proof

Define a new $r \times r$ matrix $\boldsymbol{B}$ with elements

$$\boldsymbol{B} = \{a_{ij} X_i X_j\} = \{b_{ij}\} \tag{A.9}$$

then

$$\sum_{i=1}^{r} b_{ij} = \sum_{i=1}^{r} a_{ij} X_i X_j = X_j \sum_{i=1}^{r} a_{ij} X_i \tag{A.10}$$

From Euler's theorem and equation (A.10),

$$\sum_{i=1}^{r} b_{ij} = 0 \tag{A.11}$$

Now define a diagonal matrix $\boldsymbol{E}$ with elements

$$\boldsymbol{E} = \{\delta_{ij} X_i\} \tag{A.12}$$

where δ_{ij} is the Kronecker delta. Then the matrix $\boldsymbol{B}$ is given by

$$\boldsymbol{B} = \boldsymbol{EAE} \tag{A.13}$$

Call $\boldsymbol{A}_{ij}$ the matrix that results by deleting the i row and the j column from matrix $\boldsymbol{A}$, and call $\boldsymbol{B}_{ij}$ the matrix that results by deleting the i row and the j column from matrix $\boldsymbol{B}$. Let $\boldsymbol{E}^{(i)}$ denote the matrix that results by deleting the i row and the i column from matrix $\boldsymbol{E}$. We will call C_{ij} the cofactors of matrix $\boldsymbol{A}$ and F_{ij} the cofactors of matrix $\boldsymbol{B}$. From linear algebra,

$$C_{ij} = (-1)^{i+j} |\boldsymbol{A}_{ij}| \tag{A.14}$$

$$F_{ij} = (-1)^{i+j} |\boldsymbol{B}_{ij}| \tag{A.15}$$

Also the matrices $\boldsymbol{B}_{ij}$ and $\boldsymbol{A}_{ij}$ are related by

$$\boldsymbol{B}_{ij} = \boldsymbol{E}^{(i)} \boldsymbol{A}_{ij} \boldsymbol{E}^{(j)} \tag{A.16}$$

Using equations (A.14), (A.15), and (A.16), we have

$$\begin{aligned} F_{ij} &= (-1)^{i+j} |\boldsymbol{B}_{ij}| = (-1)^{i+j} \frac{(\prod_{l=1}^{r} X_l)^2}{X_i X_j} |\boldsymbol{A}_{ij}| \\ &= \frac{(\prod_{l=1}^{r} X_l)^2}{X_i X_j} C_{ij} \end{aligned} \tag{A.17}$$

Also,

$$F_{i(j+1)} = (-1)^{i+j+1} \left| \boldsymbol{B}_{i(j+1)} \right| \tag{A.18}$$

Note that, in the $\boldsymbol{B}_{i(j+1)}$ matrix, the $j+1$ column is missing. Now if we add to the j column all the other columns in the matrix, we get a new j column with elements

$$\alpha_i = \sum_{i \neq j+1} b_{ij} = -b_{i(j+1)} \tag{A.19}$$

Equation (A.19) follows from equation (A.11). Since adding columns doesn't change the value of the determinant, we have

$$|\boldsymbol{B}_{i(j+1)}| = -|\boldsymbol{B}_{ij}| \tag{A.20}$$

Thus

$$\begin{aligned} F_{i(j+1)} &= (-1)(-1)^{(i+j+1)}|\boldsymbol{B}_{ij}| \\ &= (-1)^{(i+j)}|\boldsymbol{B}_{ij}| = F_{ij} \end{aligned} \tag{A.21}$$

From equation (A.21), it follows that all the cofactors F_{ij} of matrix $\boldsymbol{B}$ are equal. Using this result in equation (A.17), we get the identity we are after, namely equation (A.8).

Identity 3

Here we show that, if $|\boldsymbol{A}_v| = 0$, then

$$\frac{C_{i(v+1)}}{G_{ki}} = \frac{C_{(i+1)(v+1)}}{G_{k(i+1)}} \quad (i = 1, \ldots, v-1) \tag{A.22}$$

where G_{ij} are the cofactors of the $v \times v$ matrix $\boldsymbol{A}_v$.

Proof

Since the determinant of matrix $\boldsymbol{A}_v$ is zero, then the rows and columns of matrix $\boldsymbol{A}_v$ are linearly dependent. Therefore there exists a vector $\boldsymbol{w} = (W_1, \ldots, W_v)$ such that

$$\sum_{i=1}^{v} a_{ij} W_i = 0; \quad \sum_{i=1}^{v} a_{ji} W_i = 0 \tag{A.23}$$

Let G_{ij} be the cofactors of matrix $\boldsymbol{A}_v$ and C_{ij} the cofactors of matrix $\boldsymbol{A}_{v+1}$. Since $|\boldsymbol{A}_v| = 0$, we can apply identity 2. Thus we get

$$\frac{G_{ij}}{W_i} = \frac{G_{kj}}{W_k} \tag{A.24}$$

$$\frac{G_{ji}}{W_i} = \frac{G_{jk}}{W_k} \tag{A.25}$$

Let us call $\boldsymbol{G}$ the matrix whose elements are the cofactors of matrix $\boldsymbol{A}_v$, namely

$$\boldsymbol{G} = \{G_{ij}\} \tag{A.26}$$

Equations (A.24) and (A.25) imply that the rows of matrix $\boldsymbol{G}$ are proportional to one another and the columns of matrix $\boldsymbol{G}$ are also proportional to one another. Also, from equation (A.24), we can write

$$\frac{W_{i+1}}{W_i} = \frac{G_{(i+j)j}}{G_{ij}} \tag{A.27}$$

Equation (A.27) shows that the vector $\boldsymbol{w}$ is also proportional to the rows and columns of matrix $\boldsymbol{G}$. Thus since a proportionality constant doesn't affect equation (A.23) we can set $\boldsymbol{w}$ equal to any row or column of matrix $\boldsymbol{G}$; thus

$$W_i = G_{ik} = G_{ki} \tag{A.28}$$

In equation (A.28) k can assume any of the values $1, \ldots, v$. From identity 1, we have

$$C_{i(v+1)} = -\sum_{i=1}^{v} a_{(v+1)l} G_{il} \tag{A.29}$$

Using (A.25) in (A.29), we have.

$$\frac{C_{i(v+1)}}{W_i} = -\sum_{l=1}^{v} a_{(v+1)l} \frac{G_{il}}{W_i} = -\sum_{l=1}^{v} a_{(v+1)l} \frac{G_{jl}}{W_j}$$
$$= \frac{C_{j(v+1)}}{W_j} \quad (i, j = 1, \ldots, v) \tag{A.30}$$

Now using (A.28) in (A.30), we obtain equation (A.22).

Proof of Theorem 3

The proof depends upon a result in linear algebra which specfies that, if a matrix $\boldsymbol{A}$ is nonsingular, then the so-called LU decomposition can be found. This decomposes $\boldsymbol{A}$ into two matrices, a lower triangular matrix $\boldsymbol{L}$ with ones on the diagonal, and an upper triangular matrix $\boldsymbol{U}$ composed of elements that arise from Gaussian elimination on $\boldsymbol{A}$. In detail, then,

$$\boldsymbol{A} = \boldsymbol{L}\boldsymbol{U} \tag{A.31}$$

where $\boldsymbol{A} = \{a_{ij}\}$, and

$$\boldsymbol{L} = \begin{bmatrix} 1 & 0 & 0 & \cdots & 0 \\ m_{21} & 1 & 0 & \cdots & 0 \\ m_{31} & m_{32} & 1 & \cdots & 0 \\ \vdots & \vdots & \vdots & \ddots & \vdots \\ m_{n1} & m_{n2} & m_{n3} & \cdots & 1 \end{bmatrix} \tag{A.32}$$

with

$$m_{i1} = \frac{a_{i1}}{a_{11}} \quad (i = 2, \ldots, n) \tag{A.33}$$

and the m_{ij} are the multipliers in the elimination of $\boldsymbol{A}$ to the upper triangular form given by

$$\boldsymbol{U} = \begin{bmatrix} a_{11} & a_{12} & a_{13} & \cdots & a_{1n} \\ 0 & a_{22}^{(2)} & a_{23}^{(2)} & \cdots & a_{2n}^{(2)} \\ 0 & 0 & a_{33}^{(3)} & \cdots & a_{3n}^{(3)} \\ \vdots & \vdots & \vdots & \ddots & \vdots \\ 0 & 0 & 0 & \cdots & a_{nn}^{(n)} \end{bmatrix} \tag{A.34}$$

The superscripts on the elements in $\boldsymbol{U}$ denote that the elements $a_{ij}^{(k)}$ are formed during the Gaussian elimination operations in column $k - 1$. For example, the element $a_{22}^{(2)}$ equals $a_{22} - (a_{21}/a_{11})a_{12}$.

Now $|\boldsymbol{A}| = |\boldsymbol{L}||\boldsymbol{U}| = |\boldsymbol{U}|$, since the determinant of a triangular matrix is the product of its diagonal elements. Thus

$$|\boldsymbol{A}| = \prod_{k=1}^{n} a_{kk}^{(k)} \tag{A.35}$$

We want to identify the elements $a_{kk}^{(k)}$ with the principal minors of $\boldsymbol{A}$ which we denote as $|\boldsymbol{A}_k|$. For example, $|\boldsymbol{A}_2|$ is the determinant function

$$|\boldsymbol{A}_2| = \begin{vmatrix} a_{11} & a_{12} \\ a_{21} & a_{22} \end{vmatrix} \tag{A.36}$$

If $\boldsymbol{A}$ is of order 2, then

$$|\boldsymbol{A}| = a_{11}a_{22}^{(2)} \tag{A.37}$$

implying that

$$a_{22}^{(2)} = \frac{|\boldsymbol{A}|}{a_{11}} = \frac{|\boldsymbol{A}_2|}{|\boldsymbol{A}_1|} \tag{A.38}$$

where we have defined $|\boldsymbol{A}_1| = a_{11}$. If $n = 3$, then

$$|\boldsymbol{A}| = a_{11}a_{22}^{(2)}a_{33}^{(3)} = |\boldsymbol{A}_1|\frac{|\boldsymbol{A}_3|}{|\boldsymbol{A}_1|}a_{33}^{(3)} \tag{A.39}$$

implying that

$$a_{33}^{(3)} = \frac{|\boldsymbol{A}_3|}{|\boldsymbol{A}_2|} \tag{A.40}$$

Thus it can be seen by repeated applications of this approach that

$$a_{kk}^{(k)} = \frac{|A_k|}{|A_{k-1}|} \tag{A.41}$$

This result establishes that the diagonal elements of the $\boldsymbol{U}$ matrix are actually related to the determinants of principal minors of $\boldsymbol{A}$. It also shows that, if $\boldsymbol{A}$ is nonsingular, implying existence of an LU decomposition, then $|\boldsymbol{A}_k| \neq 0$ for $k = 1, \dots, n$.

Necessary condition

With this background, we are now in a position to develop the proof. We will first develop the necessity part by showing that, if a nonsingular matrix $\boldsymbol{A}$ is positive definite, then $|A_k| > 0$ for $k = 1, \dots, n$.

Since $\boldsymbol{A}$ is positive definite by assumption, this implies that $\boldsymbol{x}^{\mathrm{T}}\boldsymbol{A}\boldsymbol{x} > 0$ for any nonzero vector $\boldsymbol{x}$. Consider a vector $\boldsymbol{x}$ with its first k entries having arbitrary values and the remaining elements zero. (i.e., $x_{i>k} = 0$). Then

$$(\boldsymbol{x}, \boldsymbol{A}\boldsymbol{x}) > 0 \tag{A.42}$$

Denoting by $\boldsymbol{x}_k$ the vector formed by the first k elements of $\boldsymbol{x}$, then

$$(\boldsymbol{x}_k, \boldsymbol{A}_k\boldsymbol{x}_k) > 0 \tag{A.43}$$

where $\boldsymbol{A}_k$ is the $k \times k$ principal submatrix of $\boldsymbol{A}$. This result implies that $\boldsymbol{A}_k$ is positive definite, having all positive eigenvalues, which in turn implies that $|\boldsymbol{A}_k| > 0$. Therefore, since the size of the vector $\boldsymbol{x}_k$ allows k to be any value from 1 through n, we have shown that, in general,

$$|\boldsymbol{A}_k| > 0 \quad (k = 1, \dots, n) \tag{A.44}$$

Then, using equation (A.35), we have that $a_{kk}^{(k)} > 0$ for $k = 1, \dots, n$.

Sufficient condition

For the sufficiency part of the proof, we must establish that, if $|\boldsymbol{A}_k| > 0, k = 1, \dots, n$, then **A** itself must be positive definite.† From the previous $\boldsymbol{LU}$ decompostion of $\boldsymbol{A}$ we have that,

$$\boldsymbol{A} = \boldsymbol{LU} \tag{A.45}$$

†It is tempting to see this as a simple result since, if all the diagonal elements of $\boldsymbol{U}$ are positive—by virtue of all the $|\boldsymbol{A}_k|$ being positive, from equation (A.35)—then $|\boldsymbol{A}| = |\boldsymbol{U}| > 0$. By extension, if $|\boldsymbol{A}| > 0$, then $\boldsymbol{A}$ must be positive definite. But this latter statement is not true. The fact that the determinant of a matrix is positive does not necessarily imply that the matrix is positive definite.

with $\boldsymbol{L}$ and $\boldsymbol{U}$ given earlier in equations (A.32) and (A.34), and the elements of $\boldsymbol{L}$ defined by

$$m_{i1} = \frac{a_{i1}}{a_{11}} \quad (i = 2, \ldots, n)$$

and

$$m_{ik} = \frac{a_{ik}^{(k)}}{a_{kk}^{(k)}} \quad (i = k+1, \ldots, n) \tag{A.46}$$

To clarify our argument, let us write the predecessor of the matrix $\boldsymbol{U}$ explicity where the Gaussian elimination procedure has been carried out to the second level, namely

$$\boldsymbol{U}^{(2)} = \begin{bmatrix} a_{11} & a_{12} & a_{13} & \cdots & a_{1n} \\ 0 & a_{22}^{(2)} & a_{23}^{(2)} & \cdots & a_{2n}^{(2)} \\ 0 & a_{32}^{(2)} & a_{33}^{(2)} & \cdots & a_{3n}^{(2)} \\ \vdots & \vdots & \vdots & \ddots & \vdots \\ 0 & a_{n2}^{(2)} & a_{n3}^{(2)} & \cdots & a_{nn}^{(2)} \end{bmatrix} \tag{A.47}$$

where

$$a_{ij}^{(2)} = a_{ij} - m_{i1}a_{1j} \; (i, j = 2, \ldots, n)$$

or equivalently

$$a_{ij}^{(2)} = a_{ij} - \frac{a_{i1}a_{1j}}{a_{11}} \quad (i, j = 2, \ldots, n) \tag{A.48}$$

At this point, we invoke the symmetry of $\boldsymbol{A}$ $a_{ij} = a_{ji}$ to show that

$$a_{ij}^{(2)} = a_{ij} - \frac{a_{i1}a_{1j}}{a_{11}} = a_{ji}^{(2)} \tag{A.49}$$

since, using symmetry of $\boldsymbol{A}$,

$$a_{ji}^{(2)} = a_{ji} - m_{j1}a_{1i} = a_{ji} - \frac{a_{j1}a_{1i}}{a_{11}} = a_{ij} - \frac{a_{1j}a_{i1}}{a_{11}} \tag{A.50}$$

In general, for a symmetric matrix $\boldsymbol{A}$, we have that $a_{ij}^{(k)} = a_{ji}^{(k)}$. Using this fact with equation (A.46) yields

$$a_{ik}^{(k)} = m_{ik}a_{kk}^{(k)} = a_{ki}^{(k)}$$

This means that $\boldsymbol{U}$ itself can be decomposed as

$$\boldsymbol{U} = \boldsymbol{E}\boldsymbol{L}^{\mathsf{T}} \tag{A.51}$$

where

$$\boldsymbol{E} = \begin{bmatrix} a_{11} & 0 & 0 & 0 & \cdots & 0 \\ 0 & a_{22}^{(2)} & 0 & 0 & \cdots & 0 \\ 0 & 0 & a_{33}^{(3)} & 0 & \cdots & 0 \\ \vdots & \vdots & \vdots & \vdots & \ddots & \vdots \\ 0 & 0 & 0 & 0 & \cdots & a_{nn}^{(n)} \end{bmatrix} \tag{A.52}$$

Thus

$$\boldsymbol{A} = \boldsymbol{L}\boldsymbol{U} = \boldsymbol{L}\boldsymbol{E}\boldsymbol{L}^{\mathsf{T}} \tag{A.53}$$

and the quadratic form $(\boldsymbol{x}, \boldsymbol{A}\boldsymbol{x})$ can be written as

$$\begin{aligned}(\boldsymbol{x}, \boldsymbol{A}\boldsymbol{x}) &= (\boldsymbol{x}, \boldsymbol{L}\boldsymbol{E}\boldsymbol{L}^{T}\boldsymbol{x}) \\ &= (\boldsymbol{y}, \boldsymbol{E}\boldsymbol{y})\end{aligned} \tag{A.54}$$

with

$$\boldsymbol{y} = \boldsymbol{L}^{\mathsf{T}}\boldsymbol{x}, \quad \boldsymbol{y}^{\mathsf{T}} = \boldsymbol{x}^{\mathsf{T}}\boldsymbol{L} \tag{A.55}$$

Now,

$$(\boldsymbol{y}, \boldsymbol{E}\boldsymbol{y}) = \sum_{i=1}^{n} a_{ii}^{(i)} y_i^2 > 0 \tag{A.56}$$

provided that $a_{ii}^{i} > 0$ $(i = 1, \ldots, n)$—which is true by assumption, since

$$a_{ii}^{(i)} = \frac{|\boldsymbol{A}_i|}{|\boldsymbol{A}_{i-1}|}$$

and by premise $|\boldsymbol{A}_i| > 0$ $(i = 0, \ldots, n)$ with $|\boldsymbol{A}_0| = 1$. It follows from (A.54) and (A.56) that

$$(\boldsymbol{x}, \boldsymbol{A}\boldsymbol{x}) = \sum_{k=1}^{n} \frac{|\boldsymbol{A}_k|}{|\boldsymbol{A}_{k-1}|} y_k^2 > 0 \tag{A.57}$$

which implies that $\boldsymbol{A}$ must be positive definite. This completes the proof.

Note

The form of equation (A.41) can be generalized to include the off-diagonal elements of $\boldsymbol{U}$, that is,

$$a_{ij}^{(i)} = \frac{|\boldsymbol{A}_i^{(j)}|}{|\boldsymbol{A}_{i-1}|} \tag{A.58}$$

where $A_i^{(j)}$ is defined as the $i \times i$ matrix formed from the principal minor matrix A_i by replacing its ith column (i.e., the last column of matrix A_i) by the jth column of matrix A, truncated to the first i elements. For example,

$$A_3^{(5)} = \begin{bmatrix} a_{11} & a_{12} & a_{15} \\ a_{21} & a_{22} & a_{25} \\ a_{31} & a_{32} & a_{35} \end{bmatrix} \tag{A.59}$$

Note that

$$|A_i^{(j)}| = 0 \quad \text{for } j < i \tag{A.60}$$

since, for $j < i$, the matrix $A_i^{(j)}$ contains two identical columns.

Proof of Theorem 6

Identity 2 given earlier in this appendix provides that

$$\frac{C_{kl}}{X_k X_l} = \frac{C_{ij}}{X_i X_j} \tag{A.61}$$

for k, l, i, j taking any value $1, \ldots, r$.

In particular,

$$\frac{C_{kk}}{X_k} = \frac{C_{jk}}{X_j} \tag{A.62}$$

and

$$\frac{C_{jj}}{X_j} = \frac{C_{jk}}{X_k} \tag{A.63}$$

Therefore

$$C_{kk} = \left(\frac{X_k}{X_j}\right)^2 C_{jj} \tag{A.64}$$

Defining A^i as the matrix that results by deleting the ith row and column from matrix A, we can write equation (A.64) as

$$|A^{(k)}| = \left(\frac{X_k}{X_j}\right)^2 |A^{(j)}| \tag{A.65}$$

From equation (A.65), we can write that

$$C_{kk} = |A^{(k)}| = \left(\frac{X_k}{X_r}\right)^2 |A^{(r)}| = \left(\frac{X_k}{X_r}\right)^2 |A_{r-1}| \tag{A.66}$$

Since, by theorem 4 in this chapter, the limit of stability occurs when $|A_{r-1}| = 0$, then so will $|A^{(k)}| = 0$ for any $k = 1, \ldots, r$, by equations (A.65) and (A.66). The fact that the result extends to $(r-1) \times (r-1)$ matrices where *any* row and column can be deleted from A_r is shown by equation (A.62) which relates cross cofactors of kind C_{ij} to C_{jj}, which vanish at the stability limit.

Bibliography

References [1, 3, 4] are important texts that deal with thermodynamic stability theory. Reference [2] is the linear algebra source most used in this text.

[1] D. Chandler, *Introduction to Modern Statistical Mechanics*. New York: Oxford University Press, 1987.

[2] B. Noble and J. W. Daniel, *Applied Linear Algebra*, second ed. Englewood Cliffs, NJ: Prentice Hall, 1977.

[3] M. Modell and R. C. Reid, *Thermodynamics and Its Applications*, second ed. Englewood Cliffs, NJ: Prentice Hall, 1983.

[4] R. L. Rowlinson and F. R. Swinton, *Liquids and Liquid Mixtures*. London: Butterworths, 1982.

2 The Critical Point in Pure Fluids and Mixtures

We showed in chapter 1 that, if the following limit-of-stability condition holds,

$$\left(\frac{\partial^2 A}{\partial v^2}\right)_{T,N} = -\left(\frac{\partial P}{\partial v}\right)_T = 0 \tag{2.1}$$

then the presumption is that the system will be unstable with respect to certain fluctuations, and a phase transition will ensue at some point. However, we must also consider the possibility that *a limit-of-stability point could also be thermodynamically stable.* Such points indeed do exist and are called *critical points.* What are the conditions that pertain at such a point? We now analyze this situation in a pure fluid before generalizing the results.

2.1 The Critical Point: Pure Fluids

We now derive a set of conditions pertaining at the critical point of a pure fluid using a simple fluctuation analysis of an extensive property which is the usual route to the results as presented in Chandler [1]. Consider the following partitioning of a system into two independent volumetric subsystems:

$$\delta V = 0 = \delta V^{(1)} + \delta V^{(2)} \tag{2.2}$$

$$\delta N^{(1)} = \delta N^{(2)} = 0 \tag{2.3}$$

Note that this partitioning preserves the total volume of the system and that it involves only extensive variables, as must be the case. From our previous results on stability theory, we have that, for a stable system,

$$(\Delta A)_{T,V,N} > 0, \qquad (\delta A)_{T,V,N} = 0, \qquad (\delta^2 A)_{T,V,N} \geq 0 \tag{2.4}$$

where

$$(\delta^2 A)_{T,V,N} = \frac{1}{2}(\delta V^{(1)})^2 \left[\left(\frac{\partial^2 A}{\partial V^2}\right)^{(1)}_{T,N} + \left(\frac{\partial^2 A}{\partial V^2}\right)^{(2)}_{T,N} \right] \tag{2.5}$$

Given the thermodynamic identity

$$\left(\frac{\partial^2 A}{\partial V^2}\right)_{T,N} = -\left(\frac{\partial P}{\partial V}\right)_{T,N} \tag{2.6}$$

if $(\delta^2 A)_{T,V,N} \geq 0$, then it necessarily follows that

$$-\left[\left(\frac{\partial P}{\partial V}\right)^{(1)}_{T,N} + \left(\frac{\partial P}{\partial V}\right)^{(2)}_{T,N} \right] \geq 0 \tag{2.7}$$

Since the ordering of the subsystems is arbitrary, their indices can be interchanged without loss of generality, and so equation (2.7) must imply that $-(\partial P/\partial V)_{T,N} \geq 0$. At the stability limit, the equality holds:

$$-\left(\frac{\partial P}{\partial V}\right)_{T,N} = 0 \tag{2.8}$$

Higher-Order-Terms in the Free-Energy Expansion

We now examine the expansion of $A(T, V, N)$, which may be written in successively higher-order terms as

$$\begin{aligned}(\Delta A)_{T,V,N} = (\delta A)_{T,V,N} + (\delta^2 A)_{T,V,N} + (\delta^3 A)_{T,V,N} + (\delta^4 A)_{T,V,N} \\ + \text{higher-order terms}\end{aligned} \tag{2.9}$$

If $(\delta^2 A)_{T,V,N} = 0$, we are at a limit of stability, where equation (2.8) holds. What happens to the 3rd and higher-order terms of this expansion if we take $(\Delta A)_{T,V,N}$ to be strictly positive at this particular limit of stability? Thus, we are now interested in analyzing the situation where

$$0 < (\Delta A)_{T,V,N} = (\delta^3 A)_{T,V,N} + (\delta^4 A)_{T,V,N} + \cdots \tag{2.10}$$

From this equation we know that, if $(\Delta A)_{T,V,N} > 0$, then $(\delta^3 A)_{T,V,N} \geq 0$. Thus, for a fluctuation (represented by a partitioning of extensive variables), here defined by $\delta V^{(1)} = -\delta V^{(2)}$, we have that

$$0 \leq (\delta V^{(1)})^3 \left[\left(\frac{\partial^3 A}{\partial V^3} \right)_{T,N}^{(1)} - \left(\frac{(\partial^3 A)}{\partial V^3} \right)_{T,N}^{(2)} \right] \tag{2.11}$$

Since this equation must hold both for $\delta V^{(1)}$ positive and negative, the term in the square brackets above must be zero to satisfy this inequality. Hence

$$\left(\frac{\partial^3 A}{\partial V^3} \right)_{T,N}^{(1)} = \left(\frac{\partial^3 A}{\partial V^3} \right)_{T,N}^{(2)} \tag{2.12}$$

and, since $P = -(\partial A/\partial V)_{T,N}$, it follows that

$$-\left(\frac{\partial^2 P}{\partial V^2} \right)_{T,N}^{(1)} + \left(\frac{\partial^2 P}{\partial V^2} \right)_{T,N}^{(2)} = 0 \tag{2.13}$$

or

$$\left(\frac{\partial^2 P}{\partial V^2} \right)_{T,N}^{(1)} = \left(\frac{\partial^2 P}{\partial V^2} \right)_{T,N}^{(2)} \tag{2.14}$$

Defining $v \equiv V/N$ and substituting this into equation (2.14), we find

$$\left(\frac{\partial^2 P}{\partial v^2} \right)_{T}^{(1)} \left(\frac{1}{N^{(1)}} \right)^2 - \left(\frac{\partial^2 P}{\partial v^2} \right)_{T}^{(2)} \left(\frac{1}{N^{(2)}} \right)^2 = 0 \tag{2.15}$$

where $N^{(1)}$ and $N^{(2)}$ refer to the moles in subsystems 1 and 2, respectively. Since the ordering of subsystems is entirely arbitrary, we must have that

$$\left(\frac{\partial^2 P}{\partial v^2} \right)_{T} = 0 \tag{2.16}$$

EXERCISE 2.1

Describe the logic that leads to equation (2.16).

Equation (2.16) defines an important additional equation that applies at a stable limit of stability in a pure system. This additional condition defines a *critical point in the fluid.* At a regular first-order phase transition, it does not apply.

EXERCISE 2.2

What can we say about the magnitude of $(\partial^3 P/\partial v^3)_T$ for a system where both $(\partial P/\partial V)_T$ and $(\partial^2 P/\partial V^2)_T$ are zero?

Since we have that

$$0 < (\Delta A)_{T,V,N} = (\delta^2 A)_{T,V,N} + \delta^3 A + \delta^4 A + \cdots \tag{2.17}$$

and, by assumption,

$$\delta A = 0, \quad \delta^2 A = 0, \quad \delta^3 A = 0 \tag{2.18}$$

we need to look at the fourth-order quantity:

$$(\delta^4 A)_{T,V,N} = \frac{1}{2}(\delta V^{(1)})^4 \left[\left(\frac{\partial^4 A}{\partial V^4}\right)^{(1)}_{T,N} + \left(\frac{\partial^4 A}{\partial V^4}\right)^{(2)}_{T,N} \right] \geq 0 \tag{2.19}$$

This equation implies that, if $-(\partial^3 P/\partial V^3)_{T,N} \geq 0$ for arbitrary perturbations $\delta V^{(1)}$, then $(\Delta A)_{T,V,N} > 0$, and this would imply a stable critical point. Usually this analysis would suffice at this point. However, if $(\partial^3 P/\partial v^3)_T = 0$, then what can we say about $(\partial^4 P/\partial v^4)_T$? We repeat a fluctuation analysis to get at this question.

In this case, if the system is stable,

$$0 < \Delta A = \delta A + \delta^2 A + \delta^3 A + \delta^4 A + \delta^5 A + \cdots \tag{2.20}$$

and, by assumption, $\delta A = \delta^2 A = \delta^3 A = \delta^4 A = 0$. In this situation,

$$(\delta^5 A) \geq 0 \tag{2.21}$$

Thus $0 \leq (\partial V^{(1)})^5[(\partial^5 A/\partial V^5)^{(1)}_{T,N} - (\partial^5 A/\partial V^5)^{(2)}_{T,N}]$ and, because the ordering of subsystems is arbitrary, we must have that $(\partial^4 P/\partial v^4)_T = 0$. This relationship between the powers of $(\partial^n P/\partial v^n)$, and $(\partial^{n+1} P/\partial v^{n+1}]$, where n is even, can be generalized for any n. Obviously, for some odd value of n, the condition $-(\partial^n P/\partial v^n) > 0$ holds in order for the state to be strictly stable, namely $\Delta A(T, V, N) > 0$.

2.2 Generalization of the Results to Multicomponent Mixtures

In multicomponent systems with n components, a fluctuation analysis like that given previously is algebraically far more cumbersome. However, a compact representation of the critical-point conditions can be given using the Legendre transform. In an n-component system, we designate the r extensive variables used to characterize the stability limit in the energy representation as $X_1, \ldots, X_r$. Here, $r = n+1$ (by theorem 4 of chapter 1) and we designate as Φ^r_1 the energy function in terms of these variables. If we take the Legendre transform of this function, we have the function Φ^r_2, where the extensive variables are now $X_2, \ldots, X_r$ with associated Hessian $\mathbf{\Lambda}^r_2$. For example, in a binary fluid, if we choose $E(S, V, N_1)$, then $r = 3$ with $X_1 = S$, $X_2 = V$, $X_3 = N_1$, and $\Phi^r_2 = A(V, N_1)$. This process can be continued until we reach the Legendre function Φ^r_r which is just a function of X_r.

If we define the square $(r-i+1)\times(r-i+1)$ matrix $\mathbf{\Lambda}_i^r$ as

$$\mathbf{\Lambda}_i^r = \begin{bmatrix} \left(\dfrac{\partial^2\Phi_i^r}{\partial X_i^2}\right) & \cdots & \left(\dfrac{\partial^2\Phi_i^r}{\partial X_i \partial X_r}\right) \\ \vdots & & \vdots \\ \left(\dfrac{\partial^2\Phi_i^r}{\partial X_r \partial X_i}\right) & \cdots & \left(\dfrac{\partial^2\Phi_i^r}{\partial X r^2}\right) \end{bmatrix} \tag{2.22}$$

then the highest-order function Φ_r^r has the corresponding (1×1) "matrix" $\mathbf{\Lambda}_r^r$, and the stability limit in this representation is simply given by the condition

$$\det \mathbf{\Lambda}_r^r = \left(\frac{\partial^2\Phi_r^r}{\partial X_r^2}\right) = 0 \tag{2.23}$$

The other thermodynamic stability conditions are given by

$$\det \mathbf{\Lambda}_i^r = 0 \quad \text{for } i = 1, \ldots, r \tag{2.24}$$

with $\det \mathbf{\Lambda}_1^{r+1} = 0$ by virtue of theorem 4 in chapter 1.

Binary Mixtures: the Simplest Representation

In a binary system, for example, since $\Phi_r^r(X_r)$ is dependent upon only one extensive variable, it is easy to see by a simple fluctuation analysis in X_r alone, like that done in the previous section, that criticality in this representation is given by the equations

$$\left(\frac{\partial^2\Phi_r^r}{\partial X_r^2}\right) = 0 \quad \text{(limit-of-stability condition)} \tag{2.25}$$

$$\left(\frac{\partial^3\Phi_r^r}{\partial X_r^3}\right) = 0 \quad \text{(critical condition)} \tag{2.26}$$

which combined provide the simplest condition defining the critical point in a binary fluid. Note that here $\Phi_r^r(X_r)$ is the Gibbs free energy $G(T, P, N_1, N_2)$ with $X_r = N_1$, for example.

The expressions for the other critical-point conditions for multicomponent systems, or for pure fluids using thermodynamic potentials requiring more than one extensive variable, have been worked out [2]. These can be succinctly stated if we define the following $(r-i+1)\times(r-i+1)$ *critical-point condition matrix* $\mathbf{\Theta}_i^r$:

$$\mathbf{\Theta}_i^r \equiv \begin{bmatrix} \left(\dfrac{\partial^2\Phi_i^r}{\partial X_i^2}\right) & \cdots & \left(\dfrac{\partial^2\Phi_i^r}{\partial X_i \partial X_r}\right) \\ \vdots & \ddots & \vdots \\ \left(\dfrac{\partial^2\Phi_i^r}{\partial X_{r-1} \partial X_i}\right) & \cdots & \left(\dfrac{\partial^2\Phi_i^r}{\partial X_{r-1} \partial X_r}\right) \\ \left(\dfrac{\partial \left|\mathbf{\Lambda}_i^r\right|}{\partial X_i}\right) & \cdots & \left(\dfrac{\partial \left|\mathbf{\Lambda}_i^r\right|}{\partial X_r}\right) \end{bmatrix} \tag{2.27}$$

where for convenience we write $|\mathbf{\Lambda}_i^r| = \det \mathbf{\Lambda}_i^r$. Note that the first $r - i$ rows of this matrix are given by those of the corresponding rows in matrix $\mathbf{\Lambda}_i^r$. The remaining row is given by functions of various derivatives of the determinant of $\mathbf{\Lambda}_i^r$. If $i = 1$, and $r = n+1$ we have the matrix $\mathbf{\Theta}_1^{n+1}$, *the critical-point condition matrix* in the energy representation. Using the LU matrix decomposition technique described in the previous chapter 1, it can be shown that the various $|\mathbf{\Theta}_i^r|(= \det \Theta_i^r)$ are linked by the following identity:

$$|\mathbf{\Theta}_{i-1}^r| = \left(\frac{\partial^2 \Phi_{i-1}^r}{\partial X_{i-1}^2}\right)^2 |\mathbf{\Theta}_i^r| \quad (i = r, \ldots, 2) \tag{2.28}$$

EXERCISE 2.3

Prove equation (2.28).

Criticality can be simply given by the condition $|\mathbf{\Theta}_r^r| = 0$. Equation (2.28) implies that other *critical-point conditions* are given by

$$|\mathbf{\Theta}_j^r| = 0 \quad (j = 1, \ldots, r) \tag{2.29}$$

with

$$\mathbf{\Theta}_r^r = \left(\frac{\partial^2 \Phi_r^r}{\partial X_r^2}\right) \tag{2.30}$$

Criticality with the Helmholtz Energy Representation in Binary Mixtures

To illustrate these otherwise seemingly inscrutable results, we look at the critical-point conditions in a binary mixture. Let us express the critical-point conditions in terms of the Helmholtz energy. We have that $r = 3$ with, for example, $X_1 = S$, $X_2 = V$, $X_3 = N_1$, $\Phi_2^3 \equiv A(V, N_1)$, and $\mathbf{\Lambda}_2^3$ the Hessian matrix of Φ_2^3 defined as

$$\mathbf{\Lambda}_2^3 = \begin{bmatrix} \left(\dfrac{\partial^2 A}{\partial V^2}\right) & \left(\dfrac{\partial^2 A}{\partial V \partial N_1}\right) \\ \left(\dfrac{\partial^2 A}{\partial N_1 \partial V}\right) & \left(\dfrac{\partial^2 A}{\partial N_1^2}\right) \end{bmatrix} \tag{2.31}$$

In addition, the critical-point condition matrix is given by

$$\mathbf{\Theta}_2^3 \equiv \begin{bmatrix} A_{VV} & A_{VN_1} \\ \left(\dfrac{\partial |\mathbf{\Lambda}_2^3|}{\partial V}\right) & \left(\dfrac{\partial |\mathbf{\Lambda}_2^3|}{\partial N_1}\right) \end{bmatrix} \tag{2.32}$$

where the subscripts on A are meant to indicate partial differentiation with respect to that variable. At the critical point, both $|\mathbf{\Lambda}_2^3| = 0$ and $|\mathbf{\Theta}_2^3| = 0$. As mentioned before, conditions such as the singularity of the matrices defined in equations (2.31) and (2.32) are rarely used, since the Gibbs free-energy conditions are far simpler.

2.3 Beyond the Limit of Stability

The results derived thus far in this chapter concern the conditions that define a limit of thermodynamic stability as well as the additional conditions that define a critical point. Once the boundaries defining these conditions are traversed, the system responds by becoming unstable, thereafter forming new phases. We now analyze the thermodynamics of coexisting phases in order to arrive at results that define this "new" multiphase equilibrium situation.

The starting point once again is the energy equation as it applies to the entire system now composed of α phases. We know that

$$\delta E(S, V, N_1, \ldots, N_n) \geq 0 \quad \text{at a stable equilibrium state} \tag{2.33}$$

Expanding δE in the usual way, we have, for a situation with α phases at equilibrium, that

$$\begin{aligned}\delta E = \sum_{i=1}^{\alpha}\left(\frac{\partial E}{\partial S}\right)^{(i)}_{V,N_k} dS^{(i)} + \sum_{i=1}^{\alpha}\left(\frac{\partial E}{\partial V}\right)^{(i)}_{S,N_k} dV^{(i)} \\ + \sum_{i=1}^{\alpha}\sum_{j=1}^{n}\left(\frac{\partial E}{\partial N_j}\right)_{S,V,N_k} dN_j^{(i)} \geq 0\end{aligned} \tag{2.34}$$

The double summation in the last term of this expression reflects the fact that, in each phase, we find all of the components.

From fundamental thermodynamic relationships,

$$\left(\frac{\partial E}{\partial S}\right)^{(i)}_{V,N_k} = T^{(i)} \tag{2.35}$$

$$\left(\frac{\partial E}{\partial V}\right)^{(i)}_{S,N_k} = -P^{(i)} \tag{2.36}$$

$$\left(\frac{\partial E}{\partial N_j}\right)^{(i)}_{S,V,N_k} = \mu_j^{(i)} \tag{2.37}$$

Thus

$$\delta E = \sum_i T^{(i)} dS^{(i)} - \sum_i P^{(i)} dV^{(i)} + \sum_i \sum_j \mu_j^{(i)} \tag{2.38}$$

and since the total extensive properties are conserved, namely

$$\begin{gathered} S = \sum_i S^{(i)}, \quad V = \sum_i V^{(i)}, \quad N_j = \sum_i N_j^{(i)} \\ \text{with } dS,\ dV,\ dN_j = 0 \end{gathered} \tag{2.39}$$

we have that

$$\sum_i dS^{(i)}, \sum_i dV^{(i)}, \sum_i dN_j^{(i)} = 0 \tag{2.40}$$

Without loss of generality, we can write that

$$dS^{(1)} = -\sum_{i=2}^{\alpha} dS^{(i)} \tag{2.41}$$

$$dV^{(1)} = -\sum_{i=2}^{\alpha} dV^{(i)} \tag{2.42}$$

$$dN_j^{(1)} = -\sum_{i=2}^{\alpha} dN_j^{(i)} \tag{2.43}$$

Substituting these results in equation (2.34) leads to the result

$$\delta E = \sum_{i=2}^{\alpha}(T^{(1)} - T^{(i)})dS^{(i)} - \sum_{i=2}^{\alpha}(P^{(1)} - P^{(i)})dV^{(i)} + \sum_{i=2}^{\alpha}\sum_{j=1}^{n}(\mu_j^{(1)} - \mu_j^{(i)})dN_j^{(i)} \geq 0 \tag{2.44}$$

Since the remaining variations (i.e., $dS^{(i)}, dV^{(i)}$) are independent, the only way to guarantee that the inequality in (2.44) is preserved is that

$$T^{(1)} = T^{(i)} \tag{2.45}$$

$$P^{(1)} = P^{(i)} \tag{2.46}$$

$$\mu_j^{(1)} = \mu_j^{(i)} \tag{2.47}$$

for $i = 2, \ldots, \alpha$ and $j = 1, \ldots, n$. In other words the temperature, pressure, and chemical potential of each component in each phase are equal. This is a famous statement of stability in multiphase systems from which the celebrated *Gibbs phase rule* can be derived. It also implies that, at a stable point, $\delta E = 0$.

EXERCISE 2.4

We have just proved that, if a multiphase system is at a stable equilibrium point, then the temperature and pressure and the chemical potential of species j in each phase are equal. However, is the converse true, namely that the equality of these intensive variables implies stability?

EXERCISE 2.5

Derive the Gibbs phase rule for multiphase equilibrium systems.

2.4 Chapter Review

This chapter considered the consequences of requiring that *a limit of stability also be stable*. This is not necessarily the case, and its imposition leads to new thermodynamic conditions that define *critical points*. We presented results providing the critical-point conditions using any thermodynamic potential in terms of the Legendre transformations of the energy equation. We examined the case of critical behavior in both pure fluids and binary mixtures, leading to equations that are commonly used to describe the critical point in these systems. Finally we used stability results to derive the conditions that exist at phase equilibrium in multiphase multicomponent mixtures.

Bibliography

[1] D. Chandler, *Introduction to Modern Statistical Mechanics*. New York: Oxford University Press, 1987.

[2] M. Modell and R. C. Reid, *Thermodynamics and Its Applications*, second ed. Englewood Cliffs, NJ: Prentice Hall, 1983.

3 Thermodynamic Scaling Near the Critical Point

Thermodynamic scaling near the critical point is a signature of critical phenomena, and many useful applications of supercritical solvent fluids depend upon exploiting this behavior in some technologically interesting way. Near the critical point, many transport and thermodynamic properties show anomalous behavior which is usually linked to

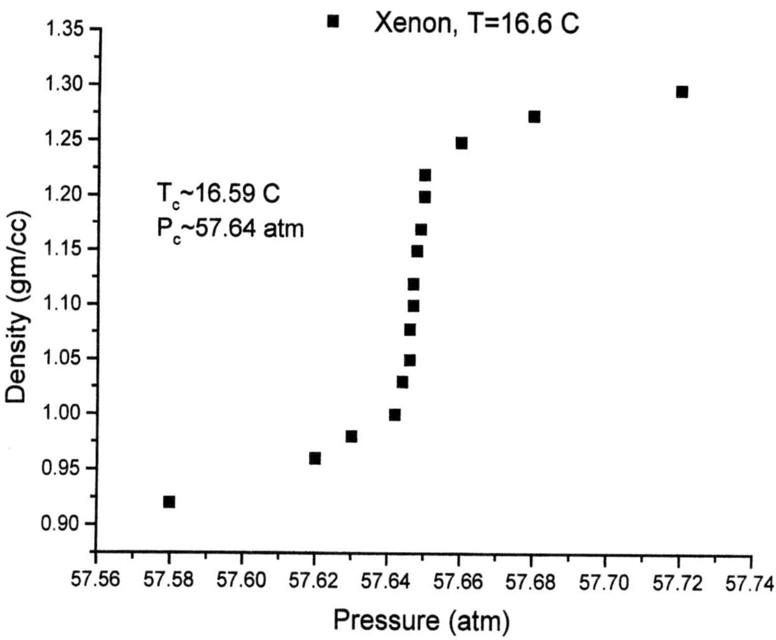

Figure 3.1 Near-critical density-pressure isotherm for xenon

the divergence of certain thermodynamic properties, such as the fluid's isothermal compressibility. In figures 3.1 and 3.2 we depict the near-critical behavior of both the density of xenon and the thermal conductivity of carbon dioxide, respectively, adapted from published data [1, 2]. The onset of what appear to be critical singularities in these properties is clearly evident in both instances. In this chapter, we focus upon the thermodynamic basis for this type of behavior.

In the theory of critical phenomena, the limiting behavior of certain thermodynamic properties near the critical point assumes special significance. In particular, properties that diverge at the critical point are of interest, and this divergence is usually described in terms of scaling laws. A typical scaling law described by Levelt-Sengers [3] takes on the form

$$|\Gamma| = A \left| \frac{\Delta q}{q_c} \right|^{\varepsilon} \tag{3.1}$$

with

$$\Delta q \equiv q - q_c$$

where Γ is the property of interest, A an amplitude factor, Δq an independent thermodynamic variable and ε an important property called a *critical exponent*. If ε is negative in this expression, then the property Γ diverges as $q \rightarrow q_c$ given that A is some finite

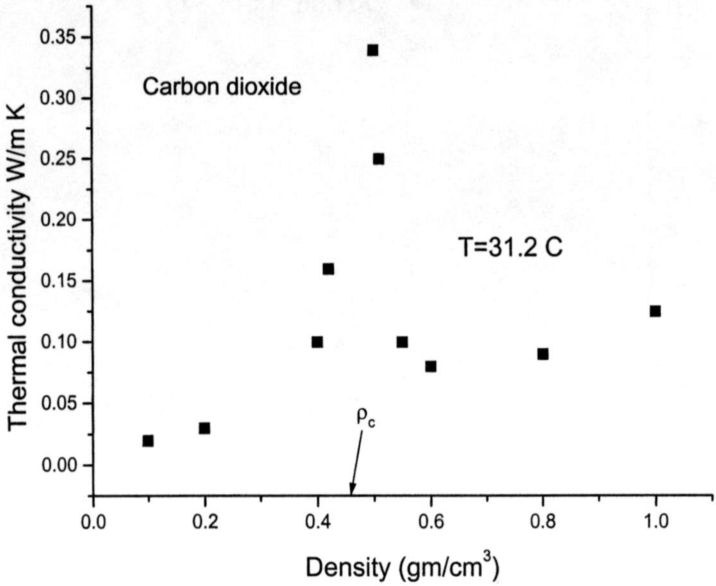

Figure 3.2 Thermal conductivity for carbon dioxide in its critical region

constant in this limit. In the case where Γ remains finite at the critical point, the scaling expression is usually modified slightly to reflect this fact:

$$\left|\frac{\Gamma - \Gamma_c}{\Gamma_c}\right| = A\left|\frac{\Delta q}{q_c}\right|^{\varepsilon} \tag{3.2}$$

It is important to realize that, in a system with more than one independent thermodynamic property, these power-law expressions are incomplete as described here. Additional thermodynamic degrees of freedom need to be specified as $\Delta q \to 0$. For example, if the power-law dependency of Γ with say temperature were being studied in a pure fluid, then one could look at an expression like equation (3.1) along either a *constant-density* or *constant-pressure* path as $T \to T_c^+$. (The notation $T \to T_c^+$ means that the temperature approaches the critical temperature from above, while $T \to T_c^-$ means that the approach is from $T < T_c$.) This is tantamount to setting a *path of approach* toward the critical point, and it may be of interest to investigate whether or not the exponent ε shows *path dependence*. We thus ask the question: does ε vary depending upon how the interrelationships among the remaining independent variables are defined as we approach the critical point?

As a simple illustration of the notion of scaling and path dependence, we look at the simple function defined by

$$f = (x^{\alpha} + y^{\beta})^2 \tag{3.3}$$

with $\alpha < \beta$. Approaching the origin along the line $y = 0$, we see that $f \to x^{2\alpha}$ as $x \to 0$, while in the direction $x = 0$, we have $f \to y^{2\beta}$ as $y \to 0$. What happens though if we approach the origin along the path $y = ax^{\gamma}$ with a some positive finite constant? In this case the scaling index for f depends upon the value of γ. For example, $f(x, ax^{\gamma}) \sim x^{2\alpha}$ if $\gamma\beta > \alpha$, but $f(x, ax^{\gamma}) \sim x^{2\gamma\beta}$ if $\gamma\beta < \alpha$.

Classical versus Nonclassical Scaling

In the theory of critical phenomena, several (somewhat idiosyncratic) terms are used to describe behavior. The *classical approach* is based upon an analytic expansion of the free energy at the critical point to develop scaling expressions for divergent thermodynamic properties. The *nonclassical* approach does not assume analyticity of the free energy at the critical point and in fact, at its core, analyzes the singular (divergent) part of the free energy using other concepts. The most common models associated with the classical approach are referred to as *mean-field* theories. These approaches explicitly ignore the role of fluctuations in determining system behavior near the critical point. Since, as we shall subsequently see, fluctuations are especially significant in the critical region, mean-field models, not surprisingly, are often inaccurate here. Notwithstanding this, the classical approach continues to be an important yardstick in the field and, given this, we now describe its use in analyzing the critical scaling behavior of fluid systems.

3.1 The Classical Equation of State, Path Dependence, and Scaling at the Critical Point

The classical model is simply one for which the system's *free energy is assumed to be analytic* at the critical point. Scaling follows from an analysis of the Taylor series expansion of the free energy around the critical point. In this section, we introduce the concept of a *critical exponent* and proceed to illustrate classical scaling ideas following the development given by Levelt-Sengers [3].

We illustrate the concept of path dependence in pure fluids by utilizing the classical expansion of the Helmholtz free energy at the critical point. A Taylor series expansion of the molar Helmholtz free energy $A(V, T)$ around the pure fluid critical point takes on the following form:

$$\begin{aligned} A(V,T) = {} & A(V_c, T_c) + A_V^c(\delta V) + A_T^c(\delta T) + A_{TT}^c(\delta T)^2/2 \\ & + A_{VT}^c(\delta V)(\delta T)/2 + A_{VVT}^c(\delta V)^2(\delta T)/6 \\ & + A_{VVVV}^c(\delta V)^4/24 + \cdots \end{aligned} \tag{3.4}$$

The subscripted indices V and T, here and elsewhere throughout this text, denote repeated partial differentiation with respect to these independent variables, and the superscript c denotes the critical value of the respective property. The variables δV, δT, and so on denote perturbations about the base expansion point, and in this case are defined by

$$\delta V \equiv V - V_c \tag{3.5}$$

$$\delta T \equiv T - T_c \tag{3.6}$$

We note that various terms in the free-energy expansion in (3.4) can, in principle, be found from an equation of state like the *van der Waals* model using standard thermodynamic identities of the form

$$A_V = -P \tag{3.7}$$

$$A_{VT} = -\left(\frac{\partial P}{\partial T}\right)_V$$

and so on. Two very important points should be made vis-a-vis the expansion shown in (3.4). The first is that the terms A^{c}_{VV} and A^{c}_{VVV} have been omitted from this equation since both of these terms are necessarily zero at the critical point, as derived in the prior chapters. Since the isothermal compressibility is given by $K_T = (VA_{VV})^{-1}$, we see that the classical expansion given above automatically satisfies the divergence in K_T mandated by thermodynamic stability requirements at the critical point. Indeed the objective of a classical analysis of this property's scaling behavior is to see how this divergence manifests itself along various thermodynamic pathways. The second point is that, since these pathways are expressed in terms of leading-order quantities in the Taylor series expansion shown in (3.4), it should not be surprising that this scaling dependence should be intimately connected to the integer exponents of the terms shown in equation (3.4). We now pursue these ideas mainly in regard to the classical scaling behavior of $K_T^{-1} = VA_{VV}$.

The Classical Scaling of the Isothermal Compressibility

The isothermal compressibility K_T defined in an earlier chapter can be related to the above free energy expansion by

$$(VK_T)^{-1} = A_{VV} \approx A^{\mathrm{c}}_{VTT}(\delta T)/3 + A^{\mathrm{c}}_{VVVV}\left(\frac{\delta V}{2}\right)^2 + \cdots \tag{3.8}$$

where the derivative A_{VV} is found from equation (3.4). We now discuss the issue concerning the choice of path in studying the classical scaling behavior of K_T.

The Isochoric Path: $V = V_{\mathrm{c}}$

In approaching the critical point, there are in principle an infinite number of paths that can be considered, for example, paths of constant density and isothermals. A constant-density path (isochoric) for example, is one for which $V = V_{\mathrm{c}}$ or $\delta V = 0$ in equation (3.8). Thus, along this path, we find from an analysis of the leading terms in (3.4) that

$$(VK_T)^{-1} \approx A^{\mathrm{c}}_{VVT}\,\delta T/3 \tag{3.9}$$

$$K_T \approx (V \cdot A^{\mathrm{c}}_{VVT})^{-1}(\delta T)^{-1}; \quad \text{for } V = V_{\mathrm{c}} \text{ (path)} \tag{3.10}$$

This expression follows the power-law template given earlier, and shows the scaling index (more commonly called the *critical exponent*) for $K_T \approx \delta T$ to be unity along this path. This critical exponent is usually designated by the symbol γ. The amplitude factor in this equation is proportional to $(V \cdot A^{c}_{VVT})^{-1}$. Divergences with exponents of 1, as is the case here, are called *strong divergences* and arise as a direct consequence of the thermodynamic stability and critical point conditions derived in prior chapters.

The Isobaric Path: $P = P_c$

Now consider an alternative *isobaric* path as we come arbitrarily close to the critical point. To define this path requires an expansion in pressure about the critical point, which can be expressed as

$$P = -A_V = -A^c_V - A^c_{VT}(\delta T) - A^c_{VVT}(\delta V)(\delta T)/3 - A^c_{VVVV}(\delta V)^3/6 + \cdots \quad (3.11)$$

Along the critical isobar, this equation reduces to

$$0 = -A^c_{VT}(\delta T) - A^c_{VVT}(\delta V)(\delta T)/3 - \frac{A^c_{VVVV}(\delta V)^3}{6} + \cdots \quad (3.12)$$

Or, in terms of δT,

$$\delta T = -\frac{A^c_{VVVV}(\delta V)^3}{6[A^c_{VT} + A^c_{VVT} \cdot \delta V/3]} \quad (3.13)$$

In the asymptotic limit as we approach the critical point,

$$\delta T = -\frac{A^c_{VVVV}(\delta V)^3}{6A^c_{VT}}; \quad P = P_c \text{ (isobaric path)} \quad (3.14)$$

This equation provides the relationship between δT and δV along the critical-pressure path. If substituted into equation (3.8) for the compressibility, it yields the scaling law for K_T along the critical isobaric path. The result is

$$(VK_T)^{-1} \approx A^c_{VVT}(\delta T)/3 + A^c_{VVVV}\frac{(\delta V)^2}{2} + \cdots \quad (3.15)$$

$$(VK_T)^{-1} \approx A^c_{VVT}(\delta T)/3 + \frac{A^c_{VVVV}}{2}\left[\frac{6A^c_{VT}}{A^c_{VVVV}}\delta T\right]^{2/3}; \quad P = P_c \quad (3.16)$$

which, in the asymptotic limit, leads to the scaling relationship

$$K_T \approx \frac{2A^{c-1/3}_{VVVV} \cdot 6^{-2/3} \cdot A^{c-2/3}_{VT}}{V_c}\delta T^{-2/3}; \quad P = P_c \quad (3.17)$$

This is a power law with exponent $-2/3$ on δT. The amplification factor is shown in parentheses in the above equation. Note that, in these two examples, the scaling exponent of K_T with temperature depends upon the path taken.

The Coexistence-Boundary Path

Another central scaling pathway concerns the path of approach toward the critical point that follows the *coexistence boundary between two phases*. This scaling law describes the way that the coexisting phase densities vary with the temperature asymptotically close to the critical point. Using a classical expansion of pressure at the critical point, it can be shown that

$$\delta V \approx \left|\frac{6A^{c}_{VVT}}{A^{c}_{VVVV}}\right|^{1/2} |\delta T|^{1/2}; \quad \text{coexisting phases} \tag{3.18}$$

where δV is either $V_{\text{gas}} - V_c$ or $V_{\text{liquid}} - V_c$. The exponent linking δV with δT in (3.18) is $\frac{1}{2}$.

Limitations with the Classical Analysis

While the classical approach is straightforward and conceptually attractive, it has limitations. The main one is that it incorrectly describes scaling exponents in real systems. A standard way of illustrating this point is to consider the classical expansion for pressure at the critical point, given by the equation

$$P = bt + 2at\eta + 4B\eta^3 + \text{higher-order terms in } \eta \tag{3.19}$$

where

$$P \equiv P - P_c, \quad t \equiv T - T_c, \quad \eta \equiv \rho - \rho_c \tag{3.20}$$

From critical-point stability conditions we know that

$$\left(\frac{\partial P}{\partial \eta}\right)_t = \left(\frac{\partial^2 P}{\partial \eta^2}\right)_t = 0 \quad (t = 0, \quad \eta = 0) \left(\frac{\partial^3 P}{\partial \eta^3}\right)_t > 0 \tag{3.21}$$

Along the coexistence envelope, it is easy to show that

$$\eta_1 = -\eta_2 = \sqrt{-at/2B} \tag{3.22}$$

where the subscripts 1 and 2 refer to the liquid and gas phases, respectively.

For $T < T_c$, we see that,

$$\frac{\rho_1 - \rho_2}{2\rho_c} = \frac{\sqrt{a(T - T_c)/2B}}{\rho_c} \tag{3.23}$$

or

$$\frac{\rho_1 - \rho_2}{2\rho_c} = A\left(1 - \frac{T}{T_c}\right)^{\frac{1}{2}}; \quad T \to T_c^- \text{ along the coexistence curve} \tag{3.24}$$

Unfortunately, density data along the coexistence curve for simple gases shows that the exponent in equation (3.24) is not valued $\frac{1}{2}$ but more of the order of 0.345 [3].

Specific-Heat Data

Another disparity between classical theory and data for real fluids involves the scaling property of C_V at the critical point. Precise data published for argon [4] showed evidence of this property's divergence, with the singularity taking the form given by

$$C_V(T) = -A^{\pm} \log \left| 1 - \frac{T}{T_c} \right| + B^{\pm} \tag{3.25}$$

The more currently accepted version of the scaling of C_V is given by the expression

$$C_V(T) \approx \frac{A^{\pm}}{\alpha}(|1 - T/T_c|^{-\alpha} - 1) + B^{\pm} \tag{3.26}$$

with $\alpha \approx 0.1$, where $A^{\pm}$ and $B^{\pm}$ are constants. From thermodynamic arguments alone, however, the classical approach predicts C_V *to be finite at the critical point*, since we have that $C_V = -T(\partial^2 A/\partial T^2)_V$. This, we now see, is a prediction at odds with experimental data.

EXERCISE 3.1

Derive a classical scaling law for the pure fluid enthalpy along the critical isochore.

3.2 The Various Critical Exponents and Their Scaling Paths

The previous discussion illustrates the concept of scaling, albeit with exponents furnished from a classical perspective. Notwithstanding the limitations seen with these results, many of the concepts introduced are relevant to a more general analysis of scaling in real systems. We now begin this more general analysis by first identifying key thermodynamic properties, their scaling exponents, and related paths of approach toward the critical point.

Exponent γ; The Critical Isochore

The exponent γ describes the divergence of the isothermal compressibility K_T along the critical isochore. The *critical isochore* in pure fluids is the direction of the co-existence curve at the critical point. We saw earlier from equation (3.10) that classically $\gamma = 1$ along this path.

The compressibility is related to a derivative of a variable by its conjugate. Quantities of this sort assume a special significance in critical scaling. They are called *susceptibilities* and can be more generally represented as second derivatives of a free energy with respect to an intensive thermodynamic variable, namely $K_T = -(1/V)(\partial V/\partial P)_{T,N} = -(1/V)(\partial^2 G/\partial P^2)_{T,N}$. In the critical-phenomena vernacular, the variable V is the system's *order parameter* with pressure P its coupling field. Both quantities feature prominently in the theory of critical phenomena.

Exponent δ; The Critical Isotherm

The exponent δ defines the asymptotic shape of the *critical isotherm* and/or *isobar*. A classical expansion for pressure around the critical point leads to the result

$$P - P_c = -\frac{A^c_{VVVV}(V - V_c)^3}{6}; \quad T = T_c \tag{3.27}$$

The exponent in equation (3.27) is designated δ and so *classically* $\delta = 3$. Note that the relationship in (3.27) above couples an extensive variable V with its conjugate coupling field P.

Exponent α; C_V along the Critical Isochore

The exponent α is used to characterize the behavior of the heat capacity along the critical isochore. The *heat capacity* is defined as

$$C_V \equiv \left(\frac{\partial E}{\partial T}\right)_{N,V} = T\left(\frac{\partial S}{\partial T}\right)_{N,V} \tag{3.28}$$

It can also be shown that

$$C_V = -T\left(\frac{\partial^2 A}{\partial T^2}\right)_V \tag{3.29}$$

We conclude from (3.29) that, classically speaking, the specific heat is finite at the critical point. If the general scaling law for C_V along this path is written as

$$C_V \propto |\delta T|^{-\alpha}; \quad V = V_c \tag{3.30}$$

this implies that, *classically*, $\alpha = 0$.

Exponent β; The Coexistence Exponent

The exponent β characterizes the *coexistence curve*. Using a classical expansion of pressure at the critical point, it can be shown that

$$\delta V \approx \left|\frac{6A^c_{VVT}}{A^c_{VVVV}}\right|^{1/2} |\delta T|^{1/2}; \quad \text{coexisting phases} \tag{3.31}$$

where δV is either $V_{gas} - V_c$ or $V_{liquid} - V_c$. The exponent in equation (3.31) is β, and thus we see that, classically, $\beta = \frac{1}{2}$. Nonclassically, the same power law applies; however, then β is different from $\frac{1}{2}$.

EXERCISE 3.2

Using the following data of Habgood and Schneider [1] for xenon at a critical temperature of 16.59°C and the scaling form given by equation (3.27), find the value of the exponent δ. Is it at variance with the classical value 3? $\rho_c = 1.099$ gm/cc, $P_c = 57.636$ atm.

ρ (gm/cc) :	1.30171	1.24892	1.19470	1.13958
P (atm) :	57.7118	57.6522	57.6380	57.6360

A comparison between classically derived exponents and their nonclassical counterparts (found from either experimental data and/or theory computer simulation) is shown in table 3.1.

We note in table 3.1 that the exponent values in nonclassical theory depend upon the dimension of the system. We describe the various systems as belonging to the *Ising critical universality class* which, broadly speaking, applies to systems with short-range interactions. These include simple fluids of the kind of interest here.

Nonclassical Approaches

Nonclassical approaches attempt to produce scaling exponents more consistent with those observed experimentally in real fluids and include the use of both computer simulation methods and advanced theories like the *renormalization group*. The *non-integer exponents* shown in table 3.1 point out the inadequacy of attempting to use an analytic expansion of the free energy around the critical point for scaling analysis. However, a simple way of attempting to force the correct scaling behavior at criticality could, for example, be achieved by proposing a free-energy expansion of the form

$$-A(V,T) = -A^c(V,T)\delta V + a_1(\delta V)^2 \delta T^\gamma + a_2(\delta V)^{\delta+1} + b_1\, \delta T\, \delta V \tag{3.32}$$

which immediately yields the result for δP:

$$\delta P = 2a_1\, \delta V(\delta T)^\gamma + a_2(\delta+1)(\delta V)^\delta + b_1\, \delta T \tag{3.33}$$

Table 3.1 Critical exponents for classical and nonclassical systems

	Classical	*Nonclassical 2-Dimensional (Ising) Universality Class*	*Nonclassical 3-Dimensional (Ising) Universality Class [3]*
γ	1	$\frac{7}{4}$	1.239
α	0	Logarithmic	0.125
β	$\frac{1}{2}$	$\frac{1}{8}$	0.326
δ	3	5	4.8
η[a]	–	$\frac{1}{4}$	0.04
ν[a]	–	1	0.63

[a] The exponents η and ν will be introduced in subsequent chapters and do not appear in classical (mean-field) theory.

For a path with $\delta T = 0$, we have that

$$\delta V \propto (\delta P)^{\frac{1}{\delta}}; \quad T = T_c \tag{3.34}$$

which is the correct scaling result if the accurate nonclassical value is used for δ. Also, along the critical isochore, we have that

$$K_T \propto \delta T^{-\gamma}; \quad V = V_c \tag{3.35}$$

which is again the correct scaling result. Note that, at the critical point, this equation also provides that

$$\left(\frac{\partial P}{\partial T}\right)_{V_c} = b_1, \quad b_1 \neq 0 \tag{3.36}$$

consistent with a finite value of the saturation pressure curve's gradient at the pure fluid critical point.

This analysis illustrates a simple way of introducing nonclassical ideas into the scaling arena and exercise 3.3 illustrates how these ideas can also be used to link the various scaling exponents using a free energy expansion of the form given by (3.32)

EXERCISE 3.3

What are the implications for the free energy if measurements of δ in (3.27) yielded a fractional scaling exponent, say $\delta = 4.8$? Show that equation (3.32) leads to the following equality for the coexistence scaling exponent: $\beta = \gamma/(\delta - 1)$.

EXERCISE 3.4

Prove equation (3.29). What does thermodynamic stability theory tell us about the value of C_V at the critical point?

EXERCISE 3.5

Prove the relationship given by equation (3.31). How would you use data to estimate a value for β?

EXERCISE 3.6

Can you think of thermodynamic results that would show that the critical exponents α, β, δ, and γ are coupled (i.e., not all independent of one another)?

3.3 Scaling in Terms of K_T, C_P, and α_P

Given the great variety of interrelationships between thermodynamic properties, we investigate the extent to which scaling for any property can be expressed in terms of a subset of a few key variables whose scaling behavior is well defined. For example, the divergence of the isothermal compressibility, whose path dependence was taken up in a previous section, is related to other thermodynamic properties. Thus a singularity in K_T at the critical point potentially generates singularities in these other properties. The three properties of fluids studied here are the constant-pressure heat capacity C_P, the isothermal compressibility K_T, and the thermal expansivity coefficient α_P. The respective definitions for C_P and α_P are

$$C_P \equiv \left(\frac{\partial H}{\partial T}\right)_{P,N} = T\left(\frac{\partial S}{\partial T}\right)_P \tag{3.37}$$

and

$$\alpha_P \equiv \frac{1}{V}\left(\frac{\partial V}{\partial T}\right)_P \tag{3.38}$$

From fundamental thermodynamic identities, we see that all these properties can be related to partial derivatives of G as follows:

$$C_P = -T\left(\frac{\partial^2 \mu}{\partial T^2}\right)_P = -TG_{TT} \tag{3.39}$$

$$K_T = -\frac{1}{V}\left(\frac{\partial^2 \mu}{\partial P^2}\right)_T = -\frac{1}{V}G_{PP} \tag{3.40}$$

$$\alpha_P = \frac{1}{V}\left(\frac{\partial^2 \mu}{\partial P \partial T}\right) = \frac{1}{V}G_{TP} \tag{3.41}$$

In addition, using the triple-product rule, we can relate all of these properties through the following identities:

$$\alpha_P = \left(\frac{\partial P}{\partial T}\right)_V K_T \tag{3.42}$$

$$\frac{C_P}{T} = V\alpha_P \Big/ \left(\frac{\partial T}{\partial P}\right)_S \tag{3.43}$$

EXERCISE 3.7

Prove (3.39)–(3.41) and use these results to develop an important identity linking all of these properties to C_V:

$$C_P = C_V + \frac{TV\alpha_P^2}{K_T} \tag{3.44}$$

or equivalently

$$C_P = C_V + T\left(\frac{\partial P}{\partial T}\right)_V^2 K_T$$

Equations (3.42)–(3.44) show how scaling properties for K_T lead directly to scaling results for C_P, α_P, and C_V—as long as we know something about the various partial derivatives in these equations. The result for C_V, from the use of equations (3.43) and (3.44), leads to the equation:

$$C_V = TV\alpha_P \left[\left(\frac{\partial P}{\partial T} \right)_S - \left(\frac{\partial P}{\partial T} \right)_V \right] \tag{3.45}$$

In order to proceed further, we need to carefully investigate the derivative terms in equation (3.45), which also arise in (3.42) and (3.43). The term $(\partial P/\partial T)_V$ at the critical point is *finite* since it is related to the slope of the fluid's vapor-pressure P–T relationship at the critical point limit. This can be seen from the *Clausius–Clapeyron* equation for the vapor-pressure curve (between phases α and β) given by

$$\left. \frac{dP}{dT} \right|_{\text{sat}} = \frac{\Delta S(T)}{\Delta V(T)} \tag{3.46}$$

where

$$\Delta S \equiv S^{\beta} - S^{\alpha} \tag{3.47}$$

and

$$\Delta V = V^{\beta} - V^{\alpha} \tag{3.48}$$

Since $dP/dT|_{\text{sat}}^{\text{c}}$ is finite at the critical point, so is $(\partial P/\partial T)_{V_{\text{c}}}$ there, because

$$\left. \frac{dP}{dT} \right|_{\text{sat}}^{\text{c}} = \left(\frac{\partial S}{\partial V} \right)_{T_{\text{c}}} = \left(\frac{\partial P}{\partial T} \right)_{V_{\text{c}}} \tag{3.49}$$

The remaining term to analyze in (3.45) is the quantity $(\partial P/\partial T)_S$, which we now look at in some detail.

Limiting Value of $(\partial P/\partial T)_S$ at the Critical Point

From standard calculus and thermodynamic identities (see the additional exercises), we can show that

$$\left(\frac{\partial P}{\partial T} \right)_V = \frac{C_P/K_T}{(\partial P/\partial T)_V VT + (C_V/\alpha_P)} \tag{3.50}$$

Since $(\partial P/\partial T)_V$ is finite at the critical point, equation (3.50) yields a matrix of possibilities for the ratios C_P/K_T and C_V/α_P there (see table 3.2).

EXERCISE 3.8

Use equation (3.50) to rationalize the results shown in table 3.2. Especially focus upon the arguments for the "yes" elements of the matrix.

Table 3.2 Matrix of possibilities for the ratios C_p/k_T and C_v/α_p at the critical point from equation (3.50)

		C_P/K_T		
		0	finite	infinite
C_V/α_P	0	No	Yes	No
	finite	No	Yes	No
	infinite	No	No	No

The results in table 3.2 show that the scaling exponents of C_P and K_T must be the same; however, the exponent for α_p must be greater than or equal to that of C_V. This is the "best" that thermodynamics can do, as subsequently discussed. We now use these results to show also that $(\partial P/\partial T)_S$ is necessarily finite at the critical point. From an earlier identity (equation (3.43)),

$$\frac{C_P}{\alpha_P} = VT\left(\frac{\partial P}{\partial T}\right)_S$$

Given that

$$\frac{C_P}{\alpha_P} = \left(\frac{C_P}{K_T}\right)\left(\frac{K_T}{\alpha_P}\right) \tag{3.51}$$

we have, using (3.42),

$$\frac{C_P}{\alpha_P} = \frac{C_P}{K_T}\left(\frac{\partial T}{\partial P}\right)_V \tag{3.52}$$

The results of table 3.2 and the finiteness of $(\partial P/\partial T)_V$ at the critical point show, from (3.52), that C_P/α_P is finite there too. This immediately proves that $(\partial P/\partial T)_S$ is finite there, from (3.43). From this we may conclude that the scaling exponents of C_P and α_P along a given path must be the same.

Table 3.2 furthermore shows that C_V/α_P can be either zero or finite at a critical point. From stability theory we have that $C_V \geq 0$ and, in the usual case where $\alpha_P \geq 0$, we then see from equation (3.45) that

$$\frac{C_V}{\alpha_P} = TV\left[\left(\frac{\partial P}{\partial T}\right)_S - \left(\frac{\partial P}{\partial T}\right)_V\right] \geq 0 \tag{3.53}$$

which implies that

$$\left(\frac{\partial P}{\partial T}\right)_S \geq \left(\frac{\partial P}{\partial T}\right)_V \tag{3.54}$$

If the inequality in (3.54) holds at the critical point, then C_V/α_P will be finite there, implying that the scaling exponents of C_V, α_P, and K_T will be identical along a given path. If, however, the equality in (3.54) holds at the critical point, then it implies that

Table 3.3 Some limiting values of thermodynamic properties at the critical point

Property	*Critical Point Value*	*Result Origin*
K_T	Infinite	Stability theory
$(\partial P/\partial T)_V$	Finite	Equation (3.49)
$(\partial P/\partial T)_S$	Finite	Equations (3.43), (3.51), (3.52)
C_P/α_P	Finite	Equation (3.52)
C_P/K_T	Finite	Equation (3.50)
C_V/α_P	Finite (or zero)	Equation (3.50)
C_V/K_T	Finite (or zero)	Equations (3.42), (3.45)
$C_V/C_P \equiv K_s/K_T$	Finite (or zero)	Equation (3.44)

C_V/α_P is zero there, so that the scaling exponent of C_V must be smaller than that of α_P and, by extension, K_T. In other words, in general, we have that

$$\alpha \leq \gamma \tag{3.55}$$

This derivation is quite general and does not depend upon classical arguments. Thus we see that, to this point, thermodynamics is not definitive on the value of the scaling exponents for K_T and C_V along the critical isochore; it merely establishes an inequality relationship between these exponents.

We summarize these results, involving limiting values of various properties at the critical point, in table 3.3.

The Use of Legendre Transforms of the Free Energy $G(T, P)$

The fact that we have seen that K_T, C_P, and α_P are related to various second-order partial derivatives of the free energy $\mu(T, P)$, equal to $G(T, P)/N$ in the case of pure fluids, is a useful result. It allows us to develop a mathematical formalism wherein all thermodynamic derivative properties can be related to these quantities, and makes possible a unified scaling analysis at the critical point. In this way, any thermodynamic property can be represented in terms of a small subset of other properties whose scaling behavior is well understood.

Given a function $\Phi(X_1, \ldots, X_N)$ and its Legendre transform $\Phi^{(1)}(I_1, X_2, \ldots, X_N)$, we prove that the following mathematical relationships hold between Φ, $\Phi^{(1)}$, and their derivatives:

$$\Phi^{(1)} = \Phi - I_1 X_1 \tag{3.56}$$

with

$$\left(\frac{\partial \Phi^{(1)}}{\partial X_i}\right)_{I_1, X_k} = \left(\frac{\partial \Phi}{\partial X_i}\right)_{I_1, X_k} - \left[\frac{\partial}{\partial X_i}(I_1 X_1)\right]_{I_1, X_k} \tag{3.57}$$

where the nomenclature $(\partial \Phi/\partial X_i)_{X_k}$ means the derivative with respect to X_i with all other Xs held constant. Now,

$$\left(\frac{\partial \Phi}{\partial X_i}\right)_{I_1, X_k} = \left(\frac{\partial \Phi}{\partial X_i}\right)_{X_k} + \left(\frac{\partial \Phi}{\partial X_1}\right)_{X_k} \left(\frac{\partial X_1}{\partial X_i}\right)_{I_1, X_k} \tag{3.58}$$

From equations (3.57) and (3.58) we see that, since

$$I_1 = \left(\frac{\partial \Phi}{\partial X_1}\right)_{X_k}$$

$$\left(\frac{\partial \Phi^{(1)}}{\partial X_i}\right)_{I_1, X_k} = \left(\frac{\partial \Phi}{\partial X_i}\right)_{X_k} \tag{3.59}$$

Using this result for example, we see that

$$\left(\frac{\partial^2 \Phi^{(1)}}{\partial I_1 \partial X_i}\right) = \left(\frac{\partial^2 \Phi}{\partial X_1 \partial X_i}\right)\left(\frac{\partial X_1}{\partial I_1}\right)_{X_k} \tag{3.60}$$

This latter result is an important one since it establishes some identities between off-diagonal second-order partial derivatives of Φ and those of $\Phi^{(1)}$, its immediate Legendre transform.

In order to complete this analysis, however, we require results for other links between all the other types of second-order partial derivatives of Φ and $\Phi^{(1)}$.

Derivatives with the Structure $(\partial^2 \Phi^{(1)}/\partial I_1^2)_{X_k}$

We have that

$$\left(\frac{\partial \Phi^{(1)}}{\partial I_1}\right)_{X_k} = \left(\frac{\partial \Phi(X_1, \ldots, X_N)}{\partial I_1}\right)_{X_k} - X_1 - I_1 \left(\frac{\partial X_1}{\partial I_1}\right)_{X_k} \tag{3.61}$$

from which follows (see the additional exercises)

$$\left(\frac{\partial^2 \Phi^{(1)}}{\partial I_1^2}\right)_{X_k} = \frac{-1}{(\partial^2 \Phi/\partial X_1^2)_{X_k}} \tag{3.62}$$

Transformations for Derivatives with the Structure $(\partial^2 \Phi^{(1)}/\partial X_i \partial X_j)_{I_1, X_k}$

From equations (1.59)–(1.61) in the first chapter, we have that

$$\left(\frac{\partial^2 \Phi^{(1)}}{\partial X_i\, \partial X_j}\right)_{I_1, X_k} = \left(\frac{\partial^2 \Phi}{\partial X_i \partial X_j}\right)_{X_k} - \left(\frac{\partial^2 \Phi}{\partial X_1 \partial X_i}\right)_{X_k}\left(\frac{\partial^2 \Phi}{\partial X_1 \partial X_j}\right)_{X_k} \Bigg/ \left(\frac{\partial^2 \Phi}{\partial X_1^2}\right)_{X_k} \tag{3.63}$$

The identities given in equations (3.60), (3.62), and (3.63) provide all the relationships between second-order partial derivatives of Φ and $\Phi^{(1)}$. (Satisfy yourself on this point.) We point out that, in this particular analysis, the variables I_i and X_i are simply thermodynamic variables involved in the Legendre exchange process. They should not be confused with the earlier use of these symbols in chapter 1 where they were used to designate intensive and extensive thermodynamic variables, respectively.

Using the Identities Starting with the Thermodynamic Potential $G(T, P, N)$

The real significance of equations (3.60), (3.62), and (3.63) is seen if the base function Φ is taken to be the free energy G, namely $\Phi(T, P, N) = G(T, P, N)$. Its immediate Legendre transform is $\Phi^{(1)}(-S, P, N)$, and $d\Phi^{(1)} = TdS + VdP + \mu dN$. In this case, we recognize that $d\Phi^{(1)} = dH(S, P, N)$, where H is the system enthalpy. We see that here $I_1 = -S$ and $X_1 = T$. From equation (3.60), for example, we could look at the property $(\partial T/\partial P)_{S,N}$ by way of example. We have that

$$\left(\frac{\partial T}{\partial P}\right)_{S,N} = \left(\frac{\partial^2 \Phi^{(1)}}{\partial S\, \partial P}\right) \tag{3.64}$$

and, from (3.60),

$$\left(\frac{\partial T}{\partial P}\right)_{S,N} = \left(\frac{\partial^2 \Phi}{\partial T \partial P}\right) \Big/ \left(\frac{\partial^2 \Phi}{\partial T^2}\right) \tag{3.65}$$

or

$$\left(\frac{\partial T}{\partial P}\right)_{S,N} = -G_{TP}/G_{TT} \tag{3.66}$$

From equations (3.39) and (3.41),

$$\left(\frac{\partial T}{\partial P}\right)_{S,N} = \frac{TV\alpha_P}{C_P} \tag{3.67}$$

which is a result consistent with that given in (3.43). Similarly, we see that

$$\left(\frac{\partial V}{\partial S}\right)_{P,N} = \left(\frac{\partial T}{\partial P}\right)_{S,N} = \frac{TV\alpha_P}{C_P} \tag{3.68}$$

From this simple example, we see that seemingly obscure properties like $(\partial T/\partial P)_{S,N}$ and $(\partial V/\partial S)_{P,N}$ have been reduced to expressions in terms of fundamental properties α_P, K_T, and C_P which in themselves have well-defined scaling relationships.

In general one might be interested in the scaling properties of partial-derivative properties defined in terms of any of the fundamental thermodynamic variables S, T, P, V, μ, and N, and these can be found using the approach just described, which we now summarize:

- Starting out with $\Phi \equiv G(T, P, N)$, systematically proceed through successive Legendre transformations until the derivative property of interest is naturally expressed in terms of the given Legendre representation (thermodynamic potential).
- Using the identities given by equations (3.60), (3.62), and (3.63) for any two successive steps between a base function and its immediate Legendre transform, chain through these successive transformations until the link with G_{TT}, G_{TP}, G_{PP} and thereby α_P, K_T, C_P is completed. This approach makes a complete scaling analysis possible to carry out systematically.

EXERCISE 3.9

Develop an expression for property $(\partial\mu/\partial P)_{S,N}$ in terms of α_P, K_T, and C_P, and show that

$$\left(\frac{\partial\mu}{\partial P}\right)_{S,N} = G_{PN} - \frac{G_{TP}G_{TN}}{G_{TT}} \tag{3.69}$$

Does $(\partial\mu/\partial P)_{S,N}$ remain finite at the critical point?

EXERCISE 3.10

Scaling of partial derivatives of entropy

It is often useful in analysis to have derivative properties of the entropy easily available. Using the approach just described, it is possible to develop all the requisite relationships between these and our basis quantities α_P, K_T, and C_P. These relationships are given by

$$\left(\frac{\partial S}{\partial V}\right)_P = \frac{C_P}{\alpha_P V T} \tag{3.70}$$

$$\left(\frac{\partial S}{\partial P}\right)_V = \frac{C_P K_T}{T\alpha_P} - V\alpha_P \tag{3.71}$$

$$\left(\frac{\partial S}{\partial T}\right)_P = C_P/T \tag{3.72}$$

$$\left(\frac{\partial S}{\partial P}\right)_T = -\alpha_P V \tag{3.73}$$

$$\left(\frac{\partial S}{\partial V}\right)_T = \frac{\alpha_P}{K_T} \tag{3.74}$$

$$\left(\frac{\partial S}{\partial T}\right)_V \equiv \frac{C_V}{T} = \frac{C_P}{T} - \frac{V\alpha_P^2}{K_T} \tag{3.75}$$

Prove the relationships given in equations (3.70)–(3.75).

3.4 The Griffiths–Wheeler Classification

Pure Fluids

In looking at the results shown in equations (3.70)–(3.75), a distinctive structure emerges as to which properties diverge and which remain finite at the critical point. A classification of these was provided by Griffiths and Wheeler [3, 5] who termed variables that are

Table 3.4 Griffiths–Wheeler classification of critical scaling in pure fluids, where the derivative structure is $(\partial A/\partial B)_C$ [3]

A	*B*	*C*	*Scaling Property*	*Example*
Density	Field	Field	Strong divergence	$\left(\frac{\partial V}{\partial P}\right)_T$
Density	Field	Density	Weak divergence	$\left(\frac{\partial S}{\partial T}\right)_V = \frac{C_V}{T}$
Density	Density	Field	Finite	$\left(\frac{\partial S}{\partial V}\right)_P = \frac{C_P}{\alpha_P V T}$
Density	Density	Density	Finite	
Field	Field	Field	Finite	$\left(\frac{\partial P}{\partial T}\right)_\mu = S$
Field	Field	Density	Finite	$\left(\frac{\partial P}{\partial T}\right)_V$
Field	Density	Field	Zero	$\left(\frac{\partial P}{\partial V}\right)_T$
Field	Density	Density	Zero	$\left(\frac{\partial T}{\partial S}\right)_V = T/C_V$

identical in each phase as *fields* while those not equal in each phase are called *densities*. For example, temperature is a field while density is a density variable. According to the Griffiths–Wheeler scheme, a derivative of a density with respect to a field (with the remaining field constant) is a strongly divergent property. A *strong divergence* is one with critical exponent greater or equal to one. A property consisting of a derivative of a density with respect to a field with another density constant shows a *weak divergence*. An example of a weakly divergent property is C_V. If we have a derivative of a density with respect to a density at constant field, or the derivative of a field with respect to a field at constant density, then the resultant property is generally finite and well-behaved at the critical point. Table 3.4 shows the Griffiths–Wheeler classification structure for a *pure fluid's critical scaling.* We point out that these are nonclassical results not dependent upon the assumption of an analytic free energy at the critical point.

3.5 The Direction of Approach to the Critical Point: The Coexistence Curve as an Eigenvector of Φ_G

From what we have seen in this chapter, the direction in field space which assumes special importance when discussing phase transitions and critical phenomena is a direction *parallel to the coexistence curve*. This direction in P–T space is given by the derivative $(\partial P/\partial T)_{v_c}$, as shown earlier in equation (3.49). We reiterate that the importance of

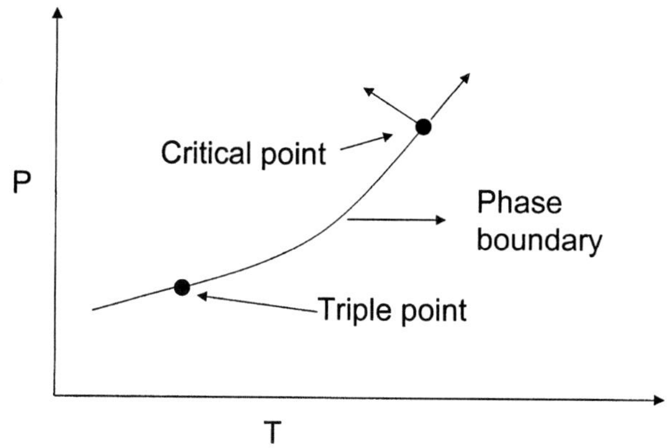

Figure 3.3 Phase boundary for a typical fluid

this direction lies in the fact that important critical exponents like α and γ are defined along this path of approach to the critical point. (There are situations, however, when the thermodynamic field variables of interest may, for some reason or another, not be P and T. In V–T space, this direction would be a vector at the critical point that is parallel to the T axis. Why?).

In $P-T$ space, the phase coexistence direction has a special relationship to Φ_G which we now discuss. In the P–T representation, stability considerations require that the free energy G be a concave function of the variables P and T. This can be expressed mathematically by the inequality

$$\boldsymbol{x}^{\mathsf{T}}\boldsymbol{\Phi}_G\boldsymbol{x} > 0 \tag{3.76}$$

where $\boldsymbol{x} \equiv \begin{bmatrix} T \\ P \end{bmatrix}$ and $\boldsymbol{\Phi}_G$ is the matrix of second derivatives of G defined by

$$\boldsymbol{\Phi}_G \equiv \begin{bmatrix} -G_{TT} & -G_{TP} \\ -G_{TP} & -G_{PP} \end{bmatrix} \tag{3.77}$$

or, using equations (3.39)–(3.41),

$$\boldsymbol{\Phi}_G \equiv \begin{bmatrix} \frac{C_P}{T} & -V\alpha_P \\ -V\alpha_P & VK_T \end{bmatrix} \tag{3.78}$$

and

$$\det \boldsymbol{\Phi}_G = \frac{VK_T C_V}{T} = \frac{VK_S C_P}{T} \tag{3.79}$$

The coexistence curve in P–T space is shown in figure 3.3.

We now show that the direction singled out by the slope of the coexistence curve at the critical point is especially significant to the matrix $\boldsymbol{\Phi}_G$ at the critical point. We look at the situation where the P–T coordinate system is rotated and examine the properties

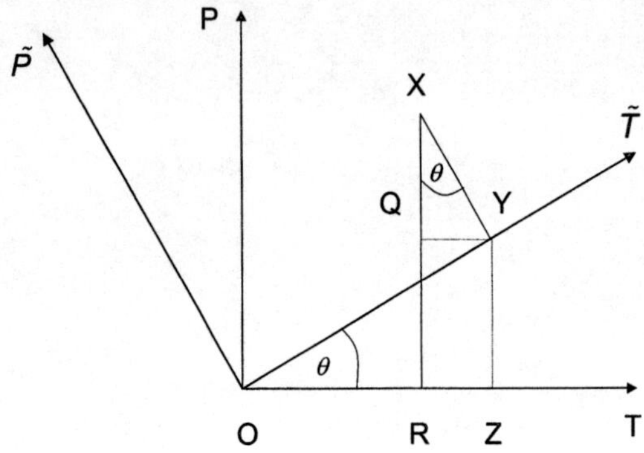

Figure 3.4 Coordinate system for the free-energy transformation

of the system in terms of this new coordinate system. A diagram of the rotation is shown in figure 3.4. The new coordinate axes, which we designate as $\tilde{\mathbf{x}} \equiv \begin{bmatrix} \tilde{T} \\ \tilde{P} \end{bmatrix}$, and the original ones are related, from simple geometrical arguments [6], by

$$T = \tilde{T}\cos\theta - \tilde{P}\sin\theta \tag{3.80}$$

$$P = \tilde{T}\sin\theta + \tilde{P}\cos\theta \tag{3.81}$$

In addition, the inner product defined by equation (3.76) becomes

$$\alpha\tilde{T}^2 + 2\kappa\tilde{T}\tilde{P} + \beta\tilde{P}^2 = c, \tag{3.82}$$

where [6]

$$\alpha = a\cos^2\theta + 2k\sin\theta\cos\theta + b\sin^2\theta \tag{3.83}$$

$$\kappa = (b-a)\sin\theta\cos\theta + k(\cos^2\theta - \sin^2\theta) \tag{3.84}$$

$$\beta = a\sin^2\theta - 2k\sin\theta\cos\theta + b\cos^2\theta \tag{3.85}$$

and we make the following associations:

$$a \equiv C_P/T \tag{3.86}$$

$$b \equiv VK_T \tag{3.87}$$

$$k \equiv -V\alpha_P \tag{3.88}$$

Using a matrix representation these equations can be expressed as,

$$\boldsymbol{x}^{\mathsf{T}}\boldsymbol{\Phi}_G\boldsymbol{x} = c \tag{3.89}$$

$$\boldsymbol{x} = \boldsymbol{S}\tilde{\boldsymbol{x}} \tag{3.90}$$

where **S**, the rotation matrix, is defined as

$$S = \begin{bmatrix} \cos\theta & -\sin\theta \\ \sin\theta & \cos\theta \end{bmatrix} \tag{3.91}$$

In the new coordinate system, the inner product (3.76) becomes

$$\tilde{x}^{\mathsf{T}}(S^{\mathsf{T}}\Phi_G S)\tilde{x} \tag{3.92}$$

Rotation matrices of the type shown in (3.91) have the property that

$$S^{\mathsf{T}} = S^{-1} \tag{3.93}$$

in which case (3.92) becomes,

$$\tilde{x}^{\mathsf{T}}(S^{-1}\Phi_G S)\tilde{x} = x^{\mathsf{T}}\Phi_G x \tag{3.94}$$

Equation (3.94) defines an important class of transformations called *similarity transformations* between the two coordinate systems x and $\tilde{x}$ [6]. The matrices $S^{-1}\Phi_G S$ and Φ_G are said to be *similar*. Of particular importance is the case where the rotation of coordinates is such that the new coordinate directions are parallel to the *eigenvectors* of the matrix Φ_G at the critical point. In this case, the matrix $S^{-1}\Phi_G S$ is diagonal, and the column vectors of S are just the eigenvectors of Φ_G.

This diagonal matrix $S^{-1}\Phi_G S$ represents the free energy of the system in terms of the new coordinate system. This observation is extremely useful in the theory of critical phenomena, as we will see in subsequent chapters dealing with finite-size scaling and the renormalization-group method. This diagonalization of Φ_G is equivalent to κ vanishing in equation (3.82), and for this to occur, the following condition must hold:

$$\tan 2\theta = \frac{2k}{a-b} \tag{3.95}$$

This equation defines the rotation such that the new coordinate directions are parallel to the eigenvectors of Φ_G. In terms of the elements of Φ_G, equation (3.95) becomes

$$\tan 2\theta = \frac{-2V\alpha_P}{\frac{C_P}{T} - VK_T} \tag{3.96}$$

which can be simplified to yield

$$\tan 2\theta = \frac{-2\left(\frac{\partial P}{\partial T}\right)_V}{\left(\frac{\partial P}{\partial T}\right)_V \left(\frac{\partial P}{\partial T}\right)_S - 1} \tag{3.97}$$

Since a trignometric identity provides that

$$\tan 2\theta = \frac{2\tan\theta}{1-\tan^2\theta} \tag{3.98}$$

we have that (3.97) becomes

$$\frac{2\tan\theta}{1-\tan^2\theta} = \frac{2\left(\frac{\partial P}{\partial T}\right)_V}{1-\left(\frac{\partial P}{\partial T}\right)_V\left(\frac{\partial P}{\partial T}\right)_S} \tag{3.99}$$

In other words, (3.99) shows that, if $(\frac{\partial P}{\partial T})_S = (\frac{\partial P}{\partial T})_V$ at the critical point (see (3.54)), then at that condition,

$$\tan\theta = \left(\frac{\partial P}{\partial T}\right)_{V_c} \tag{3.100}$$

We can show that if $\gamma > \alpha$ this result holds true (see the additional exercises), and it implies that, to diagonalize $\boldsymbol{\Phi}_G$, the original coordinates must be rotated through this angle θ at the critical point. The angular orientation of the new coordinate system q is such that the axis direction (given by $\tan\theta$) is precisely the direction of the coexistence curve at the critical point, given by $(\partial P/\partial T)_{V_c}$. These transformations play a powerful role in scaling theory, and clearly point out the importance of the coexistence direction in the theory of phase transitions.

3.6 Scaling Results from the Stable Limit of Stability Conditions

We now look for other results that might pertain at the critical point using the stable limit of stability conditions developed in chapter 2. From the Helmholtz energy stability and criticality conditions, we have that

$$-\left(\frac{\partial P}{\partial V}\right)_{T,N} = 0 \tag{3.101}$$

and

$$-\left(\frac{\partial^2 P}{\partial V^2}\right)_{T,N} = 0 \tag{3.102}$$

However, what if we choose to look at these equations in terms of $E(S, V, N)$? The limit of the stability conditions is given by the equation

$$|\boldsymbol{L}| \equiv \det \boldsymbol{L} = 0 \tag{3.103}$$

where

$$\boldsymbol{L} \equiv \begin{bmatrix} T/C_V & -\left(\frac{\partial P}{\partial S}\right)_{V,N} \\ -\left(\frac{\partial P}{\partial S}\right)_{V,N} & -\left(\frac{\partial P}{\partial V}\right)_{S,N} \end{bmatrix} \tag{3.104}$$

and the stable limit of the stability conditions is:

$$\det \boldsymbol{M} = 0, \tag{3.105}$$

where

$$\boldsymbol{M} \equiv \begin{bmatrix} T/C_V & -\left(\frac{\partial P}{\partial S}\right)_{V,N} \\ \left(\frac{\partial |L|}{\partial S}\right)_{V,N} & \left(\frac{\partial |L|}{\partial V}\right)_{S,N} \end{bmatrix} \tag{3.106}$$

From (3.104) we see that $\boldsymbol{L} = 0$ implies

$$\left(\frac{\partial P}{\partial S}\right)^2_{V,N} = -\frac{T}{C_V}\left(\frac{\partial P}{\partial V}\right)_{S,N} \tag{3.107}$$

We know from the results of chapter 1 that both C_V and $-(\partial P/\partial V)_{S,N}$ are positive in a stable phase. We also know that both C_V and $-(\partial V/\partial P)_{S,N}$ diverge at the critical point, from the results of the previous section; however, this is not evident from equation (3.107). If we look at the results for $|M| = 0$ we again find that they do not shed light on the divergence or finiteness of C_V and $-(\partial P/\partial V)_{S,N} \equiv 1/K_S V$ at the critical point. The real significance of the results developed in the previous sections is thus clear. They provide information about these divergences and the relative magnitudes of the scaling exponents.

3.7 Chapter Review

In this chapter, we addressed the important topic of scaling at the critical point from a macroscopic thermodynamic perspective. The scaling regime is the thermodynamic region in which certain properties become large, ultimately resulting in their divergence to ∞ at the critical point itself. This divergence is normally expressed in terms of scaling-law expressions containing exponents that characterize the strength of the divergence. The classical approach for investigating these scaling laws involves an analytic expansion of the free energy at the critical point from which various critical exponents can be found. The magnitudes of these classical exponents are at variance with experimental measurements that have been made, suggesting that the free energy at the critical point has a singular component which must be considered. Recognition of this result has lead to nonclassical theories of critical phenomena embodied by the renormalization-group method and various scaling theories which we consider in subsequent chapters.

The fundamental divergent thermodynamic properties in fluids are K_T, α_P, C_P, and C_V. Critical exponents were defined, with respect to scaling laws for some of these properties, and we saw that the values of these exponents depend upon the path of approach to the critical point. A general thermodynamic analysis for critical scaling in a pure fluid was developed in which it was shown how thermodynamic properties can be related to a set of three fundamental properties K_T, α_P, and C_P, whose scaling properties are well known. This approach allows a systematic analysis of scaling to be carried out for any property of interest in the fluid's critical region.

The path of approach parallel to the coexistence curve was singled out for special study, since the exponents γ and α are defined along this path for the properties K_T and C_V, respectively. We showed that the axis direction pointed out by the coexistence curve and its extension at the critical point is an eigenvector of the free-energy Hessian matrix of the thermodynamic potential $G(T, P)$ there. The other eigenvector is perpendicular to this direction; hence any path of approach can be expressed as a linear combination of these two directions (eigenvectors). These eigenvector transformations of the thermodynamic coordinates diagonalize the free-energy Hessian matrix, a result that will be useful in more advanced discussions of critical phenomena, for example, in an analysis of finite-size scaling effects in the critical region which we will consider in a subsequent chapter.

3.8 Additional Exercises

Scaling in Pure Fluids

1. The adiabatic compressibility K_S is defined as

$$K_S \equiv -\frac{1}{V}\left(\frac{\partial V}{\partial P}\right)_{S,N}$$

Is K_S divergent at the critical point?

2. On what basis do we assert that $C_V \to \infty$ at the critical point of a pure fluid? [Note: stability conditions from chapter 1 do not necessarily lead to this result.]
3. Prove the identity $K_T/K_S = C_P/C_V$ and discuss the limiting value of these ratios at the critical point.
4. Does $(\partial P/\partial T)_V$ have to be positive in all fluids? Reason your answer based upon proving the identity

$$\left(\frac{\partial P}{\partial T}\right)_V = \frac{C_P/K_T}{(\frac{\partial P}{\partial T})_V VT + (\frac{C_V}{\alpha_P})}$$

5. What does the Griffiths–Wheeler theory predict about the value of quantities like $(\partial P/\partial T)_V$ at the critical point? How could we most conveniently measure this property at the critical point?
6. Prove the following identity:

$$\left(\frac{\partial P}{\partial T}\right)_S = \left(\frac{\partial P}{\partial T}\right)_V + \frac{C_V}{K_T TV}\left(\frac{\partial T}{\partial P}\right)_V$$

What does this equation allow us to conclude about the value of the term $(\partial P/\partial T)_S$ at the critical point?

7. Prove the identities

$$\frac{C_P}{T} = V\alpha_P/(\partial T/\partial P)_S$$

$$\frac{C_V}{\alpha_P} = TV\left[\left(\frac{\partial P}{\partial T}\right)_S - \left(\frac{\partial P}{\partial T}\right)_V\right]$$

If $\alpha_P < 0$, what does this tell us about the property $(\partial P/\partial T)_S$?

8. From the thermodynamic identity

$$C_P = C_V + TV\alpha_P \left(\frac{\partial P}{\partial T}\right)_V$$

can you establish a relationship between the critical exponents α, β, and γ? [Hint: Use the scaling relationships for C_P, C_V, and α_P in this identity].

9. Is the coexistence curve for a real fluid symmetric along the critical isochore axis extended into the two-phase region? If not, what is the significance of this result? Why is the critical isochore singled out for special consideration here? If we were representing the phase diagram in terms of μ and T, what would be the equivalent direction to the critical isochore?

Properties of the Matrix Φ_G, the Hessian of the Potential $G(T, P)$

10. If a matrix $\boldsymbol{A}$ is positive definite and symmetric, we have shown (in chapter 1) that all of its eigenvalues are positive, with

$$\boldsymbol{P}^{-1}\boldsymbol{A}\boldsymbol{P} = \boldsymbol{\Lambda}$$

where $\boldsymbol{P}$ is the matrix of eigenvectors of $\boldsymbol{A}$, and $\boldsymbol{\Lambda}$ is a diagonal matrix with the eigenvalues of $\mathbf{A}$ on its diagonal. Show that these *unit eigenvectors* are orthogonal to each other, and that det $\boldsymbol{P} = 1$.

11. Prove that the determinant of $\boldsymbol{\Phi}_G$ is equal to $VK_T C_V/T = VK_S C_P/T$, and show that $\boldsymbol{\Phi}_G$ is a positive definite matrix.

12. For the (2×2) symmetric matrix

$$\mathbf{A} = \begin{bmatrix} a & k \\ k & b \end{bmatrix}$$

Show that its eigenvalues (two) are given by

$$\lambda_1 = \frac{1}{2}(a+b) + \frac{1}{2}[(a-b)^2 + 4k^2]^{1/2}$$

$$\lambda_2 = \frac{1}{2}(a+b) - \frac{1}{2}[(a-b)^2 + 4k^2]^{1/2}$$

Sketch the general shape of contours defined by the inner-product function $\boldsymbol{y}^{\mathrm{T}}\boldsymbol{A}\boldsymbol{y}$ around the origin. Discuss what happens to the shape of these functions after $\boldsymbol{A}$ is diagonalized by an eigenvector transformation. Use these results to develop expressions for the eigenvalues of $\boldsymbol{\Phi}_G$. Why do we call $\boldsymbol{\Phi}_G$ a stability matrix when it is defined in terms of *intensive* variables, while all of the stability results presented in chapter 1 are in terms of the *extensive* variables of thermodynamic potentials?

13. The matrix $\boldsymbol{\Phi}_G$ is the matrix of second derivatives of $G(T, P)$. Using the Legendre transform identities proved in this chapter transform $G(T, P) \to \bar{G}(T, \mu)$. In this case, $\bar{G}$ is the free energy with respect to scaling fields T and μ. Find the elements of $\boldsymbol{\Phi}_{\bar{G}}$ (i.e., the Hessian matrix of second derivatives of $\bar{G}(T, \mu)$.

14. What are the eigenvalues of the matrix $\boldsymbol{\Phi}_{\bar{G}}$ of the previous exercise? Prove that the coexistence direction at the critical point is an eigenvector of $\boldsymbol{\Phi}_{\bar{G}}$. In this case, moving along the coexistence direction involves a *linear combination* of μ and T—unlike many systems (like the Ising magnet) where the coexistence direction is perpendicular to the field variable H. What implications for properties such as the specific heat and susceptibility do you think this has for defining scaling relationships in fluids along this direction?
15. Find an expression for the derivative quantity

$$\left(\frac{d^2\mu}{dT^2}\right)$$

along a straight line of slope (dP/dT) in the P–T plane. If the quantities K_T and C_V are both singular at the critical point, how does this affect the property $(d^2\mu/dT^2)$ along the coexistence direction as well as a direction orthogonal to it?
16. Prove the identity given by equation (3.62) relating derivatives of the function $\Phi(X_1, X_2, \ldots, X_N)$ and its Legendre transform $\Phi^{(1)}(I_1, X_2, \ldots, X_N)$, where

$$\Phi^{(1)} \equiv \Phi - I_1 X_1$$

Experimental Data and Scaling

17. An experimentalist is trying to define the value of the exponent x in the following scaling equation of state which is to be applied in the critical region:

$$p = p_c + at + btv + cv^x$$

The most accurate apparatus available to the researcher allows for temperature and pressure measurements. Conceive of how the experiments should proceed, and find x from the following data obtained for the compressibility K_T:

K_T (bar^{-1})	2.04	3.55	4.08	6.70
t (°C)	0.075	1.010	0.006	0.001

where $t = T - T_c$. Would you expect the exponent x to be a universal one? If not, explain.
18. In their study of phase transitions of ^{4}He in a silica aeorogel, Wong and Chan [7] reported the following data for the coexistence boundary of the *gel-confined* helium:

T (K)	ρ_l (gm/cc)	ρ_g (gm/cc)
5.160	0.0820	0.0795
5.165	0.0816	0.0830
5.166	0.0815	0.0805
5.167	0.0880	0.0860

Use these data to estimate the scaling exponent β for the confined fluid. Is this exponent similar to that of the 3-d Ising model given in table 3.1? If not, are the data incorrect or can you offer other possible explanations?

19. In their measurements of compressibility in the ethane–n-heptane binary mixture, Ehrlich and Wu [8] studied the scaling of K_{Tx} with pressure at the critical composition and temperature using the scaling relationship:

$$K_{Tx} \propto \delta P^{-\lambda}; \quad x_c, T_c$$

Their results showed that λ was a function of x_c (see figure 3.5). Comment upon this. Do the results shown in table 3.1 shed light on this matter?

20. Harris and Yung [9] provide the following information for carbon dioxide near its critical point:

T (K)	ρ_l (kg/m^3)	ρ_g (kg/m^3)	P (bar)
278	898.5	114.7	39.2
288	824.4	160.7	50.2
298	713	241.9	63.5
303	593.1	343.6	71.2

Using the scaling relationship, $\rho_l - \rho_g = A(T - T_c)^\beta$, find the value of the critical temperature using a known value for β for this universality class. Compare this result to the use of the law of *rectilinear diameter* given by $\frac{1}{2}(\rho_l + \rho_g) = \rho_c + A(T - T_c)$. What is the significance of this law? [A is a constant here.]

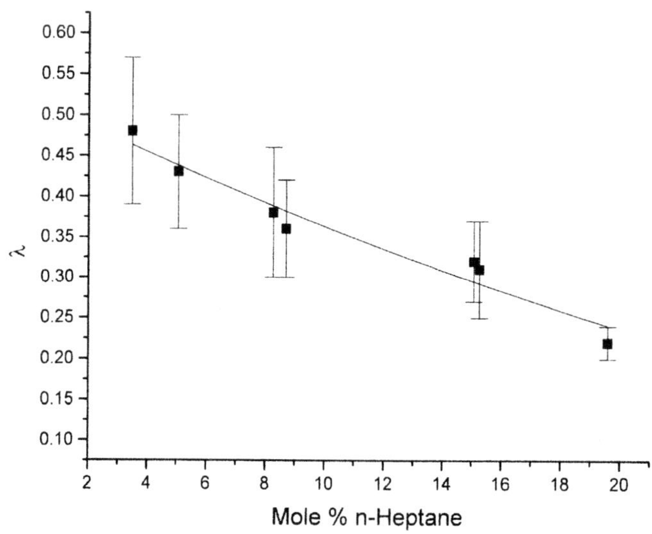

Figure 3.5 Compressibility of a hydrocarbon mixture

21. If the compressibility scaling behavior is studied using the equation of state used in exercise 17, namely

$$p = p_c + at + btv + cv^x$$

show that, along a path described by $v = t^y$, with y a constant, we have

$$K_T \propto t^{-y(x-1)} \quad \text{for} \quad y(x-1) < 1$$

$$K_T \propto t^{-1} \quad \text{for} \quad y(x-1) > 1$$

22. A correction to scaling describes the terms, beyond the first-order ones, that may be important in describing scaling behavior. For example, the scaling behavior of K_T might be of the form

$$K_T \propto \frac{1}{at^x + bt^y}$$

with $y > x$, where the term bt^y is the *first-order correction-to-scaling* term. In what circumstances do you think corrections to scaling might be important? Using the above equation how would you determine whether or not they are significant in such a system?

Bibliography

The classic references in this area are references [2, 3, 5].

[1] H. W. Habgood and W. G. Schneider, "Measurements in the critical region of Xenon," *Can. J. Chem.* vol. 32, p. 98, 1954.

[2] J. V. Sengers, "Transport processes near the critical point of gases and binary liquids in the hydrodynamic regime," *Ber. Bunsen-Ges. Phys. Chem.*, vol. 76, p. 234, 1972.

[3] J. M. H. Levelt-Sengers, "Chapter 1: Thermodynamics of solutions near the solvent's critical point," in *Supercritical Fluid Technology: Reviews in Modern Theory and Applications*, T. J. Bruno and J. F. Ely (eds.). Boca Raton, FL: CRC Press, 1991.

[4] M. I. Bagatskii, A. V. Voronel, and U. G. Gusak, "Measurements of the specific heat of Argon," *Sov. Phys. JETP* (English translation), vol. 43, p. 517, 1962.

[5] R. B. Griffiths and J. C. Wheeler, "Critical points in multicomponent systems," *Phys. Rev. A*, vol. 2, p. 1047, 1970.

[6] B. Noble and J. W. Daniel, *Applied Linear Algebra*, second ed. Englewood Cliffs, NJ: Prentice Hall, 1977.

[7] A. P. Y. Wong and M. H. W. Chan, "Liquid–vapor critical point of He in aerogel," *Phys. Rev. Lett.*, vol. 65, p. 2567, 1990.

[8] P. Ehrlich and P. C. Wu, "Isothermal compressibility in ethane–heptane mixtures," *AIChE J.*, vol. 19, p. 540, 1973.

[9] J. G. Harris and Y. Yung, "Carbon dioxide's liquid–vapor coexistence curve and critical properties as predicted by a simple molecular model," *J. Phys. Chem.*, vol. 99, p. 12022, 1995.

4 Scaling Near the Critical Point in Mixtures

The critical point of mixtures requires a more intricate set of conditions to hold than those at a pure-fluid critical point. In contrast to the pure-fluid case, in which the critical point occurs at a unique point, mixtures have additional thermodynamic degrees of freedom. They, therefore, possess a *critical line* which defines a locus of critical points for the mixture. At each point along this locus, the mixture exhibits a critical point with its own composition, temperature, and pressure. In this chapter we investigate the critical behavior of binary mixtures, since higher-order systems do not bring significant new considerations beyond those found in binaries. We deal first with mixtures at finite compositions along the critical locus, followed by consideration of the technologically important case involving dilute mixtures near the solvent's critical point. Before taking up this discussion, however, we briefly describe some of the main topographic features of the critical line of systems of significant interest: those for which nonvolatile solutes are dissolved in a solvent near its critical point.

4.1 The Critical-Line Topography in Binary Supercritical Mixtures

The critical line divides the P–T plane into two distinctive regions. The area above the line is a one-phase region, while below this line, phase transitions can occur. For example, a mixture of overall composition x_c will have a loop associated with it, like the one shown in figure 4.1, which just touches the critical line of the mixture at a unique point. The leg of the curve to the "left" of the critical point is referred to as the *bubble line*; while that to the right is termed the *dew line*. Phase equilibrium occurs between two phases at the point where the bubble line at one composition intersects the dew line; this requires two loops to be drawn of the sort shown in figure 4.1. A question naturally arises as to whether or not all binary systems exhibit continuous critical lines like that shown. In particular we are interested in the situation involving a nonvolatile solute dissolved in a supercritical fluid of high volatility.

Imagine that, at various T–P pairs along the mixture's critical locus, starting out from the pure-fluid point (i.e., the left-hand critical point in figure 4.1), we perform two types of calculation: (1) a phase-equilibrium one producing the composition of the solute in the fluid phase in equilibrium with pure solute and (2) critical-line compositions at

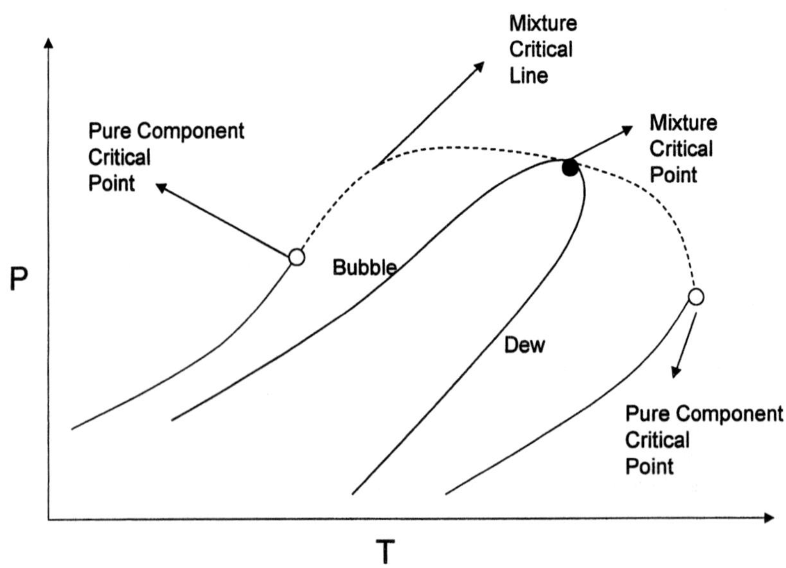

Figure 4.1 Schematic of a binary mixture critical locus

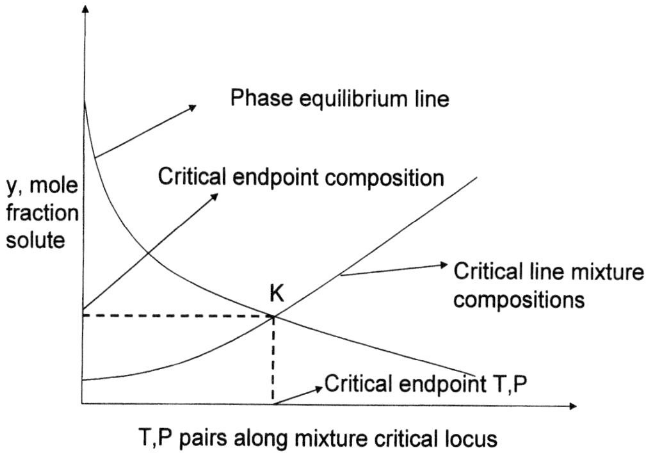

Figure 4.2 Solubility along the mixture critical locus with a critical endpoint

the given T–P pair. This phase-equilibrium problem is defined by the solution of the phase-equilibrium equation

$$\mu_1^s(T, P) = \mu_1^v(T, P, y_1^*) \tag{4.1}$$

where T and P are the temperature and pressure conditions of the system, and y_1^* is the mole fraction of solute in the fluid phase. In addition, the critical-mixture composition at (T, P) is y_1^c, where generally $y_1^* \neq y_1^c$. A plot of these compositions might look like that shown schematically in figure 4.2. At some point, both curves may intersect at the point K shown in the figure. To the right of K, we have a situation in which the critical line's composition is greater than the corresponding equilibrium composition. However, this represents an *unstable situation*, implying that the critical line does not extend beyond the point K. The point K thus represents a *critical end point*—in this case, a *lower critical end point* (LCEP). Analogous arguments, working from the opposite end of the critical line, result in another critical end point referred to as *an upper critical end point* (UCEP). At a temperature between the LCEP and the UCEP, there is no critical line, since the system will only exhibit two-phase equilibria; thus we have a system with a discontinuous critical line, commonly the case with nonvolatile solutes dissolved in supercritical fluids.

EXERCISE 4.1

Prove that critical line in figure 4.2 cannot extend beyond the point K.

4.2 Critical Stability in Binary Mixtures at Finite Compositions

We saw in an earlier chapter that, in addition to the limit-of-stability condition, critical points require that this also be a *stable limit of stability*. This latter requirement gives rise to additional equations that must hold at the mixture's critical point itself.

The most common conditions used to define the critical point in binary mixtures uses the Gibbs free energy G. Following a fluctuation analysis, similar to that followed in the pure fluid case, it is easily shown that, along the binary-mixture critical locus,

$$\left(\frac{\partial \mu_i}{\partial N_i}\right)_{T,P,N_j \neq i} = 0 \tag{4.2}$$

$$\left(\frac{\partial^2 \mu_i}{\partial N_i^2}\right)_{T,P,N_j \neq i} = 0 \tag{4.3}$$

$$\left(\frac{\partial^2 \mu_i^3}{\partial N_i^3}\right)_{T,P,N_j \neq i} > 0 \tag{4.4}$$

Based upon phase-rule considerations, this immediately shows that, for a binary mixture, there is only one intensive thermodynamic degree of freedom at its critical point.

One of the first points of interest in these systems is the relationship between material and mechanical stability conditions at the critical point. It is noteworthy how the conditions for mixtures differ from those for pure fluids.

The Relationship between the Limits of Material and Mechanical Stability along the Mixture's Critical Locus: A Classical Analysis

The critical points are located at *a limit of material stability* of the mixture given by the vanishing of the determinant of the following matrix:

$$\boldsymbol{A} = \begin{bmatrix} \left(\frac{\partial \mu_1}{\partial N_1}\right)_{T,V,N_2} & \left(\frac{\partial \mu_1}{\partial N_2}\right)_{T,V,N_1} \\ \left(\frac{\partial \mu_1}{\partial N_2}\right)_{T,V,N_1} & \left(\frac{\partial \mu_2}{\partial N_2}\right)_{T,V,N_1} \end{bmatrix} \tag{4.5}$$

where μ_i is the chemical potential of component i, N_i is the number of molecules of component i, T the temperature, and V the volume. The system is stable when $|\boldsymbol{A}| > 0$, and is at its limit of material stability when $|\boldsymbol{A}| = 0$, where $|\boldsymbol{A}|$ denotes the determinant of $\boldsymbol{A}$. From the earlier chapter on stability theory, the stability criterion $|\boldsymbol{A}| \geq 0$ can be alternatively formulated in terms of the matrix $\boldsymbol{B}$ below:

$$\boldsymbol{B} = \begin{bmatrix} \left(\frac{\partial P}{\partial \rho}\right)_{T,x_1} & -v\left(\frac{\partial P}{\partial x_1}\right)_{T,\rho} \\ -v\left(\frac{\partial P}{\partial x_1}\right)_{T,\rho} & \left(\frac{\partial^2 a}{\partial x_1^2}\right)_{T,\rho} \end{bmatrix} \tag{4.6}$$

where $v = V/N$ is the molecular volume, $a = A/N$ the molecular Helmholtz energy, $N = N_1 + N_2$ with N_i the number of molecules of component i, and ρ is the number density $N/V = v^{-1}$. These two stability criteria (given by $|\boldsymbol{A}| = 0$ and $|\boldsymbol{B}| = 0$) are related through the thermodynamic identity

$$|\boldsymbol{B}| = N^2|\boldsymbol{A}| \tag{4.7}$$

EXERCISE 4.2

Show that (4.6) leads to the following stability inequality for binary mixtures:

$$a_{xx}a_{vv} - a_{vx}^2 \geq 0$$

where the subscripted variables indicate partial differentiation.

In the pure-fluid case, both mechanical and material stability conditions are satisfied simultaneously at the critical point. We now show, at least classically, that this is generally not true for binary fluid mixtures. From equation (4.6) we can obtain a more transparent equation linking mechanical and material stability terms:

$$\left(\frac{\partial P}{\partial \rho}\right)_{T,x_1} = N^2|\boldsymbol{A}| + v^2\left(\frac{\partial P}{\partial x_1}\right)_{T,\rho}^2 \bigg/ \left(\frac{\partial^2 a}{\partial x_1^2}\right)_{T,\rho} \tag{4.8}$$

Equation (4.8) is useful since it provides a very direct connection between the terms defining the mixture's *mechanical stability limit*, namely $(\partial P/\partial \rho)_{T,x_1}$, and its material stability limit given by the vanishing of $|\boldsymbol{A}|$.

When the mixture's material stability limit is reached, $|\boldsymbol{A}| = 0$ and we see from (4.8) that the mechanical stability limit is not necessarily reached simultaneously, namely $(\partial P/\partial \rho)_{T,x_1}$ may be nonzero even though $|\boldsymbol{A}| = 0$. If this latter condition holds, then

$$\left(\frac{\partial P}{\partial \rho}\right)_{T,x_1} = v^2\left(\frac{\partial P}{\partial x_1}\right)_{T,\rho}^2 \bigg/ \left(\frac{\partial^2 a}{\partial x_1^2}\right)_{T,\rho} \tag{4.9}$$

Classically, the terms $(\partial P/\partial x_1)_{T,\rho}^2$ and $(\partial^2 a/\partial x_1^2)_{T,\rho}$ are positive along the mixture's critical line except possibly at a critical azeotrope [1] (see exercise 4.2). Except in this one circumstance, this implies that $(\partial P/\partial \rho)_{T,x_1}$ will be positive and finite along the mixture's critical line, and so the mixture will not generally be at its limit of mechanical stability.

EXERCISE 4.3

Prove that

$$\left(\frac{\partial^2 a}{\partial x_1^2}\right)_{T,\rho} = N\left[\left(\frac{\partial \mu_1}{\partial N_1}\right)_{T,V,N_2} + \left(\frac{\partial \mu_2}{\partial N_2}\right)_{T,V,N_1} - 2\left(\frac{\partial \mu_1}{\partial N_2}\right)_{T,V,N_1}\right]$$

Does the above equation ensure that, in a stable system, $(\partial^2 a/\partial x_1^2)_{T,\rho} > 0$? How about exactly at a mixture's critical point itself? [Hint: Show that, for two positive numbers a and b, their arithmetic mean is greater than or equal to their geometric mean.]

A Classical Analysis of the Critical Azeotrope

At a mixture's critical azeotrope,

$$\lim_{x_1 \to x_c} \left(\frac{\partial v}{\partial x_1} \right)_{T_c, P_c} \to \pm\infty \tag{4.10}$$

Using the thermodynamic identity

$$\left(\frac{\partial v}{\partial x_1} \right)_{T,P} = \left(\frac{\partial^2 a}{\partial x_1^2} \right)_{T,\rho} \Bigg/ \left(\frac{\partial P}{\partial x_1} \right)_{T,\rho} \tag{4.11}$$

we see that, if

$$\left(\frac{\partial^2 a}{\partial x_1^2} \right)_{T,\rho}^{c} > 0 \tag{4.12}$$

then the following conditions hold for the critical azeotrope:

$$|\boldsymbol{A}| = 0 \tag{4.13}$$

$$\left(\frac{\partial P}{\partial \rho} \right)_{T,x_1}^{c} = 0 \tag{4.14}$$

$$\left(\frac{\partial P}{\partial x_1} \right)_{T,\rho}^{c} = 0 \tag{4.15}$$

The superscript c in all of these equations here denotes a property evaluated at the mixture's critical point.

We see, therefore, that only at a mixture's critical azeotrope will both mechanical and material stability conditions be simultaneously satisfied. We reiterate that this is, strictly speaking, a classical perspective, as seen from the fact that, since

$$da = -s\,dT - P\,dv + (\mu_1 - \mu_2)dx_1 \tag{4.16}$$

the various derivatives in (4.14) and (4.15) are all related to derivatives of the free energy at the critical line, which is implicitly assumed to be well behaved there. These identities are given by

$$\left(\frac{\partial P}{\partial \rho} \right)_{T,x_1} = \frac{a_{vv}}{\rho^2} \tag{4.17}$$

and

$$\left(\frac{\partial P}{\partial x_1} \right)_{T,\rho} = -a_{vx_1} \tag{4.18}$$

We summarize these points:

- Equations (4.13)–(4.15) are necessary and sufficient conditions for critical azeotropy, as long as $(\partial^2 a/\partial x_1^2)^c_{T,\rho} > 0$.

- Except at a point of critical azeotropy, *classical analysis* predicts that the compressibility of the mixture, $K_{Tx} = -V^{-1}(\partial v/\partial P)_{T,x}$ is positive and finite at the critical point. From this it necessarily follows that both the mixture's expansion coefficient α_{Px} and its heat capacity C_{Px} also do not diverge at the critical line.

EXERCISE 4.4

Show that, in a binary mixture (composition x),

$$\left(\frac{\partial v}{\partial x}\right)_{T,P} = vK_{Tx}\left(\frac{\partial P}{\partial x}\right)_{T,v}$$

where

$$K_{Tx} \equiv -\frac{1}{v}\left(\frac{\partial v}{\partial P}\right)_{T,x} = \frac{1}{\rho}\left(\frac{\partial \rho}{\partial P}\right)_{T,x}$$

What does this identity tell you about classical predictions for the scaling of the terms K_{Tx} and $(\partial P/\partial x)_{T,v}$ as the critical azeotrope is approached?

Partial Molar Volumes along the Critical Line: A Classical Analysis

We now consider the partial molar volume, or the closely related property $(\partial v/\partial x)_{T,P}$, along the mixture's critical line in the case of mixtures that do not form critical azeotropes.

A thermodynamic identity provides that

$$\left(\frac{\partial v}{\partial x}\right)_{T,P} = -\left(\frac{\partial v}{\partial P}\right)_{T,x} \Big/ \left(\frac{\partial x}{\partial P}\right)_{T,v} \tag{4.19}$$

At finite compositions along the critical line, classical theory provides that both of the quantities on the right-hand side of (4.19) are finite; this implies that $(\partial v/\partial x)_{T,P}$ *remains finite there*. This result can be used to analyze the behavior of partial molar volumes $\bar{V}_1$ or $\bar{V}_2$ along the critical line, important properties for understanding solvation in supercritical solvents.

In a binary mixture, the following identities hold:

$$\bar{V}_1 = v + (1 - x_1)\left(\frac{\partial v}{\partial x_1}\right)_{T,P} \tag{4.20}$$

$$\bar{V}_2 = v - x_1\left(\frac{\partial v}{\partial x_1}\right)_{T,P} \tag{4.21}$$

Along the critical line with finite x_1, we have that

$$\frac{\bar{V}_1}{K_{Tx}} = \frac{v}{K_{Tx}} + \frac{1 - x_1}{K_{Tx}}\left(\frac{\partial v}{\partial x_1}\right)_{T,P} \tag{4.22}$$

$$\frac{\bar{V}_2}{K_{Tx}} = \frac{v}{K_{Tx}} + \frac{x_1}{K_{Tx}}\left(\frac{\partial v}{\partial x_1}\right)_{T,P} \tag{4.23}$$

The magnitude of the first term on the right-hand side of these equations, v/K_{Tx}, is finite along the critical line. Since the term $(\partial v/\partial x_1)_{T,P}$ given by the identity in equation (4.19) is also finite, this implies that, at finite x_1, both $\bar{V}_1$ and $\bar{V}_2$ are predicted to be finite—as are properties of the form $\bar{V}_i/K_{Tx}$.

We also have that

$$\rho \bar{V}_1 = 1 + (1 + x_1)\left(\frac{\partial P}{\partial x_1}\right)_{T,\rho} K_{Tx} \tag{4.24}$$

$$\rho \bar{V}_2 = 1 - x_1 \left(\frac{\partial P}{\partial x_1}\right)_{T,\rho} K_{Tx} \tag{4.25}$$

from which it follows that at both extremes of the mixture's critical line $\bar{V}_1^{\infty}$ and $\bar{V}_2^{\infty}$ have opposite signs in their divergences. For example, in a mixture where

$$\lim_{\substack{T\to T_{c2}\\ P\to P_{c2}\\ x_1\to 0}} \left(\frac{\partial P}{\partial x_1}\right)_{T,\rho} < 0$$

we find that

$$\lim_{\substack{T\to T_{c2}\\ P\to P_{c2}}} \bar{V}_1^{\infty} \to -\infty \tag{4.26}$$

and

$$\lim_{\substack{T\to T_{c1}\\ P\to P_{c1}}} \bar{V}_2^{\infty} \to \infty \tag{4.27}$$

The situation at a critical azeotrope is left to exercise 4.5.

EXERCISE 4.5

Show that in a system exhibiting *negative critical azeotropy*, defined as being a system where

$$\lim_{\substack{T\to T_{c2}\\ P\to P_{c2}\\ x_1\to 0}} \left(\frac{\partial P}{\partial x_1}\right)_{T,\rho} < 0 \quad \text{and} \quad \lim_{\substack{T\to T_{c1}\\ P\to P_{c1}\\ x_1\to 1}} \left(\frac{\partial P}{\partial x_1}\right)_{T,\rho} > 0$$

we have:

$$\lim_{\substack{T\to T_{c2}\\ P\to P_{c2}}} \bar{V}_1^{\infty} \to -\infty \quad \text{and} \quad \lim_{\substack{T\to T_{c1}\\ P\to P_{c1}}} \bar{V}_2^{\infty} \to -\infty$$

Can these conditions alone be used to experimentally determine the existence (or not) of critical azeotropy in a binary system?

4.3 The Nonclassical Perspective

Scaling in Binary Mixtures at Their Critical Points: The Universality Hypothesis

To explore this issue, we first consider the choice of an order parameter in these systems, since one is not preordained by any laws of physics. To do so, we need to look at the properties that necessarily diverge thermodynamically at the transition. The free energy can be written as follows:

$$dA = -SdT - pdV + \mu_1 dN_1 + \mu_2 dN_2 \tag{4.28}$$

Given that $dN = dN_1 + dN_2$, we can write the following forms for the free energies in this system:

$$dA = -SdT - pdV + (\mu_1 - \mu_2)dN_1 + \mu_2 dN \tag{4.29}$$

$$dG = -SdT + Vdp + \Delta dN_1 + \mu_2 dN \tag{4.30}$$

where we have defined

$$\Delta \equiv \mu_1 - \mu_2 \tag{4.31}$$

Thermodynamic stability theory requires that

$$\left(\frac{\partial \Delta}{\partial N_1}\right)_{T,P} = 0 \tag{4.32}$$

which implies that, at constant N,

$$\left(\frac{\partial \Delta}{\partial x_1}\right)_{T,P} = 0 \tag{4.33}$$

Furthermore, at constant composition, $dN_1 = x_1 dN$, so that, at the binary mixture's critical composition (mole fraction), we have that

$$\frac{1}{x_1^c}\left(\frac{\partial \Delta}{\partial N}\right)_{T,P} = 0 \tag{4.34}$$

which implies that

$$\left(\frac{\partial \Delta}{\partial \rho}\right)_{T,P} = 0 \tag{4.35}$$

at the mixture's critical point. We use this equation to define the *order parameter* $\rho - \rho_c$ and its coupling field Δ in binary mixtures near the vapor–liquid critical point. The stability analysis of chapter 1 can now be fully engaged, leading to Legendre transformations of the Gibbs free energy, three useful forms being

$$d\phi = -SdT - pdV - N_1 d\Delta + \mu_2 dN \tag{4.36}$$

$$d\psi = -SdT + VdP - N_1 d\Delta + \mu_2 dN \tag{4.37}$$

$$d\theta = TdS + Vdp - N_1 d\Delta + \mu_2 dN \tag{4.38}$$

From these equations and (4.29), the following derivative properties of the free energy are seen to strongly diverge at the transition:

$$\left(\frac{\partial v}{\partial P}\right)_{T,\Delta} = \left(\frac{\partial^2 \psi}{\partial P^2}\right)_{T,\Delta} = -vK_{T\Delta} \tag{4.39}$$

$$-\left(\frac{\partial x_1}{\partial \Delta}\right)_{T,P} = \left(\frac{\partial^2 \psi}{\partial \Delta^2}\right)_{T,P} \tag{4.40}$$

$$-\left(\frac{\partial s}{\partial T}\right)_{P,\Delta} = \left(\frac{\partial^2 \psi}{\partial T^2}\right)_{P,\Delta} \tag{4.41}$$

If we do another Legendre transform of equation (4.37), we get that

$$d\Gamma = -SdT + VdP - N_1 d\Delta - N d\mu_2 \tag{4.42}$$

It follows that $d\Gamma = 0$ (why?), in which case we find that

$$d\mu_2 = -sdT + vdP - x_1 d\Delta \tag{4.43}$$

By looking closely at the properties in equations (4.39)–(4.41), it is easy to appreciate the first Griffiths–Wheeler hypothesis for criticality in mixtures. This is stated as follows: if one field variable is fixed, then the binary mixture will generally behave as a one-component fluid in keeping with the idea of critical-point universality within classes of systems. Thus, a derivative of a density with respect to a field, with yet another two fields held constant, should yield a strongly diverging property.

Levelt-Sengers [2] illustrated the Griffiths–Wheeler approach using the potential μ_2 (Δ, P, T), where $\Delta \equiv \mu_2 - \mu_1$, which is immediately available from (4.43). With this function we have that

$$1 - x_2 = \left(\frac{\partial \mu_2}{\partial \Delta}\right)_{T,P}; \quad v = \left(\frac{\partial \mu_2}{\partial P}\right)_{\Delta,T}; \quad s = -\left(\frac{\partial \mu_2}{\partial T}\right)_{P,\Delta} \tag{4.44}$$

which leads to strongly diverging second derivatives:

$$\left(\frac{\partial^2 \mu_2}{\partial \Delta^2}\right)_{T,P} = -\left(\frac{\partial x_2}{\partial \Delta}\right)_{T,P} \tag{4.45}$$

$$\left(\frac{\partial^2 \mu_2}{\partial P^2}\right)_{\Delta,T} = -vK_{T\Delta} \tag{4.46}$$

Table 4.1 Nonclassical divergence characteristics at the critical point in pure fluids and binary mixtures at finite compositions

Property	*Pure Fluid*	*Mixtures*
K_T	Strongly divergent	Weakly divergent
C_P	Strongly divergent	Weakly divergent
C_V	Weakly divergent	Finite

and so on, where

$$K_{T\Delta} \equiv -\frac{1}{v}\left(\frac{\partial v}{\partial P}\right)_{T,\Delta} \tag{4.47}$$

These strong derivatives diverge with an exponent γ along the weak (coexistence) path, while along a strong path they diverge with renormalized exponent $\gamma/\beta\delta$ in complete analogy with the one-component case.

Furthering the analogy with pure fluids, we now examine the weak divergences not present in classical theory. Since, by Griffiths–Wheeler, if one field is kept constant, then the binary mixture is brought into analogy with the pure fluid; in this case, the following derivative is an example of a *weakly diverging property in binary mixtures*:

$$\left(\frac{\partial^2 \mu_2}{\partial P^2}\right)_{T,x} = \left(\frac{\partial v}{\partial P}\right)_{T,x} = -vK_{Tx} \tag{4.48}$$

This result points out a general conclusion of the nonclassical theory of mixture behavior, namely that thermodynamic properties that diverge strongly in pure fluids will diverge only weakly in binary mixtures [2]. We showed earlier that a nonazeotropic critical point in mixtures K_{Tx} is predicted to be finite by classical arguments. This result thus represents a departure between classical and nonclassical theoretical predictions of critical behavior in fluid mixtures.

Since density-like properties vary smoothly along the critical line, nonclassical theory also predicts that properties like the isochoric heat capacity of a mixture, defined as $C_{Vx} = (\partial E/\partial T)_{V,x}$, will be finite along the critical line and that, in general, properties that diverge weakly in pure fluids will not diverge in a mixture of constant composition.

These results are summarized in tables 4.1 and 4.2. The general weakening of critical divergences (along a particular path) as we move from pure fluids to mixtures at finite compositions, is shown in table 4.1. In table 4.2, a more detailed summary of mixture properties as well as a comparison of their classical and nonclassical predictions is given.

EXERCISE 4.6

Show how the results for C_{Px} in table 4.2 can be developed.

Table 4.2 Divergence characteristics of various mixture properties at finite compositions along a (nonazeotropic) mixture critical locus

Property	*Classical Theory*	*Nonclassical Theory (Griffiths–Wheeler)*
K_{Tx}	Finite	Weakly divergent
$\bar{V}_1, \bar{V}_2$	Finite	Finite
C_{Px}	Finite	Weakly divergent
C_{Vx}	Finite	Finite
$(\partial x/\partial P)_{T,V}$	Finite	Weakly divergent
$(\partial v/\partial x)_{T,P}$	Finite	Finite

Liquid–Liquid Consolute Points

For a binary mixture, represented by components 1 and 2, we have that the molar Gibbs free energy can be expressed by the equation

$$dg = -s\,dT + v\,dP + (\mu_2 - \mu_1)dx_2 \tag{4.49}$$

We thus see that the molar Gibbs free energy, at constant pressure for example, is a function of the field variable T and the density x_2; this is qualitatively similar to the pure-fluid case, where the molar Helmholtz free energy is also a function of a field and density variable. An immediate issue to address is once again the choice of a suitable *order parameter*. By analogy with the pure-fluid case, we want a density-like variable that describes the difference in the order (and free energy) between coexisting phases along the coexistence curve. In addition, it must be a choice that leads to thermodynamic divergences at the critical point involving derivatives of the order parameter and its coupling field. Given these considerations, a natural choice for this order parameter is either x_1 or x_2. Arbitrarily choosing x_2, we say that, near the consolute point, the order-parameter scaling relationship along the coexistence curve is given by the familiar result, namely

$$x_2' - x_2'' \approx t^{\beta} \tag{4.50}$$

where the primes designate the two coexisting phases, β is the universal coexistence scaling exponent, and

$$t \equiv \frac{T - T_c}{T_c} \tag{4.51}$$

Along the coexistence direction, from above the critical point, the *susceptibility* in this system, defined as

$$\chi \equiv \left(\frac{\partial^2 g}{\partial x_1^2}\right)^{-1}_{T,P} = \left(\frac{\partial x_1}{\partial \mu_1}\right)_{T,P} \tag{4.52}$$

scales as

$$\chi \approx t^{-\gamma} \tag{4.53}$$

The scaling direction to be considered in (4.53) is at constant composition with $x_2 = x_{2c}$ i.e., the critical or consolute composition for component 2. Similarly, the heat capacity at constant pressure along the same path scales as,

$$C_{P,x_c} \approx t^{-\alpha} \tag{4.54}$$

where once again the scaling laws relate the exponents α, β, and γ as follows:

$$\alpha = 2 - 2\beta - \gamma \tag{4.55}$$

4.4 An Important Case: Dilute Binary Mixtures Near the Solvent's Critical Point

A common situation of practical interest with supercritical-fluid solvents involves the case of extremely dilute solutes dissolved in a solvent near its pure critical temperature and pressure. The behavior of such mixtures near the solvent's critical point displays subtle and important differences from that found elsewhere along the critical line. Hence we examine some of the issues closely paralleling the discussion given by Levelt-Sengers [2].

The Classical Approach

The mole fraction of the solute species (component 2) is designated by the variable x and the classical Helmholtz free energy of the mixture $a(T, v, x)$ is expanded as

$$a^*(T, v, x) = a(T, v, x) - RT[x \ln x + (1 - x)\ln(1 - x)] \tag{4.56}$$

where we have subtracted the ideal-mixing term for $a(T, v, x)$ to avoid the problems associated with the terms at the condition $x \to 0$.

Expanding a^* around the solvent's critical point, we find that [2]

$$\begin{aligned} a^*(T, v, x) = {} & a^{*c} + a_v^c(\delta v) + a_T^{*c}(\delta T) + a_x^{*c} x \\ & + \cdots + a_{vT}^c(\delta v)(\delta T) + a_{vx}^c(\delta v)x \\ & + \cdots + a_{vvT}^c(\delta v)^2(\delta T)/2 + a_{vvx}^c(\delta v)^2 x/2 \\ & + \cdots + a_{vvvT}^c(\delta v)^3(\delta T)/6 + a_{vvvx}^c(\delta v)^3 x/6 \\ & + \cdots + a_{vvvv}^c(\delta v)^4/24 + \cdots \end{aligned} \tag{4.57}$$

We also have reason to make use of the thermodynamic identities linking partial molar properties for a property f (defined per mole in the following equations) to its composition derivative at constant T *and* P as follows:

$$\bar{f}_1 = f - x\left(\frac{\partial f}{\partial x}\right)_{T,P} \tag{4.58}$$

$$\bar{f}_2 = f + (1 - x)\left(\frac{\partial f}{\partial x}\right)_{T,P} \tag{4.59}$$

We now look at the divergence of the partial molar volumes in the limit $x \to 0$ at the solvent's critical point, since these are key solution thermodynamic properties in these systems, as we shall see in subsequent chapters.

Using the identity

$$\left(\frac{\partial v}{\partial x}\right)_{T,P} = -\frac{a_{vx}}{a_{vv}} \tag{4.60}$$

we differentiate equation (4.56) with respect to v; keeping the leading-order terms, we find that:

$$\begin{aligned} a_v &= a_v^* \\ &= a_v^{*\mathrm{c}} + a_{vT}^{\mathrm{c}}(\delta T) + a_{vx}^{\mathrm{c}} x \\ &\quad + \cdots + a_{vvT}^{\mathrm{c}}(\delta v)(\delta T) + a_{vvx}^{\mathrm{c}}(\delta v)x + \cdots + \frac{a_{vvvv}^{\mathrm{c}}(\delta v)^3}{6} \\ &\quad + \frac{a_{vvvv}^{\mathrm{c}}(\delta v)^2(\delta T)}{2} + \frac{a_{vvvx}^{\mathrm{c}}(\delta v)^2 x}{2} \end{aligned} \tag{4.61}$$

Differentiating (4.61) gives

$$\begin{aligned} a_{vv} &= a_{vvT}^{\mathrm{c}}(\delta T) + a_{vvx}^{\mathrm{c}} x + a_{vvvv}^{\mathrm{c}}(\delta v)^2/2 \\ &\quad + \cdots + a_{vvvT}^{\mathrm{c}}(\delta v)(\delta T) + a_{vvvx}^{\mathrm{c}}(\delta v)x \end{aligned} \tag{4.62}$$

Asymptotically near the solvent's critical point, we have that

$$a_{vx} = a_{vx}^{\mathrm{c}} + \cdots \tag{4.63}$$

Using equations (4.62) and (4.63) in (4.60), we find that:

$$\left(\frac{\partial v}{\partial x}\right)_{T,P} = -\frac{a_{vx}^{\mathrm{c}} + \cdots}{a_{vvx}^{\mathrm{c}} x + a_{vvT}^{\mathrm{c}}(\delta T) + a_{vvvv}^{\mathrm{c}}(\delta v)^2/2 + \cdots} \tag{4.64}$$

The way in which the denominator in equation (4.64) approaches zero will depend upon the path of approach to the critical point. We analyze this along the critical-line path (CRL). Since v varies linearly with x along this path, the term $(\delta v)^2$ can be ignored in equation (4.64). In addition, we use the following identity along this path:

$$\left(\frac{dT}{dx}\right)_{\mathrm{CRL}}^{\mathrm{c}} = \frac{(a_{vx}^{\mathrm{c}})^2 - a_{vvx}^{\mathrm{c}} RT_{\mathrm{c}}}{RT_{\mathrm{c}} a_{vvT}^{\mathrm{c}}} \tag{4.65}$$

in equation (4.64) which all leads to the following result, for asymptotically small x:

$$\left(\frac{\partial v}{\partial x}\right)_{T,P} = -\left(\frac{RT_{\mathrm{c}}}{a_{vx}^{\mathrm{c}}}\right) x^{-1} \quad \text{on the critical-line path} \tag{4.66}$$

Thus, from equation (4.59), the solute's partial molar volume $\bar{V}_2$ along this path diverges as x^{-1}, while from equation (4.58), $\bar{V}_1$ (for the solvent) approaches the value

$$\bar{V}_1 = v_{\mathrm{c}} + \frac{RT_{\mathrm{c}}}{a_{vx}^{\mathrm{c}}} \quad \text{on the critical-line path} \tag{4.67}$$

which is finite but not equal to v_c. Along the coexistence curve at $T = T_c$, it can also be shown [2] that

$$\left(\frac{\partial v}{\partial x}\right)_{T,P} = -\frac{a^c_{vx}}{a^c_{vvx} + 3(a^c_{vx})^2/RT_c}x^{-1} \quad \text{on the coexistence curve} \tag{4.68}$$

Once again along this path, $\bar{V}_2$ diverges as x^{-1}. Meanwhile, $\bar{V}_1$ approaches the limiting value of

$$\bar{V}_1(x=0) = \frac{v_c - a^c_{vx}}{a^c_{vvx} + 3(a^c_{vx})^2/RT_c} \tag{4.69}$$

which once again does not approach v_c at this limit. Note the differences between these results and those developed earlier for the binary critical mixture at finite compositions, where both $\bar{V}_1$ and $\bar{V}_2$ are finite along the critical line. This well illustrates the special attention required when looking at the dilute solute $x \to 0$ limit.

EXERCISE 4.7

Prove equations (4.65) and (4.68).

Osmotic Susceptibilities

These are important properties in fluid mixtures, and here we examine the osmotic susceptibilities in the limit that the solute mole fraction approaches zero (i.e., $x \to 0$ in the following equations). We look at the classical expansion of the chemical potentials in the mixture around this limit. These expansions are provided by the following identities [2]:

$$\begin{aligned}\mu_1(\text{solvent}) &= a - xa_x - va_v \\ \mu_2(\text{solute}) &= a + (1-x)a_x - va_v\end{aligned} \tag{4.70}$$

which applied to equations (4.56)and (4.57) yield the equations

$$\mu_1 = a^{*c} + a^{*c}_T(\delta T) + v_c P - a^c_{vx}(\delta v)x - a^c_{vvvv}(\delta v)^4/8 + \cdots - RTx + \cdots \tag{4.71}$$

$$\begin{aligned}\mu_2 = &- a^{*c} + a^{*c}_x + a^c_{vx}(\delta v) + (a^{*c}_T + a^{*c}_{Tx})(\delta T) + a^*_{xx}x + v_c P \\ &+ \cdots + RT\ln x + \cdots\end{aligned} \tag{4.72}$$

Recall that the osmotic susceptibilities in a binary mixture are defined by the quantities

$$RT\left(\frac{\partial x_1}{\partial \mu_1}\right)_{T,P} \quad \text{and} \quad RT\left(\frac{\partial x_2}{\partial \mu_2}\right)_{T,P} \tag{4.73}$$

and that both of these quantities necessarily approach ∞ along the critical line at finite compositions.

Table 4.3 Classical path dependence of divergences of various partial molar properties in dilute near-critical mixtures (2 represents the solute species, 1 the solvent), as $x_2 \equiv x \to 0$ (adapted from reference [2])

	Property		
Path	$\bar{V}_2$	$\bar{H}_2$	$\bar{V}_1$
P_c, T_c	$x^{\frac{-2}{3}}$	$x^{\frac{-2}{3}}$	V_c
P_c, ρ_c	x^{-1}	x^{-1}	Not V_c
T_c, ρ_c	x^{-1}	x^{-1}	Not V_c
CRL	x^{-1}	x^{-1}	Not V_c

Now consider $(\partial\mu_2/\partial x)_{T,P}$ in the limit $x \to 0$ along the critical isotherm–isobar. Using (4.89) (see the additional exercises) in (4.72), we have that

$$\left(\frac{\partial\mu_2}{\partial x}\right)_{T,P} = -|a^c_{vx}|\left|\frac{6a^c_{vx}}{a^c_{vvvv}}\right|^{1/3}\frac{x^{-2/3}}{3} + \frac{RT}{x} + a^{*c}_{xx} \tag{4.74}$$

This equation reveals that, along this path, the term RTx^{-1} dominates and $(\partial x/\partial\mu_2)_{T,P}$ approaches zero as $x \to 0$. Using the Gibbs–Duhem equation (see the additional exercises), this approach can also be used to show that the cross susceptibility

$$\left(\frac{\partial x}{\partial\mu_1}\right)_{T,P} = -\frac{1-x}{x}\left(\frac{\partial x}{\partial\mu_2}\right)_{T,P} \tag{4.75}$$

is zero along this path.

Note the apparent contradiction given by these results: equation (4.74) says that $(\partial x/\partial\mu_2)_{T,P}$ is zero at the critical point limit ($x \to 0$), while (4.53) shows that, at finite x, this quantity is finite. Hence, the susceptibility is discontinuous at this limit as a result of the conflicting demands of the critical-line conditions and the infinite-dilution requirements. All of these *classical results* are summarized in tables 4.3 and 4.4

Thus we see that, in dilute binary mixtures, both solute and solvent partial molar volumes show path-dependent behavior as the critical point is approached. In addition, the discontinuity in the value of the osmotic susceptibility for the solute species at the pure-solvent critical-point limit is evident from the results shown in table 4.4. These results have interesting implications for diffusion phenomena in near-critical fluid

Table 4.4 Classical behavior of the osmotic susceptibilities in dilute near-critical mixtures (2 represents the solute species, 1 the solvent), $x_2 \equiv x \to 0$

Property	*Critical Line, Finite x_i*	*Critical Point, $x_2 \to 0$*
$(\partial\mu_1/\partial x_1)_{T,P}$	Zero	Zero
$(\partial\mu_2/\partial x_2)_{T,P}$	Zero	Infinite
$(\partial\mu_1/\partial x_2)_{T,P}$	Zero	Zero
$(\partial\mu_2/\partial x_1)_{T,P}$	Zero	Infinite

systems, since the osmotic susceptibilities play an important role in transport processes which we discuss in chapter 12.

EXERCISE 4.8

Derive the results for $\bar{V}_1$ and $\bar{V}_2$ in table 4.3 along the T_c–P_c path.

EXERCISE 4.9

An additional way of analyzing the path dependence of the osmotic susceptibilities uses the following identities:

$$\left(\frac{\partial x}{\partial \mu_2}\right)_{T,P}^{-1} = (1-x)\left[a_{xx} - \frac{a_{vx}^2}{a_{vv}}\right] \tag{4.76}$$

$$\left(\frac{\partial x}{\partial \mu_1}\right)_{T,P}^{-1} = -x[a_{xx} - a_{vx}^2] \tag{4.77}$$

Prove these equations and use them to show (4.74).

The Nonclassical Approach: Binary Fluid Mixtures near the Critical Line

As in previous cases, the nonclassical results provide the basis for predicting the behavior of real fluid mixtures. These results often depend upon conjectures, and to the present time are much less studied for fluid mixtures than for pure fluids.

We illustrate this by considering the T_c–P_c path using a nonclassically inspired, nonanalytic expansion for the pressure, similar to that used earlier for pure fluids. In this case, we postulate that

$$P = P_c + a_1\delta T + a_2\delta V\,\delta T^{\gamma} + a_3\delta V^{\delta} + b_1 x + b_2\delta T\,x + b_3\delta V\,x \tag{4.78}$$

From this we get that

$$(V - V_c)_{\text{nonclassical}} \propto x^{1/\delta} \quad \text{on the } T_c\text{–}P_c \text{ path} \tag{4.79}$$

which implies

$$\left(\frac{\partial V}{\partial x}\right)_{T_c,P_c} \propto x^{1/\delta - 1} \quad \text{on the } T_c\text{–}P_c \text{ path} \tag{4.80}$$

This implies from equation (4.59) that

$$\bar{V}_2 \propto x^{-(1-1/\delta)} \quad \text{on the } T_c\text{–}P_c \text{ path} \tag{4.81}$$

where for δ we take the nonclassical value $\delta \approx 4.80$. This predicts that $\bar{V}_2$ has a nonclassical scaling exponent of value 0.8 along the critical isotherm–isobar in the limit of

Table 4.5 Nonclassical path dependence of divergences of partial molar properties in dilute near-critical mixtures (2 represents the solute species, 1 the solvent)

Path	$\bar{V}_2$	$\bar{H}_2$	$\bar{V}_1$
P_c, T_c	$x^{-(1-1/\delta)}$	$x^{-(1-1/\delta)}$	V_c
P_c, ρ_c	x^{-1}	x^{-1}	Not V_c
T_c, ρ_c	x^{-1}	x^{-1}	Not V_c
C_{RL}^a	x^{-1}	x^{-1}	Not V_c

[a] C_{RL} means critical-line path.

reaching the pure-solvent critical point. We note that this is once again at odds with the nonclassical result provided by the Griffiths and Wheeler result for finite compositions along the critical line.

The solvent term $\bar{V}_1$ can be analyzed from the use of equation (4.20), which shows that

$$\bar{V}_1 \propto V_c + \text{a term of order } \{x \cdot x^{(1/\delta - 1)}\} \quad \text{on the } T_c\text{–}P_c \text{ path} \tag{4.82}$$

Since the second term above is zero in the limit $x \to 0$, we have $\bar{V}_1 \to V_c$ along this path. This approach can be repeated for all paths of interest, and a summary of the main nonclassical results for partial molar volumes is given in table 4.5 adapted from [2].

The Osmotic Susceptibilities as $x \to 0$

The osmotic susceptibilities have interesting values at the $x \to 0$ limit for the dilute solute, component 2. At that limit, we must have for the free energy that

$$\lim_{x\to 0} \{g = \mu_1 - \Delta x\} \to \text{finite} \tag{4.83}$$

Since the $\lim_{x\to 0} \Delta \to \infty$, where $\Delta \equiv \mu_1 - \mu_2$, this must imply that

$$\Delta \propto x^{-1} \quad \text{on the } T_c\text{–}P_c \text{ path} \tag{4.84}$$

From the identity

$$-\left(\frac{\partial \Delta}{\partial \rho}\right)_{T,P} = \left(\frac{\partial x}{\partial \rho}\right)_{T,P} \left[\left(\frac{\partial \mu_2}{\partial x}\right)_{T,P} + \frac{x}{x_1}\left(\frac{\partial \mu_2}{\partial x}\right)_{T,P}\right] \tag{4.85}$$

and the use of (4.84) we find that:

$$\left(\frac{\partial \Delta}{\partial x}\right)_{T,P} \propto x^{-2} \quad \text{on the } T_c\text{–}P_c \text{ path} \tag{4.86}$$

with

$$\lim_{x\to 0}\left(\frac{\partial \mu_2}{\partial x}\right)_{T,P} \propto x^{-2} \tag{4.87}$$

and

$$\lim_{x\to 0}\left(\frac{\partial \mu_1}{\partial x}\right)_{T,P} \propto x^{-1} \tag{4.88}$$

These particular results prove very useful in studying diffusion in near-critical binary mixtures [3].

EXERCISE 4.10

Prove equations (4.87) and (4.88).

To conclude this section, in table 4.6, we provide a summary showing the conjectured results for the nonclassical scaling properties of many of the most important properties in binary mixtures as the critical point is approached, both at the finite-composition limit and at the limit of infinite solute dilution ($x \to 0$). The results of this chapter point out the subtle, but significant, differences in these two circumstances.

4.5 Chapter Review

In this chapter we analyzed the conditions defining critical behavior in mixtures. Classical analysis reveals that, in contrast to the pure-fluid case, the mixture compressibility does not vanish at a (nonazeotropic) critical point, which implies

Table 4.6 Limiting values for some fundamental solution properties of binary mixtures in the critical region predicted by nonclassical theory; x represents the solute mole fraction for component 2

Finite Compositions on the Critical Line; x Refers to Solute Mole Fraction	*Property Limiting Value, Type of Scaling, Coexistence Direction*
$\left(\frac{\partial \mu_1}{\partial x_1}\right)_{T,P}, \left(\frac{\partial \mu_2}{\partial x_2}\right)_{T,P}, \left(\frac{\partial \mu_1}{\partial x_2}\right)_{T,P}, \left(\frac{\partial \mu_2}{\partial x_1}\right)_{T,P}$	Strong zero, $\propto t^{\gamma}$
$\left(\frac{\partial v}{\partial x}\right)_{T,P}$	Finite
$\left(\frac{\partial P}{\partial x}\right)_{T,v}$	Weak zero, $\propto t^{\alpha}$
$\left(\frac{\partial P}{\partial v}\right)_{T,x}$	Weak zero, $\propto t^{\alpha}$
In the Limit $x \to 0$	*T_c–P_c Direction*
$\left(\frac{\partial \mu_2}{\partial x}\right)_{T,P}$	$\infty, \propto x^{-2}$
$\left(\frac{\partial \mu_1}{\partial x}\right)_{T,P}$	$\infty, \propto x^{-1}$
$\left(\frac{\partial P}{\partial x}\right)_{T,v}$	Finite
$\left(\frac{\partial v}{\partial x}\right)_{T,P}$	Diverges, $\propto x^{-\left(1-\frac{1}{\delta}\right)}$

that neither do other related properties, such as the constant pressure specific heat. Nonclassical results, however, predict a weak singularity in K_{Tx} at the critical line in keeping with general results from these theories which state that properties that diverge strongly in pure fluids do so only weakly in mixtures. We presented the main results for scaling behavior in fluid mixtures, most of which flow from the Griffiths–Wheeler hypothesis about the universal nature of critical phenomena in fluid systems.

The special case of dilute mixtures near a pure-solvent critical point was analyzed in great detail, given the technological importance of such mixtures. These cases show the discontinuity in osmotic susceptibilities at the critical-line extremes, as well as path-dependent scaling behavior for various fundamental thermodynamic properties of interest. While many of the results in this chapter may appear arcane at first reading, many of them are extremely useful to have at one's command in succeeding chapters. The reader is thus advised to patiently master the concepts presented herein.

4.6 Additional Exercises

Basic Thermodynamic Analysis

1. In a binary mixture, prove the following identities:

$$\mu_1 = a - xa_x - va_v$$

$$\mu_2 - \mu_1 = a_x$$

$$\left(\frac{\partial a}{\partial x}\right)_{T,v} = \left(\frac{\partial g}{\partial x}\right)_{T,P}$$

where the subscripts are meant to signify partial differentiation with respect to the given variable and a and g are molar free energies.

2. Find the relationship between the osmotic susceptibility, defined as:

$$\left(\frac{\partial x_1}{\partial \mu_1}\right)_{T,P}$$

with the system's free energy g.

3. The following identity was previously proved:

$$v = \bar{V}_j + x_i \left(\frac{\partial v}{\partial x_i}\right)_{T,P}$$

Devise a credible experiment that would allow one to measure partial molar properties using this equation.

4. Prove the following identity in a binary mixture:

$$-\left(\frac{\partial \Delta}{\partial \rho}\right)_{T,P} = \left(\frac{\partial x_i}{\partial \rho}\right)_{T,P}\left[\left(\frac{\partial \mu_i}{\partial x_i}\right)_{T,P} + \frac{x_i}{x_j}\left(\frac{\partial \mu_i}{\partial x_i}\right)_{T,P}\right]$$

Use it to discuss the scaling behavior of the osmotic susceptibilities as the finite-composition critical point is approached.

Properties along the Critical Line

5. Explain how you would distinguish between the following derivatives in a binary mixture:

$$\left(\frac{\partial P}{\partial x}\right)^{c}_{T,\rho}, \quad \left(\frac{\partial P}{\partial x}\right)^{c}_{T,\mathrm{CXS}}, \quad \left(\frac{\partial P}{\partial x}\right)^{c}_{T,\mathrm{CRL}}$$

where CXS means along the coexistence line and CRL along the critical line.

6. Show that

$$\left(\frac{\partial P}{\partial T}\right)^{c} = \left(\frac{dP}{dT}\right)^{c}_{\mathrm{CXS}}$$

in the limit as $x \to 0$.

7. Prove that, along the critical isotherm-isobar, from a pressure expansion, $P = -a_v$, about the critical point, we find that

$$(\delta v)^3 \approx -\left(\frac{6a^{c}_{vx}}{a^{c}_{vvvv}}\right) x \quad \text{on the critical isotherm–isobar} \qquad (4.89)$$

The relationship in (4.89) is useful for studying the divergence of various properties along the T_c–P_c path, an important case being that for the osmotic susceptibilities.

8. What are the critical-point conditions in a binary mixture at the Vapor–liquid equilibrium (VLE) critical point (finite composition) in terms of the Helmholtz energy representation?

9. Along the critical line, show that the following identity holds:

$$\left.\frac{dP}{dT}\right|^{c}_{\mathrm{CRL}} = \left(\frac{\partial P}{\partial T}\right)^{c}_{x} + \left(\frac{\partial P}{\partial x}\right)^{c}_{T} \cdot \left.\frac{dx}{dT}\right|_{\mathrm{CRL}}$$

10. What does thermodynamics tell us about the values of the derivative properties

$$\left(\frac{\partial P}{\partial T}\right)^{c}_{x,\mathrm{CXS}}, \quad \left(\frac{\partial P}{\partial x}\right)^{c}_{T,\mathrm{CXS}}$$

evaluated at the binary mixture critical point?

11. Show that the following identity holds [2]:

$$\lim_{x\to 0}\left[\left(\frac{\partial P}{\partial x}\right)_{T,\rho}\right]^{c} = \left.\frac{dP}{dx}\right|^{c}_{\mathrm{CRL}} - \left(\frac{\partial P}{\partial T}\right)^{c}_{x,\mathrm{CXS}} \cdot \left.\frac{dT}{dx}\right|^{c}_{\mathrm{CRL}}$$

Does this imply that $\lim_{x\to 0}[(\partial P/\partial x)_{T,\rho}]^{c}$ is finite as $x \to 0$? We note that, at finite compositions along the critical line, this property is predicted (nonclassically) to show a weak zero.

Bibliography

The material in this chapter is based mainly upon the following references.

[1] F. M. Munoz and E. H. Chimowitz, "Critical phenomena in mixtures, part 1: thermodynamic theory for the binary critical azeotrope," *J. Chem. Phys.,* vol. 99, p. 5438, 1993.

[2] J. M. H. Levelt-Sengers, "Chapter 1: Thermodynamics of solutions near the solvent's critical point," in *Supercritical Fluid Technology: Reviews in Modern Theory and Applications*, T. J. Bruno and J. F. Ely (eds.). Boca Raton, FL: CRC Press, 1991.

[3] S. De, Y. Shapir, and E. H. Chimowitz, "Diffusion in dilute binary fluids confined in porous structures near the solvent critical point," *J. Chem. Phys.*, vol. 119, p. 1035, 2003.

5 Solvation in Supercritical Fluids

The use of supercritical fluids as solvent media is driven mainly by the need to reduce the use of organic and halogenated solvents in chemical processes. In the future, one of the main aims of research in this area will be to supplant organic solvent use in many of these processes with solvents such as supercritical carbon dioxide, environmentally a much more acceptable alternative.

One of the most common engineering requirements in this area is the need to predict solubility, and other thermodynamic behavior, in high-pressure mixtures where the solvent is close to its critical point and contains nonvolatile solute species of large molecular weight present in small amounts. In this chapter, we address this problem focusing upon solvation in organic solid–supercritical fluid systems which are among the most technologically interesting. The extension of the analyses presented here to situations where the condensed phase may be a mixture of miscible liquids, for example, is straightforward and left to a problem in the additional exercises.

5.1 Solubility Analysis along the Phase Envelope

The Binary Case

At thermodynamic equilibrium, the equality of chemical equilibrium for solute species i in both phases is expressed by the equation $\mu_i^{\mathrm{s}}(T, P) = \mu_i(T, P, y_i)$, where we have assumed negligible solubility of the solvent in the solid phase. To distinguish the solid phase properties from those in the solvent fluid phase, we designate the former using the superscript s, while y_i denotes the mole fraction of species i in the fluid phase. Along the phase envelope, $d\mu_i^{\mathrm{s}} = d\mu_i$, and expanding these derivatives in terms of the thermodynamic variables, T, P, and y_i leads to the equation

$$\left(\frac{\partial \mu_i^{\mathrm{s}}}{\partial T}\right)_P dT + \left(\frac{\partial \mu_i^{\mathrm{s}}}{\partial P}\right)_T dP = \left(\frac{\partial \mu_i}{\partial T}\right)_{P,y_i} dT + \left(\frac{\partial \mu_i}{\partial P}\right)_{T,y_i} dP + \left(\frac{\partial \mu_i}{\partial y_i}\right)_{T,P} dy_i \tag{5.1}$$

Both isobaric and isothermal changes in the system are of interest. At constant pressure, for example, $dP = 0$ and so

$$\left(\frac{\partial \mu_i}{\partial y_i}\right)_{T,P} dy_i = \left[\left(\frac{\partial \mu_i^{\mathrm{s}}}{\partial T}\right)_P - \left(\frac{\partial \mu_i}{\partial T}\right)_{P,y_i}\right] dT \tag{5.2}$$

or

$$\left(\frac{\partial y_i}{\partial T}\right)_{p,\sigma} = \frac{(\partial \mu_i^{\mathrm{s}}/\partial T)_P - (\partial \mu_i/\partial T)_{P,y_i}}{(\partial \mu_i/\partial y_i)_{T,P}} \tag{5.3}$$

where σ denotes a derivative quantity evaluated along the phase envelope. Using standard thermodynamic identities in equation (5.3) leads to the equation:

$$\left(\frac{\partial y_i}{\partial T}\right)_{p,\sigma} = \frac{\bar{H}_i(T, P, y_i) - H_{i,\mathrm{pure}}(T, P)}{T(\partial \mu_i/\partial y_i)_{T,P}} \tag{5.4}$$

EXERCISE 5.1

Prove equation (5.4).

Since

$$\left(\frac{\partial \mu_i}{\partial y_i}\right)_{T,P} = RT\left[\frac{1}{y_i} + \left(\frac{\partial \ln \hat{\phi}_i}{\partial y_i}\right)_{T,P}\right] \tag{5.5}$$

where $\hat{\phi}_i$ is the fugacity coefficient of i in the supercritical phase, the following equation relates the solubility of i with T along the phase envelope:

$$\left(\frac{\partial \ln y_i}{\partial T}\right)_{P,\sigma} = \frac{\bar{H}_i(T, P, y_i) - H_{i,\text{pure}}(T, P)}{RT^2[1 + y_i(\partial \ln \hat{\phi}_i/\partial y_i)_{T,P}]} \tag{5.6}$$

For a pure nonvolatile solute,

$$H_{i,\text{pure}}(T, P) = H_i^\circ(T) - \Delta H_i^{\text{S}} + v_i^{\text{S}} P \tag{5.7}$$

where H_i° is the enthalpy in the ideal gas state, ΔH_i^{S} the heat of sublimation at temperature T, and v_i^{S} the pure solid phase molar volume. Substituting (5.7) into (5.6) leads to the following equation for solubility changes with T along the phase envelope:

$$\left(\frac{\partial \ln y_i}{\partial T}\right)_{P,\sigma} = -\frac{(H_i^\circ - \bar{H}_i) - (\Delta H_i^{\text{S}} - v_i^{\text{S}} P)}{RT^2[1 + y_i(\partial \ln \hat{\phi}_i/\partial y_i)_{T,P}]} \tag{5.8}$$

By a similar analysis the change of composition in the fluid phase with pressure along the equilibrium surface can be established as

$$\left(\frac{\partial \ln y_i}{\partial P}\right)_{T,\sigma} = \frac{(v_i^{\text{S}} - \overline{V}_i)}{RT[1 + y_i(\partial \ln \hat{\phi}_i/\partial y_i)_{T,P}]} \tag{5.9}$$

EXERCISE 5.2

Prove (5.9).

Since the denominators in equations (5.8) and (5.9) are positive, from considerations of thermodynamic stability theory like those given in chapter 1, the signs of the numerators in these expressions determine the sign of the solubility changes $(\partial \ln y_i/\partial T)_{P,\sigma}$ and $(\partial \ln y_i/\partial P)_{T,\sigma}$ with respect to T and P, respectively. It is clear that these depend upon the behavior of the solute partial molar properties $\bar{V}_i$ and $\bar{H}_i$ in the fluid phase.

Partial Molar Properties in Supercritical Mixtures

We now examine these terms more closely using their infinite-dilution limits, a convenient and often used reference condition in these very dilute solutions. For the solute, designated as species 1, the following thermodynamic identity for $\bar{V}_1^\infty$ can be derived:

$$\bar{V}_1^\infty = \rho^{-1} K_T \left[\left(\frac{\partial P}{\partial N_1}\right)_{T,V,N_2}^\infty N\right] \tag{5.10}$$

where ∞ designates the infinite-dilution limit, K_T is the solvent system's isothermal compressibility, ρ its density and $N = N_1 + N_2$. We note that all of the other properties besides K_T in equation (5.10) are finite at the critical point, a result that follows from the scaling-analysis results given in chapter 3 and 4. Furthermore, the following thermodynamic identity relates the properties $\bar{H}_i^\infty$ and $\bar{V}_i^\infty$:

$$\bar{H}_i^\infty = T\left(\frac{\partial P}{\partial T}\right)_\sigma \bar{V}_i^\infty + \text{a finite constant} \tag{5.11}$$

where $(\partial P/\partial T)_\sigma$ is a pure-solvent property evaluated at its critical point.

EXERCISE 5.3

Prove the identities given by equations (5.10) and (5.11), and show that the property $(\partial P/\partial T)_\sigma$ is finite and positive at the solvent's critical point.

From (5.10) and (5.11), we see that $\bar{H}_i^\infty$ and $\bar{V}_i^\infty$ are closely related (see exercise 5.3) and that, since K_T diverges to infinity at the critical point, so do both $\bar{V}_i^\infty$ and $\bar{H}_i^\infty$. Since both ρ and N are positive, the sign of the quantity $(\partial P/\partial N_1)^\infty_{T,V,N_2}$, which can be either positive or negative, determines the sign of the divergence of both $\bar{V}_i^\infty$ and $\bar{H}_i^\infty$ to either $+\infty$ or $-\infty$ at the solvent's critical point. This has implications for the form of the solubility data in the solvent's critical region, as we now discuss.

Shape of the y–P Isotherm

Equations (5.8) and (5.9) show that large solubility changes should be evident as the solvent's near-critical region is approached. In the majority of systems of technological interest, $\bar{V}_1^\infty$ shows negative divergence at the solvent's critical point (why?), and there will in general be three distinct data regimes in the y–P isotherm, at slightly supercritical temperatures:

1. At low pressures corresponding to an ideal gas, the term $(\partial P/\partial N_1)_{T,V,N_2}$ is positive, and thus both $\bar{V}_1^\infty > 0$ and $\bar{V}_1^\infty > v_1^s$. From (5.9), $(\partial y_1/\partial P)_{T,\sigma}$ will be negative, showing a low-pressure turning condition where $\bar{V}_1^\infty = v_1^s$.
2. As the density increases, however, attractive forces begin to dominate the fluid structure. The term $(\partial P/\partial N_1)_{T,V,N_2}$ represents the change in pressure observed by replacing a small solvent molecule with a larger solute one possessing "greater attractive forces." If it becomes negative at some density, this leads to negative values for both $\bar{V}_i^\infty$. This implies that the quantity $(\partial y_1/\partial P)_{T,\sigma}$ will be positive, with the largest solubility change occurring in the solvent's near-critical region. This is usually the most interesting thermodynamic region for designing separation processes, since effective solvent recovery can be achieved with modest reductions in pressure.
3. At very high densities, where repulsive forces are dominant, we expect the addition of a solute molecule to cause $(\partial P/\partial N_1)^\infty_{T,V,N_2}$ to be positive. At these conditions, the system compressibility is much smaller, giving rise to modest positive values in both $\bar{V}_i^\infty$ and $\bar{H}_i^\infty$, and a solubility maximum may exist where $\bar{V}_1^\infty = v_1^s$. Beyond this turning-point condition, the solubility starts to decrease with increasing pressure.

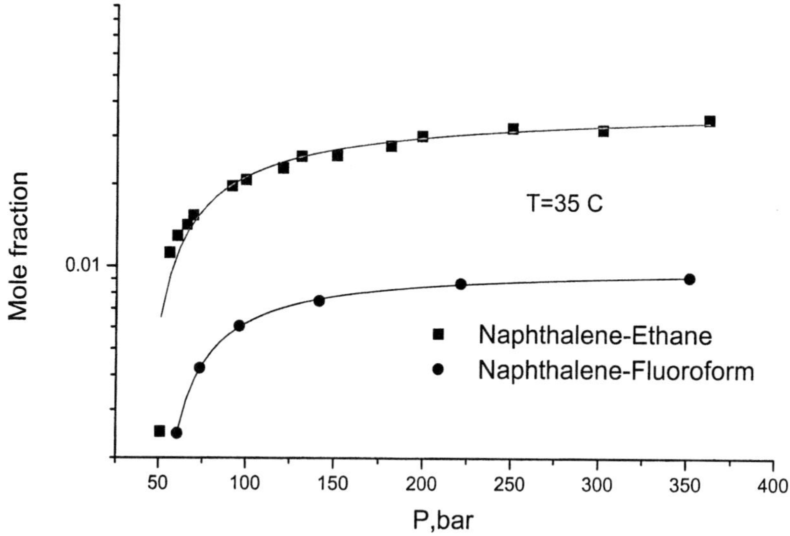

Figure 5.1 Solubility isotherms in supercritical binary mixtures

Figure 5.1 presents typical solubility data for naphthalene in two different solvents with reasonably similar critical properties [1] (ethane: $T_c = 32.2°C$, $P_c = 48.8$ bar, $\rho_c = 0.0067$ mol/cm^3; and fluoroform: $T_c = 26.1°C$, $P_c = 49.5$ bar, $\rho_c = 0.0075$ mol/cm^3). The rapid increase of solubility in the critical region is clear as the critical pressure is exceeded. Also evident is the stronger affinity of ethane for the solute species.

Experimental Data for $\bar{V}_i^\infty$ in Binary Mixtures

Several sets of data for partial molar properties in the near-critical region have been published. Some typical results are shown in figure 5.2 for the partial molar volume of naphthalene in supercritical carbon dioxide [2]—the large decrease in the partial molar volume of naphthalene in the critical region predicted by (5.10) is evident. The density of carbon dioxide at its critical point is approximately 0.0106 mole/cc, while $T_c = 31.3°C$, $P_c = 73.8$ bar. Figure 5.3 shows the compressibility cross-plotted with the solute's partial molar volume and, the linear correlation between these two quantities points to the constancy of the term $\rho^{-1}[(\partial P/\partial N_1)^\infty_{T,V,N_2} N]$ (see equation (5.10)) in this system over wide density ranges, a result that proves to be useful in representing data in these general systems, as we will soon discuss.

EXERCISE 5.4

In their study, Eckert et al. [2] found the solute property $\bar{V}_1^\infty$ from

$$\bar{V}_1^\infty(T, P) = v_{2,\text{pure}}(T, P) - \left(\frac{\partial v}{\partial y_2}\right)^\infty_{T,P} \quad (5.12)$$

Discuss the origin and validity of this equation for measuring $\bar{V}_1^{\infty}$. Measurements of the last term in equation (5.12), in addition to a knowledge of the specific volume (or density) dependence of the solvent on temperature and pressure, are used to yield $\bar{V}_1^{\infty}$ at the conditions of interest.

One may also consider the isobaric form of y–T solubility data in the critical region. These data are closely related to the retrograde effect in supercritical fluid solutions, which has been proposed as the basis for the design of selective schemes for chemical separations [3]. We now consider this phenomenon in some detail.

5.2 Retrograde Phenomena in Supercritical Mixtures

We start with an analysis of the single-solute case. The governing equation is (5.8), with the *retrograde region* defined as being where the quantity $(\partial y/\partial T)_{P,\sigma}$, or equivalently $(\partial \ln y/\partial T.)_{P,\sigma}$, is negative. The sign of the derivative $(\partial \ln y_i/\partial T)_{P,\sigma}$ is determined solely by the sign of the numerator term, which is dominated by the quantity $(H_i^{\circ} - \bar{H}_i) - \Delta H_i^{s}$. These enthalpy quantities are shown schematically in figure 5.4, which presents the *residual partial molar enthalpy* $H_i^0 - \bar{H}_i$ at constant pressure, with the solute *heat of sublimation* (vaporization) ΔH_i^{s}, which is generally constant over reasonable temperature ranges, versus temperature. Figure 5.4 illustrates equation (5.8) with the hatched area showing the retrograde region for the solute. In this figure, the temperatures T_{L} and T_{H} at which $(\partial \ln y_i/\partial T)_{P,\sigma} = 0$ correspond to limiting temperatures for the retrograde region along this isobar. At temperatures less that T_{L} or greater than T_{H}, the solute is not retrograde at this pressure. In addition, the pressure P^*, at which

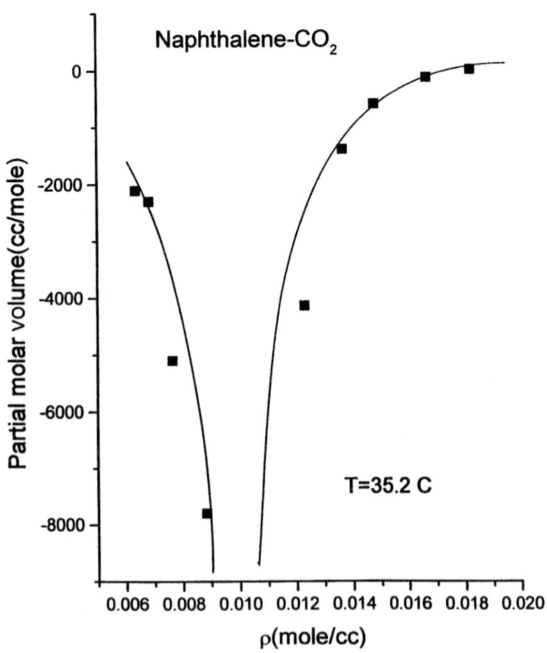

Figure 5.2 Partial molar volume of naphthalene in carbon dioxide in the critical region

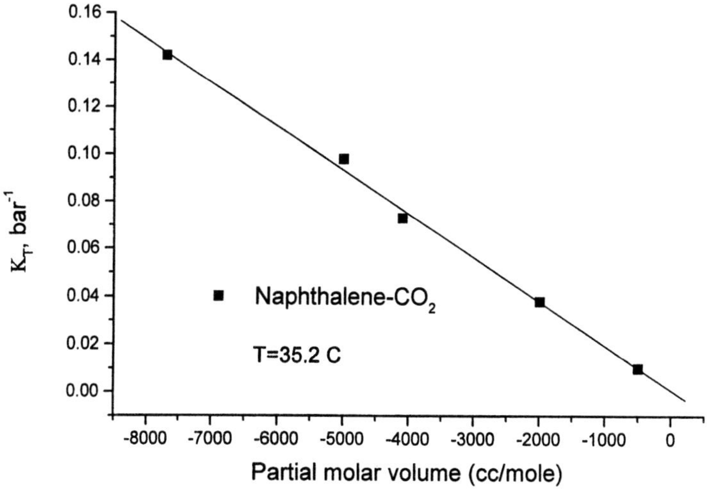

Figure 5.3 Relationship of compressibility to partial molar volume in naphthalene–carbon dioxide mixtures in the critical region

point $H_i^\circ - \bar{H}_i = \Delta H_i^s$ at a temperature T^*, defines an important limiting condition. At pressures above P^*, retrograde behavior does not exist over any temperature interval.

A companion scheme for figure 5.4 is shown in figure 5.5. At pressures close to the critical pressure of the pure solvent, solubility isobars exhibit both a maximum and minimum, corresponding to temperatures T_L and T_H, respectively, in figure 5.4. Once again, in the temperature region $T_L - T_H$, we have $(\partial \ln y_i / \partial T)_{P,\sigma} < 0$, and the solute is retrograde. The situation where the locus of inflection points (see inflection in figure 5.5) intersects an isobar at a point where $(\partial \ln y_i / \partial T)_{P,\sigma} = 0$ represents the limiting isobar at P^*, referred to in figure 5.4. We surmise from this discussion that retrograde behavior

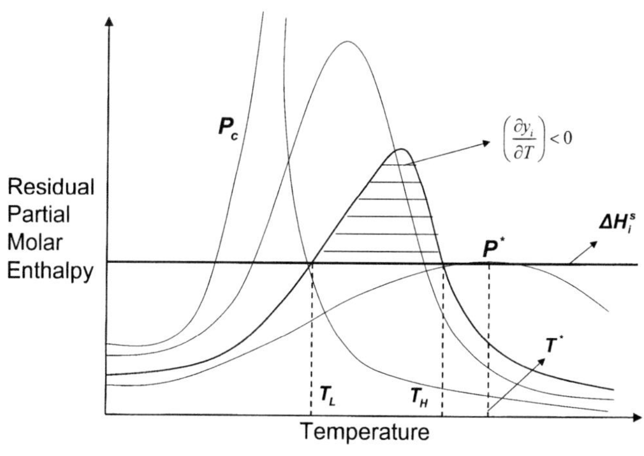

Figure 5.4 Schematic of residual partial molar enthalpy functions in the retrograde region

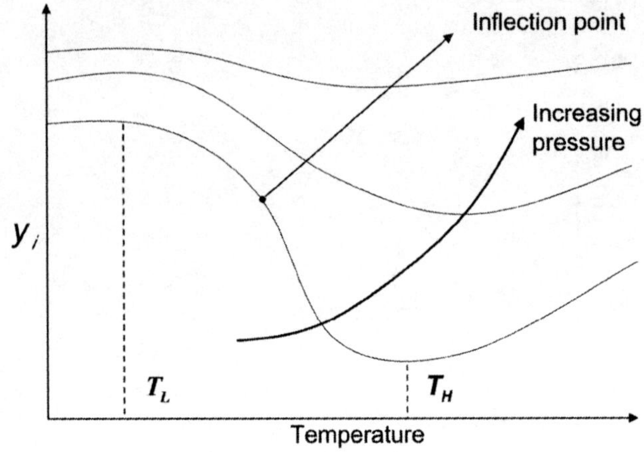

Figure 5.5 Schematic of solubility isobars in the retrograde region

is not observed at either very low or high pressures but rather at conditions in the fluid's critical region.

Crossover Regions in Dilute Binary Supercritical Mixtures

The utility of retrograde behavior for designing chemical separations depends upon the existence of *crossover phenomena* in these mixtures. A common feature of data for single solutes dissolved in a supercritical fluid is the existence of what has been termed a *crossover pressure* [3]. This is a pressure around which isotherms at various near-critical temperatures tend to converge. This absolute pressure region, for a specific solute, can vary considerably depending on the solvent's critical pressure. For 2,3-dimethylnaphthalene in ethylene (critical pressure = 50.4 bars), for example, the crossover pressure is approximately 120 bars, while it is closer to 145 bars in carbon dioxide (critical pressure = 73.8 bars).

A set of data for a mixture of 1,10-decanediol and carbon dioxide showing this behavior is shown in figure 5.6. Below the crossover pressure, an isobaric *increase* in temperature causes a solubility *decrease*, so that the solute is retrograde. Above the crossover pressure, the opposite effect occurs. This behavior can be understood by considering two opposing effects of temperature on solubility. The vapor pressure of the solid solute always increases with temperature, while the density of the fluid (related to its solvent power) usually decreases. Below the crossover pressure where the compressibility is larger, the density effect dominates, and the solubility decreases with increasing temperature. At pressures above the crossover pressure the vapor-pressure effect predominates, and in this case the solubility increases with temperature.

An important question is whether or not crossover points persist in binary and higher-order mixtures? Interestingly, this is often the case [4]. Given the existence of discrete crossover pressures for each component in such a system, we define the *crossover region*. This is a thermodynamic regime where one component is retrograde while the

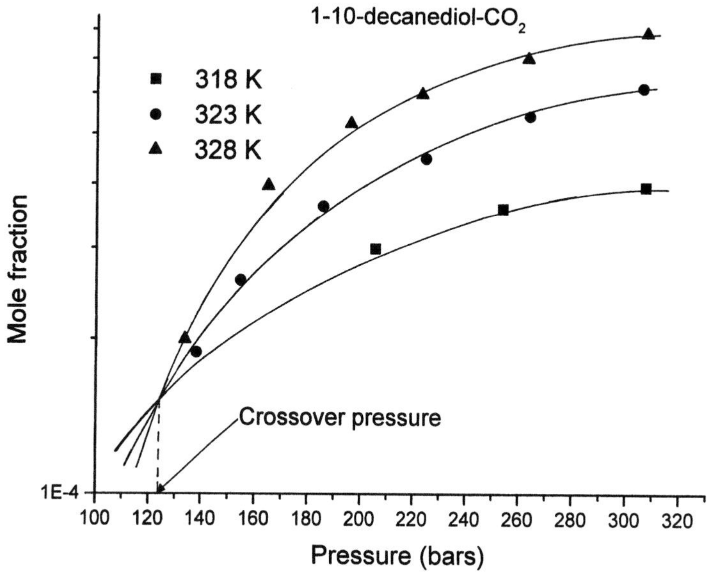

Figure 5.6 Solubility isotherms in binary supercritical mixtures showing a crossover pressure

other is not, as depicted schematically in figure 5.7. From a separations viewpoint, the implications of behavior like that depicted in figure 5.7 is interesting and can be used to design "perfectly" selective separations techniques. Figure 5.8 can be used to illustrate this graphically. Consider two compounds dissolved in a supercritical solvent with crossover pressures, assumed to be distinct for each component in the gas phase, shown at pressures P_1^* and P_2^*. We focus upon the middle pressure regime between P_1^* and P_2^*. At a pressure P_0, intermediate to P_1^* and P_2^*, consider that the gas phase is initially at a temperature T_H (with $T_H > T_L$) and subsequently cooled to a temperature T_L. For component 2, this is analogous to following a process path shown as a to b in figure 5.8. Along this path, component 2 is retrograde and so its solubility potential in the gas phase *increases* while the solubility of component 1 (which is not retrograde along the path c–d) *decreases*. Therefore, if a mixture of components 1 and 2 is initially extracted at a pressure P_0 and temperature T_H, and subjected to a decrease in temperature to T_L, pure component 1 will be forced to precipitate out of the gas phase as a pure species. This process has been studied and shown to yield pure components from a binary solute mixture. Figures 5.9 and 5.10 show the experimental crossover regions relevant to the process [4].

What is not clear is how wide the crossover regions in a given system might be. For example, at a pressure P_0 in a system's crossover pressure region, is there a reasonable temperature interval over which one component is retrograde while the others are not? The larger this interval, the better the prospect for designing an efficient separation process. In multicomponent solute systems, simple graphical representations like those just used for the binary case are not as useful; a more sophisticated mathematical description of the phenomenon is required which also demonstrates how to formally include solute–solute interactions in the fluid-phase analysis.

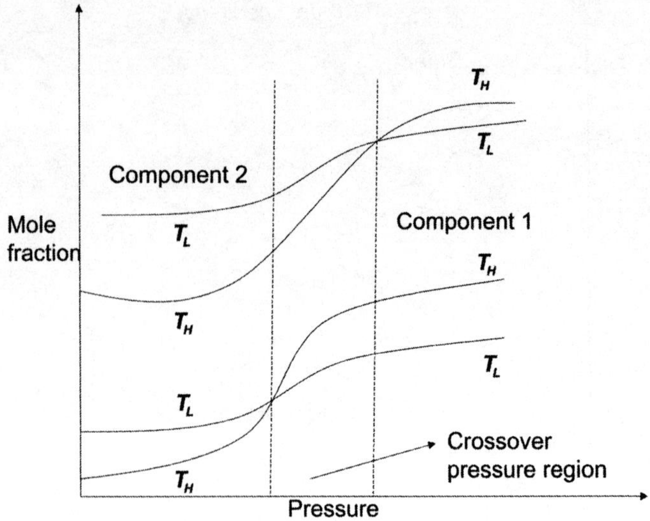

Figure 5.7 Schematic of a crossover region in a binary supercritical fluid mixture

Conditions for Retrograde Behavior in Multicomponent Mixtures

The equations for the variation of solubilities with temperature in multicomponent mixtures are given by the following matrix equations:

$$\boldsymbol{By} = \boldsymbol{c} \tag{5.13}$$

where

$$\boldsymbol{B} = \boldsymbol{I} + \boldsymbol{A} \tag{5.14}$$

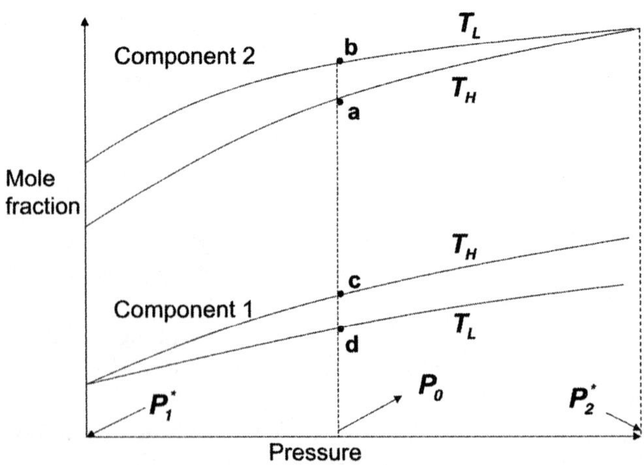

Figure 5.8 Crossover process schematic

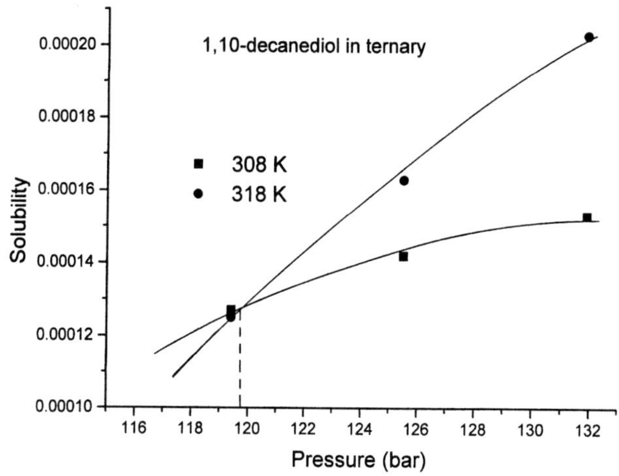

Figure 5.9 Crossover pressure in a ternary mixture

with $\boldsymbol{I}$ the identity matrix. The matrix $\boldsymbol{A}$ is defined as $\boldsymbol{A} \equiv \{a_{ij}\}$ with elements

$$a_{ij} = y_i \left(\frac{\partial \ln \hat{\phi}_i}{\partial y_j} \right)_{T,P,y_{k \neq j,N}} \tag{5.15}$$

The vector $\boldsymbol{c}$ is defined as

$$\boldsymbol{c} \equiv \{c_i\} \tag{5.16}$$

where

$$c_i \equiv -\frac{H_i^\circ - \bar{H}_i - (\Delta H_i^{\mathrm{s}} - v_i^{\mathrm{s}} P)}{RT^2} \tag{5.17}$$

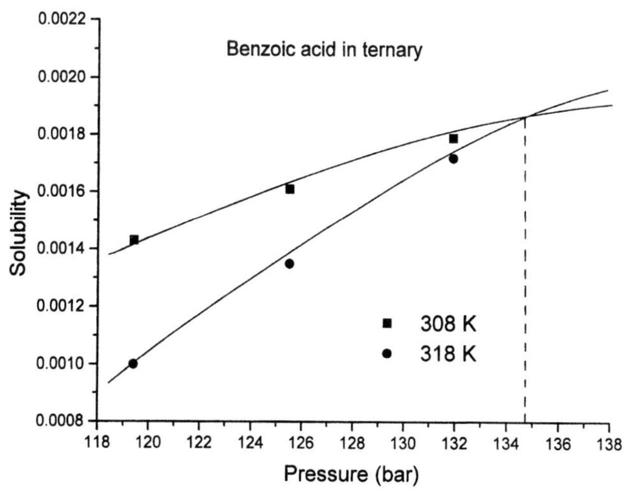

Figure 5.10 Crossover pressure in a ternary mixture

The vector $\mathbf{y}$ of derivatives represents the solubility change of each component with temperature along the phase envelope, and is defined as

$$\mathbf{y} \equiv \begin{bmatrix} \left(\dfrac{\partial \ln y_1}{\partial T}\right)_{P,\sigma} \\ \left(\dfrac{\partial \ln y_2}{\partial T}\right)_{P,\sigma} \\ \vdots \\ \left(\dfrac{\partial \ln y_{N-1}}{\partial T}\right)_{P,\sigma} \end{bmatrix} \tag{5.18}$$

For a single solute i dissolved in a pure supercritical fluid, equations (5.13)–(5.18) reduce to equation (5.8).

For an N-component phase ($N - 1$ solutes in a supercritical fluid), the matrix $\boldsymbol{B}$ in equation (5.13) has a structure corresponding to

$$\boldsymbol{B} = \begin{bmatrix} y_1\left(\dfrac{\partial \ln \hat{\phi}_1}{\partial y_1}\right)_{T,P} + 1 & y_2\left(\dfrac{\partial \ln \hat{\phi}_1}{\partial y_2}\right)_{T,P} & \cdots & y_{N-1}\left(\dfrac{\partial \ln \hat{\phi}_1}{\partial y_{N-1}}\right)_{T,P} \\ y_1\left(\dfrac{\partial \ln \hat{\phi}_2}{\partial y_1}\right)_{T,P} & y_2\left(\dfrac{\partial \ln \hat{\phi}_2}{\partial y_2}\right)_{T,P} + 1 & \cdots & y_{N-1}\left(\dfrac{\partial \ln \hat{\phi}_2}{\partial y_{N-1}}\right)_{T,P} \\ \vdots & \vdots & \ddots & \vdots \\ y_1\left(\dfrac{\partial \ln \hat{\phi}_{N-1}}{\partial y_1}\right)_{T,P} & y_2\left(\dfrac{\partial \ln \hat{\phi}_{N-1}}{\partial y_2}\right)_{T,P} & \cdots & y_{N-1}\left(\dfrac{\partial \ln \hat{\phi}_{N-1}}{\partial y_{N-1}}\right)_{T,P} + 1 \end{bmatrix} \tag{5.19}$$

A general solution to equation (5.13) can be obtained using Cramer's rule, given by

$$\left(\frac{\partial \ln y_i}{\partial T}\right)_{P,\sigma} = \frac{|\boldsymbol{B}_i|}{|\boldsymbol{B}|} \tag{5.20}$$

where $|\boldsymbol{B}_i|$ is the determinant of matrix $\boldsymbol{B}_i$ formed by replacing the ith column of $\boldsymbol{B}$ by the vector $\boldsymbol{c}$; here $|\boldsymbol{B}|$ represents the determinant of the matrix $\boldsymbol{B}$.

The matrix $\boldsymbol{B}$ can also be expressed as

$$\boldsymbol{B} = \begin{bmatrix} \dfrac{y_1}{RT}\left(\dfrac{\partial \mu_1}{\partial y_1}\right)_{T,P} & \dfrac{y_2}{RT}\left(\dfrac{\partial \mu_1}{\partial y_2}\right)_{T,P} & \cdots & \dfrac{y_{N-1}}{RT}\left(\dfrac{\partial \mu_1}{\partial y_{N-1}}\right)_{T,P} \\ \dfrac{y_1}{RT}\left(\dfrac{\partial \mu_2}{\partial y_1}\right)_{T,P} & \dfrac{y_2}{RT}\left(\dfrac{\partial \mu_2}{\partial y_2}\right)_{T,P} & \cdots & \dfrac{y_{N-1}}{RT}\left(\dfrac{\partial \mu_2}{\partial y_{N-1}}\right)_{T,P} \\ \vdots & \vdots & \ddots & \vdots \\ \dfrac{y_1}{RT}\left(\dfrac{\partial \mu_{N-1}}{\partial y_1}\right)_{T,P} & \dfrac{y_2}{RT}\left(\dfrac{\partial \mu_{N-1}}{\partial y_2}\right)_{T,P} & \cdots & \dfrac{y_{N-1}}{RT}\left(\dfrac{\partial \mu_{N-1}}{\partial y_{N-1}}\right)_{T,P} \end{bmatrix} \tag{5.21}$$

which can be further decomposed into the product of two matrices given by

$$\boldsymbol{B} = \frac{N_t}{RT}\boldsymbol{UV} \tag{5.22}$$

with

$$\boldsymbol{U} = \begin{bmatrix} \left(\frac{\partial \mu_1}{\partial N_1}\right)_{T,P} & \left(\frac{\partial \mu_1}{\partial N_2}\right)_{T,P} & \cdots & \left(\frac{\partial \mu_1}{\partial N_{N-1}}\right)_{T,P} \\ \left(\frac{\partial \mu_2}{\partial N_1}\right)_{T,P} & \left(\frac{\partial \mu_2}{\partial N_2}\right)_{T,P} & \cdots & \left(\frac{\partial \mu_2}{\partial N_{N-1}}\right)_{T,P} \\ \vdots & \vdots & \ddots & \vdots \\ \left(\frac{\partial \mu_{N-1}}{\partial N_1}\right)_{T,P} & \left(\frac{\partial \mu_{N-1}}{\partial N_2}\right)_{T,P} & \cdots & \left(\frac{\partial \mu_{N-1}}{\partial N_{N-1}}\right)_{T,P} \end{bmatrix} \tag{5.23}$$

and

$$\boldsymbol{V} = \begin{bmatrix} \frac{y_1}{1-y_1} & 0 & \cdots & 0 \\ 0 & \frac{y_2}{1-y_2} & \cdots & 0 \\ \vdots & \vdots & \ddots & \vdots \\ 0 & 0 & \cdots & \frac{y_{N-1}}{1-y_{N-1}} \end{bmatrix} \tag{5.24}$$

N_j denotes the number of moles of species j and N_t the total number of moles of all the solute species. Equations (5.22)–(5.24) show that $\boldsymbol{B}$ can be *decomposed into the product of two positive definite matrices*. A theorem in linear algebra [6] states that, for any two $(n \times n)$ matrices $\boldsymbol{X}$ and $\boldsymbol{Y}$, $\det \boldsymbol{XY} = \det \boldsymbol{X} \det \boldsymbol{Y}$. From this result it follows that $\det \boldsymbol{B}$ in equation (5.22) must be positive (see the additional exercises at the end of this chapter). As a result, the sign of the quantity $(\partial \ln y_i/\partial T)_{P,\sigma}$ in equation (5.20) is determined solely by the sign of the numerator determinant $|\boldsymbol{B}_i|$ in equation (5.20). The sign of this determinant thus provides both a necessary and sufficient condition for the retrograde behavior of a solute i in a multicomponent system. In figure 5.11, we show the function $|\boldsymbol{B}_i|$ calculated for 1,10-decanediol in the 1,10-decanediol – benzoic acid – carbon dioxide system at 120 bar. It predicts a range of temperature (T_L to T_H) for which 1,10-decanediol is not retrograde.

5.3 Density Dependence of Isothermal Solubility Data in the Critical Region

A notable feature of solubility data in supercritical fluids is the linear appearance of the data when density is used as the thermodynamic state variable [7]. We examine the origin of this behavior and show that it leads to very useful results for experimentally determining various thermodynamic properties in supercritical mixtures.

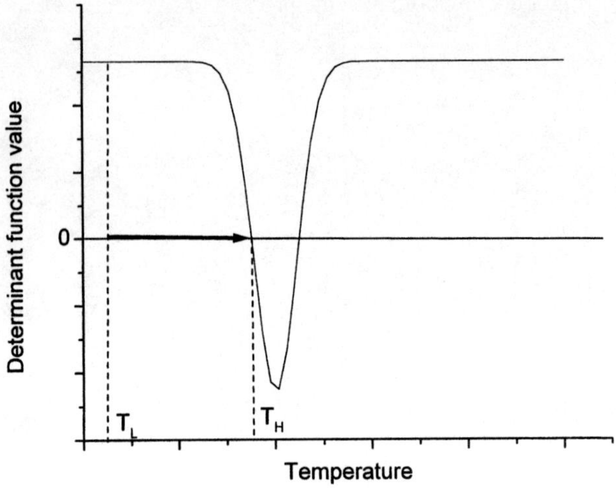

Figure 5.11 Schematic showing crossover temperature region

Equation (5.9) can be rearranged to yield

$$\left(\frac{\partial \ln y_i}{\partial \ln \rho}\right)_{T,\sigma} = \frac{v_i^{\mathrm{s}} - \bar{V}_i}{RTK_T[1 + y_i(\partial \ln \hat{\phi}_i/\partial y_i)_{T,P}]} \tag{5.25}$$

where K_T is the isothermal compressibility of the fluid phase and $\hat{\phi}_i$ the fugacity coefficient of the solute in that phase. Resorting to the infinite-dilution limit, we may reduce equation (5.25) to the following form (see exercise 5.5):

$$\left(\frac{\partial \ln y_i}{\partial \ln \rho}\right)_{T,\sigma}^{\infty} = \frac{v_i^{s} - \bar{V}_i^{\infty}}{RTK_T} \tag{5.26}$$

EXERCISE 5.5

Prove the identity

$$\lim_{y_i \to 0}\left[y_i\left(\frac{\partial \ln \hat{\phi}_i}{\partial y_i}\right)_{T,P}\right] = 0 \tag{5.27}$$

In the near-critical region $K_T \gg v_i^{\mathrm{s}}$. From equation (5.26), we see that the relationship between $\ln y_i$ and $\ln \rho$ given by that equation will then be determined largely by the variation of the term $\bar{V}_i^{\infty}/K_T$ with density along an isotherm. The results in figure 5.3 exemplify that this term is virtually constant over wide density ranges, as has been observed in a great variety of solute–solvent systems; this constancy will lead to linearity in the data.

EXERCISE 5.6

Try and develop a form of equation (5.25) in terms of reduced solvent properties defined as $T_{\mathrm{r}} \equiv T/T_{\mathrm{c}}$ and $\rho_{\mathrm{r}} \equiv \rho/\rho_{\mathrm{c}}$. What do you conclude from this analysis?

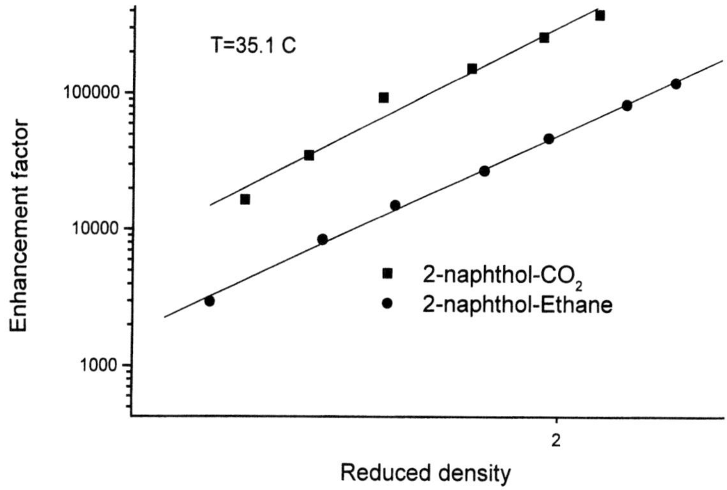

Figure 5.12 Enhancement-factor data in the critical region of a binary supercritical mixture

Enhancement Factors

The data in dilute supercritical mixtures are also often represented in terms of the solute's *enhancement factor* E [1]. This property is defined as

$$E \equiv y_i/(P_i^{\text{vap}}/P) \tag{5.28}$$

from which it can be seen that E represents the ratio of the actual solubility to the ideal gas solubility at the same thermodynamic conditions. Using this definition we see that:

$$\left(\frac{\partial \ln E_1}{\partial \ln \rho}\right)_{T,\sigma} \approx -\frac{\bar{V}_1^{\infty}}{RTK_T} \tag{5.29}$$

In figure 5.12, we show a set of typical enhancement-factor data [1] for a nonvolatile solute in two different supercritical solvents with virtually identical critical temperatures; the linearity over the density range is striking and is once again a consequence of the invariance of the property $\bar{V}_i^{\infty}/K_T$ over a wide density range. The advantage of an enhancement-factor representation such as that seen in figure 5.12 is that it vividly illustrates the different solvent (chemical) affinities for a given solute, since vapor-pressure effects are absent at the same solvent temperature—for the same solute species.

5.4 Data Modeling with Engineering Equations of State

In the last section, we showed how the prediction of a few key properties can lead to useful macroscopic thermodynamic equations for representing solubility behavior in dilute supercritical mixtures. In this section, we investigate standard approaches for modeling experimental data in these systems with equations of state. We begin by examining the idea of *corresponding states*, which attempts to express the thermodynamic behavior of fluids and their mixtures relative to their critical points.

In its macroscopic form one way of stating the corresponding states hypothesis for pure fluids is that *all substances, at the same reduced conditions of temperature and density (volume), have the same reduced pressures.* In other words, if the relevant thermodynamic variables are scaled (reduced) with respect to their critical values, the relationship describing the P–v–T behavior for all fluids becomes universal. This intuitive powerful idea has permeated much of model development in applied molecular thermodynamics. Here we further explore this idea, initially dealing with pure fluids, followed by consideration of mixtures.

The Corresponding-States Concept: Pure Fluids

The celebrated *van der Waals* (VdW) equation of state is a convenient form to illustrate these ideas, since it obeys the corresponding states principle. This equation of state is given by the model

$$P = \frac{RT}{v-b} - \frac{a}{v^2} \tag{5.30}$$

with

$$T_c = \frac{8a}{27Rb} \tag{5.31}$$

$$P_c = \frac{a}{27b^2} \tag{5.32}$$

$$v_c = 3b \tag{5.33}$$

The relationships (5.31)–(5.33) are found from solving the critical-point conditions $(\partial P/\partial v)_{T_c} = 0$ and $(\partial^2 P/\partial v^2)_{T_c} = 0$. (Note that $z_c = 3/8$ for this model, the same for all substances). Rearranging the VdW equation of state in terms of the reduced variables $T_r \equiv T/T_c$, $P_r \equiv P/P_c$, and $v_r \equiv v/v_c$ leads to the equation

$$\left(P_r + \frac{3}{v_r^2}\right)\left(v_r - \frac{1}{3}\right) = \frac{8}{3}T_r \tag{5.34}$$

This equation obviously captures the corresponding-states principle.

EXERCISE 5.7

Prove equation (5.34), where the variables T, v, and P have been reduced (scaled) accordingly. Show that this model gives rise to the following universal equations for the following properties:

$$P_c K_T = -\frac{1}{v_r}\left(-\frac{24T_r}{(3v_r-1)^2} + \frac{6}{v_r^3}\right)^{-1} \tag{5.35}$$

$$T_{\rm c}\alpha_{\rm p} = -\frac{1}{v_{\rm r}}\left(\frac{8}{3v_{\rm r}-1}\right)\left(-\frac{24T_{\rm r}}{(3v_{\rm r}-1)^2}+\frac{6}{v_{\rm r}^3}\right)^{-1} \tag{5.36}$$

Even in pure fluids, however, agreement between a corresponding-states model and experimental results will usually only be approximate. Improving the accuracy of these corresponding-state equations of state has thus been a staple of molecular-thermodynamics research in chemical engineering for decades.

Some of the concepts advanced for doing so have been quite successful, particularly the introduction of 3-parameter corresponding-states equations of state such as Soave's modification of the Redlich–Kwong (RK) model [8] (see exercise 5.8).

EXERCISE 5.8

Show that the *Redlich–Kwong equation of state* (presented below) predicts that the reduced saturation-pressure curve for all components, when plotted against reduced temperature, will be a universal function of reduced temperature.

This equation of state is given by

$$P = \frac{RT}{v-b} - \frac{a}{v(v+b)T^{1/2}}$$

How would you overcome this limitation with this model? Compare your ideas with those provided by Soave.

Corresponding States: Mixtures

The corresponding-states idea carries through to mixtures with little trouble. If one establishes critical parameters for a mixture, with which to scale thermodynamic variables such as temperature, density, and pressure, then the idea of a universal relationship can still be derived for the system's, P–v–T behavior. In this case, a mixture of given composition is treated as a *pseudopure fluid* with the result that these approaches are often referred to as *one-fluid* theories. The pseudopure fluid then obeys a universal P–v–T equation in terms of a new set of reduced parameters related to specific mixture properties.

We now illustrate this approach using so-called *mixing rules* for the mixture parameters $a_{\rm m}$ and $b_{\rm m}$ in the VdW model which implicitly provide these reduced parameters. They utilize the pure-component parameters a_i and b_i of the mixture's constituent species to provide "equivalent" a and b parameters for the mixture. These mixing rules are given by the equations

$$a_{\rm m} = \sum_i \sum_j x_i x_j a_{ij} \tag{5.37}$$

$$b_{\rm m} = \sum_i x_i b_i \tag{5.38}$$

and are often referred to as the *VdW 1 mixing rules*. They indirectly provide a pseudopure-fluid critical temperature, pressure, and volume for a mixture of given composition given by the equations

$$T_{\rm cm} = \frac{8a_{\rm m}}{27Rb_{\rm m}} \tag{5.39}$$

$$P_{\rm cm} = \frac{a_{\rm m}}{27b_{\rm m}^2} \tag{5.40}$$

$$v_{\rm cm} = 3b_{\rm m} \tag{5.41}$$

If we use (5.37)–(5.38) in the VdW equation of state, we end up with the universal equation (5.34) which now represents the mixture properties in terms of these reduced properties.

EXERCISE 5.9

It is instructive to examine other solution properties to ascertain whether or not they are universalized by the one-fluid approach as described here. For example, the solute partial molar volume in a dilute binary system of solute (i) and fluid (j) has been shown to be an important and useful property in previous chapters. It is given by the thermodynamic identity

$$\bar{V}_i^\infty = \lim_{N_i \to 0} -\frac{(\partial P/\partial N_i)_{T,V,N_j}}{(\partial P/\partial V)_{T,N}} \tag{5.42}$$

$$N = N_i + N_j \tag{5.43}$$

Using the VdW model with mixture rules given by equations (5.42) and (5.43) shows that

$$\left(\frac{\bar{V}_i^\infty}{v_{\rm cj}}\right) = \frac{-\frac{8T_{\rm r}}{3v_{\rm r}-1} - \frac{8T_{\rm r}}{(3v_{\rm r}-1)^2}\frac{b_i}{b_j} + \frac{6a_{ij}}{v_{\rm r}^2 a_{jj}}}{-\frac{24T_{\rm r}}{(3v_{\rm r}-1)^2} + \frac{6}{v_{\rm r}^3}} \tag{5.44}$$

where $v_{\rm cj}$ is the solvent critical volume. Clearly this is *not* a universal equation.

The accuracy of this approach for predicting mixture P–v–T behavior can be evaluated as follows: Choose a set of P–v–T conditions from mixture data, calculate the one-fluid results using equations (5.34) and (5.37)–(5.38), and compare with the actual conditions of the data. In general, such an approach will yield quantitatively disappointing results, and it is essential to use experimental data to enhance the accuracy of such models as we now discuss.

The Use of Experimental Data with Equation-of-State Models

Notwithstanding the limitations of the cubic equations of state mentioned thus far, they continue to be widely used for the calculation of thermodynamic properties in supercritical fluid systems. The ability to do these calculations is an important requirement for designing supercritical fluid chemical processes, since many of these properties are not easily measurable. The usual approach taken for modeling fluid-phase properties has generally used modifications of these equations of state. These adaptations incorporate at least a third parameter, like the acentric factor, that extends the capability of these models beyond the usual corresponding-states limitations.

There are three such equations widely used for these purposes, the van der Waals (VdW) equation, Soave's modification of the Redlich–Kwong equation [8], and the Peng–Robinson model [9]. These models are basically of the same genre, rooted in the VdW approach. The advantage of all of these models is that they can easily be fitted to experimental data and subsequently used for design purposes. Since these equations of state are cubic in density (volume), the numerical techniques required for manipulating these them are also much simplified. Furthermore, except for the very-near-critical region, they work quite satisfactorily—a view based upon their long-standing successful use in many diverse solute–fluid systems. Here we demonstrate how one of these models, the Peng–Robinson equation, may be used for modeling supercritical fluid mixtures, in conjunction with experimental solubility data.

The central task is to use solubility data in the fluid phase to fit a parameter(s) in the equation-of-state model. Solid–fluid solubility data are chosen here, since they are the most common type of data readily available. It is customary to assume that the fluid-phase solubility in the solid is negligible, and under these circumstances the defining equation for y_i, the mole fraction of solute i in the fluid phase, is given by the equation

$$y_i = \frac{P_i^{\text{sat}}}{P\hat{\phi}_i} \exp[(P - P_i^{\text{sat}})v_i^{\text{s}}/RT] \tag{5.45}$$

where y_i denotes the gas-phase mole fraction of component i, P_i^{sat} is the vapor pressure of i, P the total system pressure, $\hat{\phi}_i$ the fugacity coefficient of i in the fluid phase, v_i^{s} the pure component solid molar volume, T the temperature, and R the gas constant. The equation of state is used to estimate the vapor-phase fugacity coefficient.

For the Peng–Robinson equation of state given by

$$P = \frac{RT}{v - b} - \frac{a(T)}{v(v + b) + b(v - b)} \tag{5.46}$$

$\hat{\phi}_i$ is given by

$$\ln \hat{\phi}_i = \frac{b_i}{b}(Z - 1) - \ln(Z - B) - \frac{A}{2^{1.5}B} 2 \sum_j (1 - k_{ij}) \frac{(a_i a_j)^{1/2}}{a} y_i - \frac{b_i}{b} \ln \frac{Z + 2.414B}{Z - 0.414B} \tag{5.47}$$

where the compressibility Z of the mixture, satisfies

$$Z^3 - (1 - B)Z^2 + (A - 3B^2 - 2B)Z - (AB - B^2 - B^3) = 0 \tag{5.48}$$

with

$$A = \frac{aP}{(RT)^2} \tag{5.49}$$

$$B = \frac{bP}{RT} \tag{5.50}$$

$$a_i = \Omega_a \frac{R^2 T_{ci}^2}{P_{ci}} [1 + \kappa_i (1 - T_{ri}^{1/2})]^2 \tag{5.51}$$

$$b_i = \Omega_b \frac{R T_{ci}}{P_{ci}} \tag{5.52}$$

$$\kappa_i = 0.37464 + 1.54226\omega_i - 0.26992\omega_i^2 \tag{5.53}$$

and

$$\Omega_a = 0.45724, \quad \Omega_b = 0.0778 \tag{5.54}$$

The mixture parameters a and b are given by the mixing rules

$$a = \sum_i \sum_j y_i y_j a_i^{1/2} a_j^{1/2} (1 - k_{ij}) \tag{5.55}$$

$$b = \sum_i y_i b_i \tag{5.56}$$

The symbols T_{ci}, P_{ci}, and ω_i in these equations refer to the critical temperature, critical pressure, and acentric factor of the *pure species i*, respectively. The term T_{ri} refers to the reduced temperature of component i: $T_{ri} = T/T_{ci}$. The *binary interaction parameter* k_{ij} in equation (5.55) is introduced to account for interactions between species i and j, with $k_{ii} = 0$ and $k_{ij} = k_{ji}$. It is a purely empirical factor that is evaluated by regressing equation (5.45) against experimental data.

For a ternary system consisting of two solutes (components 2 and 3) in a supercritical fluid solvent (component 1), the following k_{ij} factors must be determined: k_{12} from binary data for components 1 and 2, k_{13} from binary data for components 1 and 3, and k_{23} from ternary data for the gas phase. In the data regression, an objective function is minimized, usually in the *least-squares* sense. An example of such a function is given by

$$\Phi(\mathbf{y}, T, P) = \sum_{i=1}^{m} (y_i^{\text{model}} - y_i^{\text{data}})^2 \tag{5.57}$$

where y_i^{model} is found from equation (5.45) in conjunction with equations (5.46)–(5.56). The y_i^{data} are the experimentally measured solubilities at given thermodynamic conditions (m represents the total number of data points used).

The objective of the regression analysis is to minimize the function Φ with respect to the interaction parameters k_{ij}—in this way, we achieve a so-called "best fit" of the model to the data. An example of the results of this procedure to work with, for the system of 1,10-decanediol with carbon dioxide, is given in the additional exercises for data at three temperatures, over a wide pressure range (120–310 bar). In this case, the single binary interaction parameter k_{12} (representing the solute–fluid interaction) was relatively constant with respect to temperature, but this need not be the case. In cases where this temperature dependence is strong, it can be accommodated by fitting to a simple function of temperature. Thus, one resorts to empiricism, with the payoff being equations that do a very reasonable job of fitting experimental data.

Finally, an additional point of concern may be that, for the solute species, the respective pure-component critical properties may not be known experimentally. In this case, one usually resorts to estimation methods [10], and any inaccuracies in these estimated critical properties are absorbed into the k_{ij} factors. This makes extreme precision in the determination of these critical properties unnecessary. Once the equation of state has been fitted to the available data, it can then be used for a variety of calculations of the system's thermodynamic properties. These may range from solubility calculations, at conditions for which data are unavailable, to more exotic types of thermodynamic properties that are difficult, if not impossible, to measure directly.

EXERCISE 5.10

Develop the analogous result to equation (5.45) where the condensed phase consists of a binary mixture of miscible liquids.

5.5 The Infinite-Dilution Reference Condition

Thus far in this chapter, we have focused upon the important case of dilute mixtures in the vicinity of the solvent's critical point. The infinitely dilute solute reference state is often used to describe mixture thermodynamics in these systems, and we now discuss the accuracy of this widely adopted approximation, often referred to as *Henry's law*, especially in the critical region.

With this approach, the solute fugacity in the fluid phase can be found from the relationship

$$\hat{f}_1 = x_1 k_1 \tag{5.58}$$

with x_1 being the solute mole fraction in the phase and k_1 the Henry's-law constant defined as

$$k_1(T, P) \equiv \lim_{x_1 \to 0} \frac{\hat{f}_1}{x_1} = P \lim_{x_1 \to 0} \frac{\hat{f}_1}{x_1 P} = P\hat{\phi}_1^{\infty} \tag{5.59}$$

In (5.59), $\hat{\phi}_1^{\infty}$ is the solute's fugacity coefficient in the infinite-dilution limit and P the system pressure. In the case where the condensed phase is treated as pure solute, the Henry's law prediction of solubility in the supercritical phase is given by

$$x_{i,\mathrm{HL}} = \frac{P_1^{\mathrm{S}}}{k_1} = \frac{P_1^{\mathrm{S}}}{P\hat{\phi}_1^{\infty}} \tag{5.60}$$

where P_1^S is the solute vapor pressure; here the Poynting correction, usually a small factor, has been neglected. By construction, equation (5.60) does not account for solute–solute interactions in the solution since we are at the $x \to 0$ limit. Incorporating these effects can be systematically done, however, by expanding $\hat{\phi}_1$ in a Taylor series about the infinite-dilution limit [11, 12]. The Taylor expansions of $\ln \hat{\phi}_1$ and $\ln \hat{\phi}_2$ around $x_1 = 0$ in a binary mixture are given by

$$\ln \hat{\phi}_1 = \ln \hat{\phi}_1^{\infty} + \sum_{n=1}^{\infty} \frac{1}{n!} \left(\frac{\partial^n \ln \hat{\phi}_1}{\partial x_1^n} \right)_{T,P,x_1 \to 0} x_1{}^n \tag{5.61}$$

$$\ln \hat{\phi}_2 = \ln \hat{\phi}_{2,\text{pure}} + \sum_{n=1}^{\infty} \frac{1}{n!} \left(\frac{\partial^n \ln \hat{\phi}_2}{\partial x_1^n} \right)_{T,P,x_1 \to 0} x_1{}^n \tag{5.62}$$

Using the Gibbs–Duhem equation, it is straightforward to show that the coefficients with the summation sign in equation (5.61) are related to those in equation (5.62). For example,

$$\lim_{x_1 \to 0} \left(\frac{\partial \ln \hat{\phi}_2}{\partial x_1} \right)_{T,P} = 0 \tag{5.63}$$

$$\lim_{x_1 \to 0} \left[\frac{1}{n!} \left(\frac{\partial^{n+1} \ln \hat{\phi}_2}{\partial x_1^{n+1}} \right)_{T,P} + \frac{1}{(n-1)!} \left(\frac{\partial^n \ln \hat{\phi}_1}{\partial x_1^n} \right)_{T,P} - \frac{1}{(n-1)!} \left(\frac{\partial^n \ln \hat{\phi}_2}{\partial x_1^n} \right)_{T,P} = 0 \right] \tag{5.64}$$

Using (5.60)–(5.63) leads to the following expansions for $\ln \hat{\phi}_1$ and $\ln \hat{\phi}_2$ around the infinite-dilution reference condition:

$$\begin{aligned} \ln \hat{\phi}_1 &= \ln \hat{\phi}_1^{\infty} - K_{11} x_1 + \text{higher-order terms} \\ \ln \hat{\phi}_2 &= \ln \phi_{2,\text{pure}} + \tfrac{1}{2} K_{11} x_1^2 + \text{higher-order terms} \end{aligned} \tag{5.65}$$

where

$$K_{11} = - \left(\frac{\partial \ln \hat{\phi}_1}{\partial x_1} \right)_{T,P}^{\infty} \tag{5.66}$$

These expansions are valid everywhere except exactly at the critical point of the solvent itself where K_{11} diverges (see the results in chapter 4). Thus, at all other conditions, the first-order correction terms to be the quantities $\hat{\phi}_1^{\infty}$ and $\hat{\phi}_{2,\text{pure}}$ are given by the equations

$$\begin{aligned} \hat{\phi}_1 &= \hat{\phi}_1^{\infty} \exp(-K_{11} x_1) \\ \hat{\phi}_2 &= \hat{\phi}_{2,\text{pure}} \exp\left(\tfrac{1}{2} K_{11} x_1^2\right) \end{aligned} \tag{5.67}$$

From (5.67) in (5.60), we have that

$$\frac{x_1}{x_{1,\mathrm{HL}}} = \exp(K_{11}x_1) \tag{5.68}$$

with

$$K_{11} = \frac{v}{kT}\left(\frac{\partial P}{\partial x_1}\right)_{T,\rho}^{\infty^2} K_T > 0 \tag{5.69}$$

Equations (5.68) and (5.69) predict that, in the critical region, Henry's law will always underestimate the true solubility, possibly quite significantly, since K_T can be large here. Extension to higher-order solutions follows analogously (see exercise 5.11).

EXERCISE 5.11

A case of interest is that of a binary mixture consisting of two solutes in a supercritical solvent. In this case, the first-order corrections for both solutes (here designated as species 1 and 2) are given as,

$$\begin{aligned} \hat{\phi}_1 &= \hat{\phi}_1^{\infty} e^{-K_{11}x_1} e^{-K_{12}x_2} \\ \hat{\phi}_2 &= \hat{\phi}_2^{\infty} e^{-K_{22}x_2} e^{-K_{12}x_1} \end{aligned} \tag{5.70}$$

Prove these results and discuss their implications for describing synergistic solvation effects in the critical region, that is, the effects of solute–solute interactions on their mutual solubilities compared to the pure solute case at identical pressures and temperatures [13].

5.6 Chapter Review

The discussion in this chapter was devoted to an analysis of the solubility behavior typical of low-vapor-pressure organic solutes, of large molecular weight, dissolved in a solvent near its critical point. These types of mixture are commonly encountered in technological applications using supercritical fluids, ranging from chemical separations to the production of microparticles, by rapid expansion of a solvent near its critical point.

The equations describing both isobaric and isothermal solubility behavior were developed in detail. The thermodynamic origins as to why the fluid density alone proves to be such a useful variable for describing solvation phenomena, particularly in the critical region, were explicated. Retrograde solubility behavior in these systems was also discussed in detail and shown to be a signature occurrence for dilute mixtures in the solvent's critical region. It results in the existence of pressure crossover regions for a mixture, which can, in principle, be exploited to design highly selective chemical separations of low-vapor-pressure organic compounds.

Finally, the use of equations of state to model the solubility behavior in supercritical mixtures was described. We focused upon the use of cubic equations of state since these are the most common models used in engineering applications in the field.

5.7 Additional Exercises

Basic Thermodynamic Analysis

1. Three important solution properties in dilute supercritical binary mixtures are

$$K_{Tx} \equiv -\frac{1}{v}\left(\frac{\partial v}{\partial P}\right)_{T,x}, \quad \left(\frac{\partial P}{\partial x}\right)_{T,v}, \quad \left(\frac{\partial v}{\partial x}\right)_{T,P}$$

Near the critical point when $x \to 0$, we have the following limiting results:

$$K_{Tx} \equiv -\frac{1}{v}\left(\frac{\partial v}{\partial P}\right)_{T,x} \text{ diverges,} \quad \left(\frac{\partial P}{\partial x}\right)_{T,v} \text{ is finite,} \quad \left(\frac{\partial v}{\partial x}\right)_{T,P} \text{ diverges}$$

These conditions hold in general, that is, both classically and nonclassically. On what basis do we assert these results?

2. If v is the molar volume of a binary mixture of two species i and j, prove the following thermodynamic identity:

$$v = \bar{V}_j + x_i\left(\frac{\partial v}{\partial x_i}\right)_{T,P}$$

where $\bar{V}_j$ is the partial molar volume of species j in the solution. How would you use this equation to devise an experiment to measure the solute property $\bar{V}_j^{\infty}$ in a dilute supercritical binary mixture?

3. At a binary mixture critical line (with $x_i, x_j > 0$) are the properties $\bar{V}_j$ and $\bar{V}_i$ finite?

4. From the identity

$$\bar{H}_i^{\infty} = \frac{5kT}{2} + \int_0^P \left[\bar{V}_1^{\infty} - T\left(\frac{\partial \bar{V}_1^{\infty}}{\partial T}\right)_P\right] dP$$

one can in principle use $\bar{V}_j^{\infty}$ data to estimate $\bar{H}_i^{\infty}$ data in dilute supercritical mixtures. Prove this identity and discuss the feasibility of this approach.

5. How would you use $\bar{V}_j^{\infty}$ data for a solute to estimate the property

$$\left(\frac{\partial P}{\partial x}\right)_{T,v}$$

in the limit as $x \to 0$.

6. By defining the solute mole fraction $x_1 \equiv N_1/N_t$ do a classical (Taylor series) expansion of pressure about the solvent's critical point, where $P \equiv P(x_1, T, \rho)$. Derive the *classical conditions* for which the term $\bar{V}_1^{\infty}/K_T$ is independent of density at a given near-critical temperature. [Hint: use the identity linking $(\partial P/\partial x)_{T,v}$ and $\bar{V}_1^{\infty}/K_T$.]

7. Prove equation (5.9), namely that the variation of solubility with pressure along the phase envelope in a binary mixture of solid with supercritical fluid is given by

$$\left(\frac{\partial \ln y_i}{\partial P}\right)_{T,\sigma} = \frac{v_i^s - \bar{V}_i}{RT[1 + y_i(\partial \ln \hat{\phi}_i/\partial y_i)_{T,P}]}$$

8. If the independent variables in the previous question's equation are chosen to be T and ρ, is it possible for $(\partial \ln y_i/\partial \ln \rho)_{T,\sigma}$ to change sign along the phase envelope in the vicinity of the solvent's critical region? (By this is meant modest density changes at temperatures close to the solvent's T_c)
9. The molecular energy and entropy of transfer from a pure solid solute into a supercritical fluid phase (designated v) are defined as

$$\Delta S_{\text{transfer}} = \bar{S}_i^v - S_{i,\text{pure}}(T, P), \quad \Delta H_{\text{transfer}} = \bar{H}_i^v - H_{i,\text{pure}}(T, P)$$

In the retrograde solubility region, the enthalpy of transfer is negative. What does this and the entropy-of-transfer equation tell you about the qualitative microscopic structure of the fluid phase in this region?
10. *Attractive* solutes are those for which $\bar{V}_1^\infty$ and $\bar{H}_1^\infty$ tend to minus infinity at the solvent's critical point. At a point where $\bar{H}_1^\infty \ll 0$, what can we say about the pressure dependence of solubility?
11. What is the thermodynamic condition for a solute to exhibit a crossover temperature? A crossover temperature is the temperature at which two solubility isobars at pressures P_L and P_H cross.
12. Show that, if a solute in a supercritical fluid is retrograde, then to a good approximation we have that

$$\bar{H}_i(T, P) < H_i^{\text{solid}}(T, P^{\text{sat}}(T))$$

If the zero-enthalpy state is the ideal-gas condition, show that in this case a necessary but not sufficient condition for retrograde solute behavior is given by the inequality $\bar{H}_i(T, P) < 0$.
13. *Repulsive* (positive) solutions are defined as those for which $\bar{V}_1^\infty$ and $\bar{H}_1^\infty \to +\infty$ at the solvent's critical point. Sketch solubility isobars for systems of this sort. Outline the design of a crossover process for separating a mixture of two solutes where one is attractive and the other repulsive. Is it possible to split such mixtures into pure constituent components with this process? What would be the most probable types of chemical mixtures to display this behavior?
14. Does the crossover pressure occur at a unique point in a given solute–solvent system? in other words, do the solubility isotherms *for all temperatures* cross at a unique pressure?
15. Prove equation (5.22) and show that both matrices on the right-hand side of that equation are positive definite. (Recall that *positive definite* matrix is defined as being one in which the product $\boldsymbol{y}^{\mathsf{T}}\boldsymbol{A}\boldsymbol{y}$ is positive for all nonzero vectors $\boldsymbol{y}$.)

Numerical Use of an Equation of State

16. Using the *Peng–Robinson* model described in this chapter and the k_{ij} parameters given below for the ternary system 1:10 decanediol (1) – benzoic acid (2) – CO_2

T, K	k_{13}	k_{12}	k_{23}
308	0.14	–0.14	0.018
318	0.14	0.14	0.01

published by Chimowitz and Pennisi [3], determine the extent of the crossover pressure region for this mixture at the temperatures shown.
If we repeat this set of calculations using the *infinite-dilution reference state* for calculating the crossover regions, how do these results compare with the finite-concentration results done in the first case?

17. If the pair intermolecular potential for species $i-j$ can be written in the general form $u(r_{ij}) = \varepsilon\phi(r_{ij}/\sigma)$, where σ is the collision diameter, show that the configuration integral [14]

$$Z_N \equiv \int \cdots \int e^{-\beta U_N} dr_1 \cdots d\mathbf{r}_N$$

where $U_N = \sum_{i<j}^{N-1} u(r_{ij})$, can be expressed in terms of a universal function with reduced temperature and density defined by $T^* \equiv kT/\varepsilon$ and $\rho^* \equiv \rho\sigma^3$. What is the equation of state of the fluid in terms of this universal function? If you proved this result, you've developed a thermodynamic model that is a corresponding-states equation from microscopic considerations—referred to as *microscopic* corresponding states.

18. The divergence of $\bar{V}_i^{\infty}$ at the critical point is result of long-range correlations at the critical point. Some properties, however, remain finite and quite unremarkable even when the fluid phase is very close to its critical point. One such property is the *solute chemical potential*. This can be demonstrated by the following problem: Prove that, in a binary solute(1)–solvent(2) system,

$$\bar{H}_1^{\mathrm{r}} = \left(\frac{\partial H^{\mathrm{r}}}{\partial N_1}\right)_{T,V,N_2} + \bar{V}_1\left(\frac{\partial H}{\partial V}\right)_{T,N_1,N_2}$$

and

$$\bar{S}_1^{\mathrm{r}} = \left(\frac{\partial S^{\mathrm{r}}}{\partial N_1}\right)_{T,V,N_2} + \bar{V}_1\left(\frac{\partial S}{\partial V}\right)_{T,N_1,N_2} - k$$

where H^{r} is the residual enthalpy of the solute, S^{r} is its residual entropy, and k is Boltzmann's constant.

Since $H = U + PV + \frac{3}{2}NkT$, where U is the system's potential energy, show that

$$\bar{H}_1^{\mathrm{r}} = \left(\frac{\partial U}{\partial N_1}\right)_{T,V,N_2} + \bar{V}_1\left[P + \left(\frac{\partial U}{\partial V}\right)_{T,N_1,N_2}\right] - kT$$

and that

$$\bar{H}_1^{\mathrm{r}} = \left(\frac{\partial U}{\partial N_1}\right)_{T,V,N_2} + T\bar{V}_1\left(\frac{\partial S}{\partial V}\right)_{T,N_1,N_2} - kT$$

The divergent parts of $\bar{H}_1^{\mathrm{r}}$ and $\bar{S}_1^{\mathrm{r}}$ are only those terms with $\bar{V}_1$ in them (why?). Define these terms as

$$\Delta S_1^{\mathrm{D}} = \bar{V}_1 \left(\frac{\partial S}{\partial V}\right)_{T,N_1,N_2}$$

and

$$\Delta H_1^{\mathrm{D}} = T\bar{V} \left(\frac{\partial S}{\partial V}\right)_{T,N_1,N_2} = T\Delta S_1^{\mathrm{D}}$$

Since the reduced chemical potential of the solute is given by

$$\mu_1^{\mathrm{r}} \equiv \bar{H}_1^{\mathrm{r}} - T\bar{S}_1^{\mathrm{r}}$$

it is clear that, in this equation, the divergent parts of $\bar{H}_1^{\mathrm{r}}$ and $\bar{S}_1^{\mathrm{r}}$ cancel exactly. Hence, μ_1^{r} does not diverge at the critical point and in fact is well-behaved in that vicinity, a result that proves useful in computer simulation as discussed in a subsequent chapter.

Bibliography

The solubility properties of supercritical fluids are described in references [1, 2, 7, 11], the crossover effect and retrograde phenomena in references [3, 4, 5], cubic equations of state in references [8, 9], and the infinite-dilution reference state in references [11, 12, 13].

[1] W. J. Schmitt and R. C. Reid, "Solubility of monofunctional organic solids in chemically diverse supercritical fluids," *J. Chem. Eng. Data*, vol. 31, p. 204, 1986.

[2] C. A. Eckert, D. H. Ziger, K. P. Johnston, and S. Kim, "Solute partial molar volumes in supercritical fluids," *J. Phys. Chem.*, vol. 90, p. 2738, 1986.

[3] E. H. Chimowitz and K. J. Pennisi, "Process synthesis concepts for supercritical gas extraction in the crossover region," *AIChE J.*, vol. 32, p. 1665, 1986.

[4] F. D. Kelley and E. H. Chimowitz, "Experimental data for the crossover process in a model supercritical system," *AIChE J.*, vol. 35, p. 981, 1989.

[5] E. H. Chimowitz, F. D. Kelley, and F. M. Munoz, "Analysis of retrograde behavior and the cross-over effect in supercritical fluids," *Fluid Phase Equilib.*, vol. 44, p. 23, 1988.

[6] B. Noble and J. W. Daniel, *Applied Linear Algebra*, second ed. Englewood Cliffs, NJ: Prentice Hall, 1977.

[7] S. K. Kumar and K. P. Johnston, "Modeling the solubility of solids in supercritical fluids with density as the independent variable," *J. Supercrit. Fluids*, vol. 1, p. 15, 1988.

[8] G. Soave, "Equilibrium constants from a modified Redlich–Kwong equation of state," *Chem. Eng. Sci.*, vol. 27, p. 1197, 1972.

[9] D. Y. Peng and D. B. Robinson, "A new two-constant equation of state," *Ind. Eng. Chem. Fund.*, vol. 15, p. 59, 1976.

[10] R. C. Reid, J. M. Prausnitz, and T. K. Sherwood, *The Properties of Gases and Liquids*, third ed. New York: McGraw Hill, 1977.

[11] P. G. Debenedetti and S. K. Kumar, "Infinite dilution fugacity coefficients and the general behavior of dilute binary systems," *AIChE J.*, vol. 32, p. 1253, 1986.

[12] F. M. Munoz, T. W. Li, and E. H. Chimowitz, "Henry's law and synergism in dilute near-critical mixtures: theory and simulation results," *AIChE J.*, vol. 41, p. 3899, 1995.

[13] F. M. Munoz and E. H. Chimowitz, "Critical phenomena in mixtures, part 2: Synergistic and other effects near mixture critical points," *J. Chem. Phys.*, vol. 99, p. 5450, 1993.

[14] D. A. McQuarrie, *Statistical Mechanics*. New York: Harper & Row, 1976.

6 Supercritical Adsorption

In this chapter, we discuss adsorption phenomena in supercritical systems, a situation that occurs in many application areas in chemical-process and materials engineering. An example of a commercial application in this area, which has achieved wide acceptance as a tool in analytical chemistry, is supercritical fluid chromatography (SFC). Not only is SFC a powerful technique for chemical analysis, but it also is a useful method for measuring transportive and thermodynamic properties in the near-critical systems.

In the next section, we analyze adsorption-column dynamics using simple dynamic models, and describe how data from a chromatographic column can be used to estimate various thermodynamic and transport properties. We then proceed to discuss the effects of proximity to the critical point on adsorption behavior in these systems. The closer the system is to its critical point, the more interesting is its behavior.

6.1 Mathematical Model of Adsorption Dynamics

For very dilute solute systems, like those considered here, the energy balance is often ignored to a first approximation; this leads to a simple set of mass-balance equations defining transport for each species. These equations can be developed to various levels of complexity, depending upon the treatment of the adsorbent (stationary phase). The conceptual view of these phases can span a wide range of possibilities ranging from completely nonporous solids (fused structures) to porous materials with complicated ill-defined pore structures. Given these considerations, it is customary to make the following assumptions in the development of a simple model of adsorber-bed dynamics:

1. The stationary and mobile phases are continuous in the direction of the flow, with the fluid phase possessing a flat velocity profile ("plug" flow).
2. The porosity of the stationary phase is considered constant irrespective of pressure and temperature conditions (i.e., it is incompressible).
3. The column is considered to be radially homogeneous, leading to a set of equations with one spatially independent variable, representing distance along the column axis.
4. The dispersion term in the model equation represents the combined effects of molecular diffusion and dispersion due to convective stirring in the bed. These effects are combined into an effective phenomenological dispersion coefficient, considered to be constant throughout the column.
5. Transport is assumed to occur through a process whereby solute adsorbs onto the stationary phase and then diffuses into the particle.

The model that emerges from these assumptions is now developed and proves to be useful and quite robust. A shell balance over a small section of the column leads to the following equation describing the fixed-bed adsorber dynamics for a single solute:

$$-\varepsilon\left(\frac{\partial(uc)}{\partial x}\right)+\varepsilon\frac{\partial}{\partial x}\left(D\frac{\partial c}{\partial x}\right)=\frac{\partial}{\partial t}(\varepsilon c+(1-\varepsilon)q) \tag{6.1}$$

where u is the velocity of the supercritical fluid's mobile phase, c the bulk-phase concentration of solute, q its concentration in the adsorbent phase, x the distance in the axial direction, ε the bed porosity, and D the effective dispersion coefficient of the solute.

EXERCISE 6.1

Prove equation (6.1).

In order to solve this partial differential equation, appropriate *boundary conditions* are required to be specified. The bed is initially considered to be free of any solute, and so

$$c(x,0)=0 \tag{6.2}$$

In the case of an *impulse solute input into the bed*, relevant to chromatographic-column experiments, for example, we have that

$$c(0, t) = \delta(t)$$

For *continuous* operation of the bed, as would be the case in a supercritical fluid adsorption process for remediating an incoming gas stream from a dilute organic pollutant species, this boundary condition could be written as

$$c(0,\ t) = c_o \tag{6.3}$$

where c_o is the concentration of the contaminant species in the inlet stream. Another boundary condition used at the end of the bed is the so-called *Kucera condition* [1]:

$$\frac{\partial c(L, t)}{\partial x} = 0 \tag{6.4}$$

Equation (6.1) and its attendant boundary conditions define the basic conservation equations in the system; one set of such equations is written for each of the species in the system. At this point, however, the manner in which the adsorbed-phase concentration q is determined remains unspecified.

Model Simplification: The Local Equilibrium Model

A common way of simplifying these equations is to assume that the solute concentrations in the fluid and stationary phases are locally at equilibrium. This formulation of the model is thus referred to as the *local equilibrium model.* It has the advantage of not requiring a detailed description of the stationary phase's structure, or transport-rate expressions for solute mass transfer onto the adsorbent surface. This model is exemplified by the following transport equation:

$$-D\frac{\partial^2 c}{\partial x^2} + \frac{\partial (uc)}{\partial x} + \frac{\partial}{\partial t}[(1 + k)c] = 0 \tag{6.5}$$

where the term k is often referred to as the solute's *capacity factor*, which is related to its *adsorption coefficient* K_{eq} through the definition

$$k \equiv \frac{(1 - \varepsilon)}{\varepsilon} K_{eq} = K_o K_{eq} \tag{6.6}$$

where we define the solute adsorption coefficient (equilibrium constant) K_{eq} as:

$$K_{eq} \equiv \frac{\text{solute concentration per volume of stationary phase at equilibrium}}{\text{solute concentration per volume of mobile phase at equilibrium}}$$

that is, $K_{eq} = q/c$ and $K_o \equiv (1 - \varepsilon)/\varepsilon$. This model leads to a particularly simple relationship between the equilibrium constant in the model and measurements of the *solute's residence time* available from chromatographic experiments, for example. Even with this model, however, there are further useful simplifications to consider.

Case 1: The Omission of Axial Dispersion ($D = 0$)

The analyses that follow often use the Laplace transform of a function. The Laplace transform of the function $c(t, x)$ applied to the t variable is defined as

$$\mathrm{L}\{c(t, x)\} = \hat{c}(s, x) = \int_0^\infty c(t, x)e^{-st}\, dt \tag{6.7}$$

With $D = 0$, and assuming u constant, the Laplace transform of equation (6.5) yields

$$\frac{\partial \hat{c}(s, x)}{\partial x} = -\frac{(1 + k)s\hat{c}(s, x)}{u} \tag{6.8}$$

This is a first-order differential equation in x which can be integrated, resulting in the equation

$$\hat{c}(s, x) = \hat{c}(s, 0) \exp\left[\frac{-(k + 1)sx}{u}\right] \tag{6.9}$$

The term $\hat{c}(s, 0)$ can be found from the Laplace transform of the boundary condition $c(t, 0)$. For a *square pulse input* of the solute species, with the width of the pulse being $b - a$ and height c_0, the solution in the s–x plane is

$$\hat{c}(s, x) = \frac{c_0}{s}\left[\exp\left(-sa - \frac{(1 + k)sx}{u}\right) - \exp\left(-sb - \frac{(1 + k)sx}{u}\right)\right] \tag{6.10}$$

In the t–x plane, the Laplace inverse of this equation yields

$$c(t, x) = c_0\left[H\left(t - a - \frac{(1 + k)x}{u}\right) - H\left(t - b - \frac{(1 + k)x}{u}\right)\right] \tag{6.11}$$

which is an equation for a square pulse of width $b - a$ with a *time offset* of $(1 + k)x/u$.

The information held in these equations is more readily accessed using the mathematical concept of a moment. The *first moment* t of a time-dependent function $c(t)$ is defined as

$$\bar{t} = \frac{\int_0^\infty c(t)t\, dt}{\int_0^\infty c(t)dt} \tag{6.12}$$

For students of statistics, we recognize that first moments provide the mean of a function. Applying this definition to the function $c(t, x)$ in (6.11) leads to the *mean residence time* $\bar{t}$ of the species in the column, with the input being a rectangular pulse of width $b - a$. The result is

$$\bar{t} = \frac{b + a}{2} + \frac{(1 + k)L}{u} \tag{6.13}$$

where L is the length of column traveled by the solute from the injection point to the detector position. For an infinitesimally small pulse with $a = 0$ and $b \ll 1$,

$$\bar{t} = \frac{(1 + k)L}{u} \tag{6.14}$$

The term L/u in (6.14), denoted by the symbol t_o, is the residence time of an *unretained species* in the column. The relationship given by (6.14) is a simple yet practically important result. It shows that, if one establishes t_o experimentally, then measuring the mean residence time of the species in the column allows one to establish k for that species under the thermodynamic conditions pertinent to the experiment. From equation (6.6), we have that k itself is simply related to K_{eq}, the adsorption constant of the species. We now inquire about the consequences of including dispersion effects in the model formulation: how does this change these results?

Case 2: $D \neq 0$, Dispersion Effects Included

The resultant transformed equation for the column is now given by

$$D\frac{\partial^2 \hat{c}}{\partial x^2} - u\frac{\partial \hat{c}}{\partial x} = (1+k)\hat{c}(s,x) \tag{6.15}$$

which is a familiar second-order homogeneous equation of $\hat{c}(s,x)$. For a pulse input, standard differential-equation manipulations provide the following solution for $\hat{c}(s,x)$:

$$\hat{c}(s,x) = c_o\left[\frac{e^{-sa}}{s} - \frac{e^{-sb}}{s}\right]\exp\left[\left(\frac{xu}{2D}\right) - \frac{x}{2}\left(\left(\frac{u}{d}\right)^2 + \frac{4(1+k)s}{D}\right)\right]^{1/2} \tag{6.16}$$

which can be inverted to yield [2]

$$c(t,x) = \frac{c_o\alpha}{2\sqrt{\pi}}\exp\left(\frac{ux}{2D}\right)(I_a(t,x) - I_b(t,x)) \tag{6.17}$$

where

$$\alpha \equiv \frac{x(1+k)^{1/2}}{D^{1/2}} \tag{6.18}$$

$$I_\theta \equiv \int_0^{t-\theta} \exp\left(\frac{-\beta\tau - 0.25\alpha^2\tau^{-1}}{\tau^{-1.5}}\right) d\tau \tag{6.19}$$

and

$$\beta \equiv \frac{u^2}{4D(1+k)} \tag{6.20}$$

EXERCISE 6.2

Prove the results given in (6.15)–(6.20).

Clearly, this is a more complicated solution than the previous case where D was ignored. However, it is still an analytical result with the potential for further manipulation. Instead of working with the complicated time-domain function $c(t,x)$ to find $\bar{t}$, it is possible to often simplify the analysis of data from these systems by once again exploiting the following relationships between *moments of the dynamic response function* $c(t)$ and the function $\hat{c}(s,x)$.

Mathematical Moments of the Response Function

The following well-known mathematical identity defines the relationship between the various moments of the response function $c(t, x)$ and $\hat{c}(s, x)$:

$$\int_0^\infty t^k c(t)dt = (-1)^k \lim_{s \to 0} \frac{\partial^k \hat{c}(s)}{ds^k} \quad \text{for all integers } k \tag{6.21}$$

The order of the *moment of the function* $c(t)$ in equation (6.21) is the integer k. Moments to any order can thus be found from either the time-domain or the Laplace-domain forms of the response functions. Experimental data can be used to calculate the left-hand integrals to any order in k in equation (6.21), and if the Laplace function of the model is available, as it usually is, then model parameters can be evaluated from these data (see the additional exercises). This is a powerful and widely used technique for measuring thermodynamic and transport properties using chromatographic methods as now discussed in more detail.

EXERCISE 6.3

Prove the result given in equation (6.21).

Residence Time in the $D \neq 0$ Model

The first moment of $\hat{c}(s, x)$, found from differentiating equation (6.16) and applying (6.21), yields

$$\bar{t} = \frac{b + a}{2} + \frac{(1 + k)x}{u} \tag{6.22}$$

which is *identical* to the result for the model formulation with $D = 0$ shown in (7.14). In other words, once transport is modeled with equation (6.5), the results used for finding thermodynamic adsorption coefficients from the data are independent of whether or not axial dispersion effects are included in formulating the local equilibrium model. The effects of including D show up in the second moments of the data, which are only related to the spread of the species concentration profile about its mean value (see the additional exercises).

EXERCISE 6.4

Using the transport equation and its Laplace representation (equations (6.15) and (6.16)) develop the expression for the second moment of the concentration's dynamic function $c(t)$ using (6.21). Explain how you would use data for $c(t)$ to measure D, stating which other parameters are needed.

Notwithstanding its simplicity, the local equilibrium model described here has been widely used for data analysis using chromatographic techniques with supercritical mixtures at high pressures [3]. It is reasonable, however, to ask how a more complicated model, assuming finite-rate mass transfer between both phases and/or a more complicated picture of the adsorbent phase, might impact data analysis in these systems. We now consider this and related questions.

Mass Transfer into a Stationary Phase Consisting of Porous Particles

We consider the model formulation discussed by Suzuki and Smith [4] for a bed of uniform spherical porous particles with each particle having radius R. The solute is transported to the particle surface from where it diffuses into the interior. The basic solute conservation equation is similar to that considered before except that now the term accounting for solute accumulation in the particle is an expression for finite-rate mass transfer, leading to the solute transport equation:

$$D\frac{\partial^2 c}{\partial x^2} - u\frac{\partial c}{\partial x} - \frac{3(1-\varepsilon)}{\varepsilon R}N_{\mathrm{o}} = \frac{\partial c}{\partial t} \tag{6.23}$$

Here N_{o} is the solute flux from the mobile phase to the particle surface. The term factor $3(1-\varepsilon)/R$ in equation (6.23) arises because of the surface–volume ratio of spherical particles. Using a conventional first-order mass-transfer equation, this flux can be expressed as,

$$N_{\mathrm{o}} = k_{\mathrm{f}}[c - c_{\mathrm{p}}(R)] = D_{\mathrm{e}}\left(\frac{\partial c_{\mathrm{p}}}{\partial r}\right)_{r=R} \tag{6.24}$$

where $c_{\mathrm{p}}(R)$ is the solute concentration on the surface of the particle, D_{e} the effective solute diffusivity into the particle interior, and k_{f} the external mass transfer coefficient with units of velocity (L/t). The conservation equation for the particle itself is given by

$$D_{\mathrm{e}}\left(\frac{\partial^2 c_{\mathrm{p}}}{\partial r^2} + \frac{2}{r}\frac{\partial c_{\mathrm{p}}}{\partial r}\right) - N_{\mathrm{p}} = \beta\frac{\partial c_{\mathrm{p}}}{\partial t} \tag{6.25}$$

where β is the internal particle porosity and N_{p} the rate of disappearance of the solute onto the solid surface within the particle. Using a first-order kinetic equation for adsorption, the expression for N_{p} is

$$N_{\mathrm{p}} \equiv \frac{\partial n_{\mathrm{p}}}{\partial t} = k_{\mathrm{a}}\left(c_{\mathrm{p}}(r) - \frac{q(r)}{K_{\mathrm{eq}}}\right) \tag{6.26}$$

where K_{eq} is the solute equilibrium adsorption coefficient linking solid and gas-phase concentrations within the particle, and where k_{a} is the adsorption-rate constant within the particle. This system of equations for a square pulse input of solute was solved by Kubin [5] leading to the first moment for the species given by

$$\bar{t} = \frac{L}{u}\left[1 + \frac{1-\varepsilon}{\varepsilon}\beta\left(1 + \frac{K_{\mathrm{eq}}}{\beta}\right)\right] \tag{6.27}$$

Comparing this to the earlier result for the simpler nonporous model, we see that both results are similar in spite of the present model's much greater complexity. With $\beta = 0$, the above case reduces to the result given in (6.14), as it should be where transport to the pore interior was neglected. We note from equation (6.27) that the equilibrium constant K_{eq} can still be established from solute residence-time data with this more complicated model.

The second moment σ_2 of these equations yields the following results

$$\sigma_2 = \frac{2L}{u}\left[A_1 + \frac{D[(1+A_o)^2 + A_o]^2}{u^2}\right] \tag{6.28}$$

where

$$A_o = \frac{1-\varepsilon}{\varepsilon}(\beta + K_{eq}) \tag{6.29}$$

and

$$A_1 = \frac{1-\varepsilon}{\varepsilon}\left[\frac{K_{eq}^2}{k_a} + (\beta + K_{eq})^2\left(\frac{1}{5D_e} + \frac{1}{k_f R}\right)\frac{R^2}{3}\right] \tag{6.30}$$

We note that both diffusion coefficients D and D_e appear in these equations as well as the mass-transfer coefficients k_f and k_a.

EXERCISE 6.5

By finding the second moments of equations (6.23)–(6.26), prove the results given in equations (6.28)–(6.30).

The expressions (6.28)–(6.30) contain a number of empirical parameters, some of which (e.g., K_{eq}) can be found independently from first-moment data. In addition, D and D_e can be found from these equations if the mass-transfer rates are considered to be high enough to neglect the terms $1/k_f R$ and K_{eq}^2/k_a. Since mass-transfer coefficients are large in the near-critical region, this approximation has often been adopted [6], leading to the results

$$\frac{\sigma_2(1+A_o)^2}{2\bar{t}^2}L = A_3 u + \frac{D}{u}(1+A_o)^2 \tag{6.31}$$

where

$$A_3 = \frac{R^2(1-\varepsilon)(\beta + K_{eq})^2}{15\varepsilon D_e} \tag{6.32}$$

From an analysis of the second moments of the data, a plot of $\sigma_2(1+A_o)^2 L/2\bar{t}^2$ versus u will provide estimates of D_e (from the slope) and D/u (from the intercept).

EXERCISE 6.6

How would one find mass-transfer coefficients from an analysis of chromatographic data using equations (6.30)–(6.32) to model the data?

Given the ability to measure thermodynamic and transport properties in these systems, we now investigate interesting and unique effects of proximity to the critical point, on solute adsorption when dissolved in a near-critical solvent.

6.2 Equilibrium Adsorption Coefficients in the Critical Region

One of the major characteristics of adsorption in the critical region is the occurrence of either *endothermic or exothermic adsorption* depending upon the fluid phase's proximity to the critical point [3]. In the engineering literature, the equality of fugacities is often used to describe phase equilibrium in chemical systems. Adopting this approach, we now analyze this problem in a binary system.

The differential changes in the fugacities of the two components, denoted a and b in the two phases, are given by the relationships:

$$d \ln f_{\mathrm{b}}^{\mathrm{m}} = d \ln f_{\mathrm{b}}^{\mathrm{s}} \tag{6.33}$$

and

$$d \ln f_{\mathrm{a}}^{\mathrm{m}} = d \ln f_{\mathrm{a}}^{\mathrm{s}} \tag{6.34}$$

In these and subsequent equations, b refers to the solute species and a to the solvent, and the superscripts m and s refer to the mobile and stationary phases, respectively. The intensive thermodynamic variables for the mobile phase are y_{b}, T, and P, while we choose θ_{a}, θ_{b}, T, and P in describing the stationary phase; the symbol θ_i refers to the fractional coverage (composition) of species i in the adsorbed phase and, in general $\theta_{\mathrm{a}} + \theta_{\mathrm{b}} \leq 1$. Thus the mobile-phase fugacity functions in (6.33) and (6.34) can be expanded in terms of y_{b}, T, and P, while the stationary-phase functions are expanded in terms of θ_{a}, θ_{b}, T, and P. This approach yields the following set of equations:

$$\begin{aligned} &-\left[\frac{\bar{H}_{\mathrm{a}}^{\mathrm{m}} - H_{\mathrm{a}}^{\mathrm{o}}}{RT^2} + \frac{\bar{H}_{\mathrm{a}}^{\mathrm{s}} - H_{\mathrm{a}}^{\mathrm{o}}}{RT^2}\right] dT + \left[\frac{\bar{V}_{\mathrm{a}}^{\mathrm{m}} - V_{\mathrm{a}}^{\mathrm{s}}}{RT}\right] dP \\ &+ \left(\frac{\partial \ln f_{\mathrm{a}}^{\mathrm{m}}}{\partial y_{\mathrm{b}}}\right)_{p,T} dy_{\mathrm{b}} - \left(\frac{\partial \ln f_{\mathrm{a}}^{\mathrm{s}}}{\partial \theta_{\mathrm{a}}}\right)_{p,T,\theta_{\mathrm{b}}} d\theta_{\mathrm{a}} - \left(\frac{\partial \ln f_{\mathrm{a}}^{\mathrm{s}}}{\partial \theta_{\mathrm{b}}}\right)_{p,T,\theta_{\mathrm{a}}} d\theta_{\mathrm{b}} = 0 \end{aligned} \tag{6.35}$$

for the solvent, and

$$\begin{aligned} &-\left[\frac{\bar{H}_{\mathrm{b}}^{\mathrm{m}} - H_{\mathrm{b}}^{\mathrm{o}}}{RT^2} + \frac{\bar{H}_{\mathrm{b}}^{\mathrm{s}} - H_{\mathrm{b}}^{\mathrm{o}}}{RT^2}\right] dT + \left[\frac{\bar{V}_{\mathrm{b}}^{\mathrm{m}} - V_{\mathrm{b}}^{\mathrm{s}}}{RT}\right] dP \\ &+ \left(\frac{\partial \ln f_{\mathrm{b}}^{\mathrm{m}}}{\partial y_{\mathrm{b}}}\right)_{p,T} dy_{\mathrm{b}} - \left(\frac{\partial \ln f_{\mathrm{b}}^{\mathrm{s}}}{\partial \theta_{\mathrm{a}}}\right)_{p,T,\theta_{\mathrm{b}}} d\theta_{\mathrm{a}} - \left(\frac{\partial \ln f_{\mathrm{b}}^{\mathrm{s}}}{\partial \theta_{\mathrm{b}}}\right)_{p,T,\theta_{\mathrm{a}}} d\theta_{\mathrm{b}} = 0 \end{aligned} \tag{6.36}$$

for the solute, where $\bar{H}$ and $\bar{V}$ (as is the convention used in this text) refer to the partial molar enthalpies and volumes respectively; here $^{\circ}$ denotes the ideal-gas state.

Variation of k_{b} with T and P

We consider P and y_{b} constant in equations (6.35) and (6.36), which can then be rearranged to solve for $d\theta_{\mathrm{a}}$. The approximation that the mobile-phase composition remains constant (i.e., $dy_{\mathrm{b}} \approx 0$) is rationalized here based upon the fact the solute concentration in these systems is extremely dilute. Using equations (6.35) and (6.36), it is easily shown (see exercise 6.7) that the temperature derivative of the solute capacity factor under these conditions is given by

$$\left(\frac{\partial \ln k_{\text{b}}}{\partial T}\right)_{P,y_{\text{b}}} = \frac{C_{\text{b}}(\bar{H}_{\text{b}}^{\text{s}} - \bar{H}_{\text{b}}^{\text{m}}) + C_{\text{a}}(\bar{H}_{\text{a}}^{\text{s}} - \bar{H}_{\text{a}}^{\text{m}})}{RT^2} + \alpha_{\text{m}} \tag{6.37}$$

The pressure derivative follows analogously, the result being

$$\left(\frac{\partial \ln k_{\text{b}}}{\partial P}\right)_{T,y_{\text{b}}} = \frac{C_{\text{b}}(\bar{V}_{\text{b}}^{\text{m}} - \bar{V}_{\text{b}}^{\text{s}}) + C_{\text{a}}(\bar{V}_{\text{a}}^{\text{m}} - \bar{V}_{\text{a}}^{\text{s}})}{RT} - K_{T\text{m}} \tag{6.38}$$

The coefficients C_{a} and C_{b} are related to properties of the stationary phase, while α_{m} and $K_{T\text{m}}$ are the volume expansivity and isothermal compressibility of the mobile phase, respectively. The equations for C_{a} and C_{b} are given by

$$C_{\text{a}} = -\left(\frac{\partial \ln f_{\text{b}}^{\text{s}}}{\partial \theta_{\text{a}}}\right)_{p,T,\theta_{\text{b}}} \bigg/ \theta_{\text{b}}\Delta \tag{6.39}$$

$$C_{\text{b}} = \left(\frac{\partial \ln f_{\text{a}}^{\text{s}}}{\partial \theta_{\text{a}}}\right)_{p,T,\theta_{\text{b}}} \bigg/ \theta_{\text{b}}\Delta \tag{6.40}$$

with

$$\Delta = \left(\frac{\partial \ln f_{\text{b}}^{\text{s}}}{\partial \theta_{\text{b}}}\right)_{p,T,\theta_{\text{a}}} \left(\frac{\partial \ln f_{\text{a}}^{\text{s}}}{\partial \theta_{\text{a}}}\right)_{p,T,\theta_{\text{b}}} - \left(\frac{\partial \ln f_{\text{b}}^{\text{s}}}{\partial \theta_{\text{a}}}\right)_{p,T,\theta_{\text{b}}} \left(\frac{\partial \ln f_{\text{a}}^{\text{s}}}{\partial \theta_{\text{b}}}\right)_{p,T,\theta_{\text{a}}} \tag{6.41}$$

Thus we see that the determinantal function Δ is a stability coefficient, which is positive for a stable phase (see chapter 1).

EXERCISE 6.7

Recall from the definition of the capacity factor given in (6.6) for component i that it can be expressed as

$$k_i \equiv \frac{\theta_i}{y_i} \cdot \frac{1}{K_0} \cdot \frac{\rho^{\text{s}}}{\rho^{\text{m}}}$$

where K_0 is related to the column porosity, ρ^{s} is the molar adsorption-site density of the adsorbent phase, and ρ^{m} is the molar density of the fluid phase. Using this definition for k_i for the solute and solvent, respectively, prove equations (6.37)–(6.38) from (6.33) and (6.34). In order to do this, one assumes that K_0 and ρ^{s} are constants—why is this a good assumption?

In order to proceed further, one requires a model for describing the fugacity terms in the stationary phase appearing in (6.37) and (6.38) with more easily recognizable thermodynamic forms. One may, in abstract terms, view this stationary phase as an incompressible 3-dimensional lattice structure. If the celebrated *Langmuir* model [7] is used to describe this phase (see exercise 6.8), the terms in equations (6.39) and (6.40) can be derived quite simply. Using this approach the temperature dependence of the capacity factor is given by the equation [3]:

$$\left(\frac{\partial \ln k_{\mathrm{b}}}{\partial T}\right)_{P,y_{\mathrm{b}}} = (1-\theta_{\mathrm{b}})\frac{\bar{H}_{\mathrm{b}}^{\mathrm{s}} - \bar{H}_{\mathrm{b}}^{\mathrm{m}}}{RT^2} - \theta_{\mathrm{a}}\frac{\bar{H}_{\mathrm{a}}^{\mathrm{s}} - \bar{H}_{\mathrm{a}}^{\mathrm{m}}}{RT^2} + \alpha_{\mathrm{m}} \tag{6.42}$$

It should be noted that equation (6.42) includes terms for both the solute and supercritical solvent species. Both enthalpy terms in this equation can be grouped into a term that is sometimes referred to as the *enthalpy of transfer*, denoted by $\Delta H_{\mathrm{b}}^{\mathrm{trans}}$. For L other adsorbing dilute solute species, this term is given by

$$\Delta\bar{H}_{\mathrm{b}}^{\mathrm{trans}} = \bar{H}_{\mathrm{b}}^{\mathrm{s}} - \bar{H}_{\mathrm{b}}^{\mathrm{m}} - \sum_{i=1}^{L}\theta_i(\bar{H}_i^{\mathrm{s}} - \bar{H}_i^{\mathrm{m}}) \tag{6.43}$$

Thus equation (6.37) can be rewritten as

$$\left(\frac{\partial \ln k_{\mathrm{b}}}{\partial T}\right)_{P,y_{\mathrm{b}}} = \frac{\Delta\bar{H}_{\mathrm{b}}^{\mathrm{trans}}}{RT^2} + \alpha_{\mathrm{m}} \tag{6.44}$$

EXERCISE 6.8

The Langmuir model for adsorption considers an M-site lattice with N adsorbing species; each site can accept one adsorbed molecule. The statistical-mechanical canonical partition function in this system is given by

$$Q(T, M, N_1, N_2, \ldots, N_c) = \Omega(T, M, \underline{N})\prod_{i=1}^{c} q_i^{N_i}$$

where Ω is the *configurational degeneracy* of a particular lattice energy state and N_i the number of adsorbed molecules of species i.

Here q_i is the partition function for an individual adsorbed molecule, a function of temperature only often defined as

$$q_i = q_i(T_0)e^{-\beta U_i}$$

where U_i is the configurational energy of the surface-bound species. Show that Ω is given by the equation

$$\Omega = \frac{M!}{\prod_{i=1}^{c} N_i!(M - \sum_{j=1}^{c} N_j)!}$$

Find the chemical potential and fugacity of an adsorbed solute i using this model and verify equation (6.42) using (6.37)–(6.41)

[Hint: $(\partial \ln Q/\partial N_i)_{T,M,N_k} = -\beta\mu_i^{\mathrm{s}}$ and $\mu_i^{\mathrm{s}} - \mu_i^{\mathrm{o}} = RT \ln \frac{f_i^{\mathrm{s}}}{f_i^{\mathrm{o}}}$ where ^{o} represents the ideal-gas state.]

Retrograde Adsorption in Supercritical Systems

We now highlight the important role played by critical phenomena in the fluid phase and how it determines behavior in these systems based upon the analysis just presented. We discuss the analogy between retrograde phenomena in solubility phenomena, described in chapter 5, and similar effects that occur in supercritical adsorption systems.

Equation (6.42) can be reformulated, as follows, in terms of configurational enthalpy functions, which represent the property's departure from the ideal-gas state:

$$\left(\frac{\partial \ln k_{\mathrm{b}}}{\partial T}\right)_{P,y_{\mathrm{b}}} = \frac{1-\theta_{\mathrm{b}}}{RT^2}[(\bar{H}_{\mathrm{b}}^{\mathrm{s}} - H_{\mathrm{b}}^{\mathrm{o}}) - (\bar{H}_{\mathrm{b}}^{\mathrm{m}} - H_{\mathrm{b}}^{\mathrm{o}})] - \theta_{\mathrm{a}}[(\bar{H}_{\mathrm{a}}^{\mathrm{s}} - H_{\mathrm{a}}^{\mathrm{o}}) - (\bar{H}_{\mathrm{a}}^{\mathrm{m}} - H_{\mathrm{a}}^{\mathrm{o}})] + \alpha_{\mathrm{m}} \tag{6.45}$$

The incompressible-stationary-phase functions $\bar{H}_i^{\mathrm{s}} - H_i^{\mathrm{o}}$ in this equation are analogous to the usual heats of adsorption for the solute species. The mobile-phase functions $\bar{H}_i^{\mathrm{m}} - H_i^{\mathrm{o}}$, however, are the only ones that are strongly affected by proximity to the critical point in the system. In a binary solution for example, there are two of these functions, the dominant one being that of the solute species (why?). In fact, if the infinite-dilution limit of equation (6.45) were taken, then the solute enthalpy function is the only one that diverges at the critical point of the solvent; the solvent's enthalpy function remains finite there. This suggests that the determining factor in equation (6.45) is the solute enthalpy term given by

$$H_b^{\mathrm{o}} - \bar{H}_b^{\mathrm{m}} - \Delta H_{\mathrm{b}}^{\mathrm{ads}} \tag{6.46}$$

where for convenience we have defined $\Delta H_{\mathrm{b}}^{\mathrm{ads}}$ as

$$\Delta H_{\mathrm{b}}^{\mathrm{ads}} = H_{\mathrm{b}}^{\mathrm{o}} - \bar{H}_{\mathrm{b}}^{\mathrm{s}} \tag{6.47}$$

We now use the analogy with retrograde solubility behavior developed in the previous chapter.

We showed there that, if a pure solute is in contact with a supercritical fluid, retrograde solvation occurs when

$$H_{\mathrm{b}}^{\mathrm{o}} - \bar{H}_{\mathrm{b}}^{\mathrm{m}} > \Delta H_{\mathrm{b}}^{\mathrm{sub}} \tag{6.48}$$

where $\Delta H_{\mathrm{b}}^{\mathrm{sub}}$ is the heat of sublimation of the solute at the given temperature. In dilute systems, θ_{b} in equation (6.45) will be small, with the result that, in the critical region of the fluid phase, equation (6.45) can be very well approximated by

$$\left(\frac{\partial \ln k_{\mathrm{b}}}{\partial T}\right)_{P,y_{\mathrm{b}}} \cong \frac{H_{\mathrm{b}}^{\mathrm{o}} - \bar{H}_{\mathrm{b}}^{\mathrm{m}} - \Delta H_{\mathrm{b}}^{\mathrm{ads}}}{RT^2} + \alpha_{\mathrm{m}} \tag{6.49}$$

We have retained only the terms that can show critical-point divergences in equation (6.45) in formulating equation (6.49); it is a key result, since it suggests that

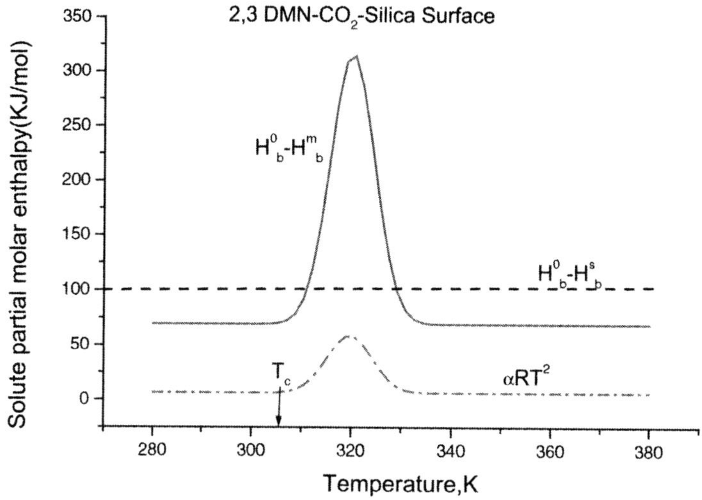

Figure 6.1 Fluid-phase partial molar enthalpy function in a system of supercritical fluid with adsorbent

the variation of the solute capacity factor (or equivalently its adsorption coefficient) with temperature depends upon the relative magnitude of the terms

$$\frac{H_{\mathrm{b}}^{\mathrm{o}} - \bar{H}_{\mathrm{b}}^{\mathrm{m}} - \Delta H_{\mathrm{b}}^{\mathrm{ads}}}{RT^2} \text{ and } \alpha_{\mathrm{m}} \tag{6.50}$$

In most systems of interest, α_{m} in equation (6.49) will be positive, exhibiting divergence at the solvent's critical point, as discussed in chapter 2. Hence the sign of the derivative in equation (6.49) will be closely related to the sign of

$$\frac{H_{\mathrm{b}}^{\mathrm{o}} - \bar{H}_{\mathrm{b}}^{\mathrm{m}} - \Delta H_{\mathrm{b}}^{\mathrm{ads}}}{RT^2} \tag{6.51}$$

A plot of the enthalpy functions in equation (6.51) for a typical solute–solvent system is shown in figure 6.1. The numerator of (6.51) is positive if

$$H_{\mathrm{b}}^{\mathrm{o}} - \bar{H}_{\mathrm{b}}^{\mathrm{m}} > \Delta H_{\mathrm{b}}^{\mathrm{ads}} \tag{6.52}$$

In this case, it can be seen that there is an immediate analogy between this result, which applies for adsorption, and that for solvation given in chapter 5. The inequality in (6.52) has been referred to as the *retrograde adsorption inequality* [3] which can be used to illustrate adsorption behavior in the following circumstances:

1. When the inequality in (6.52) is satisfied, the solute's adsorption coefficient will increase, possibly quite dramatically with increasing temperature.
2. At a temperature more remote from the critical one, this inequality is likely not to be satisfied since the enthalpy term decays strongly away from the critical point of the system. If the negative values induced by the violation of this inequality are large enough to overcome the value of α_{m}, then the derivative quantity will be

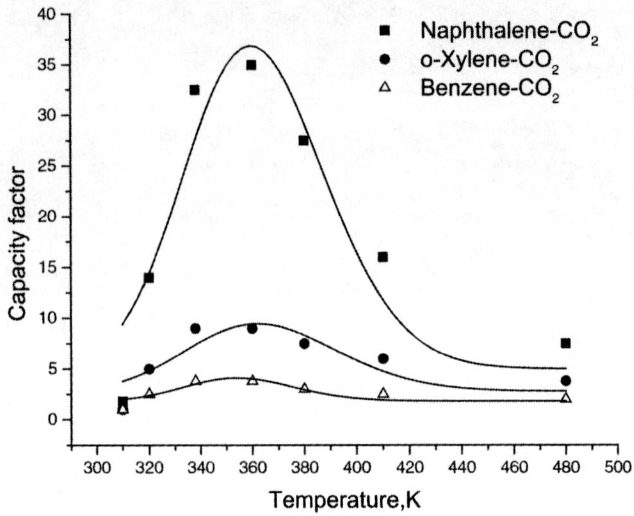

Figure 6.2 Capacity-factor isobars measured using supercritical fluid chromatography

negative; in this case, k_b will start to decline with increasing temperature. Data [2] showing this behavior is given in figure 6.2. It is evident that this picture describes two adsorption regimes in near-critical mixtures: (1) an *endothermic region* where adsorption coefficients increase with temperature, and (2) an *exothermic regime* where the opposite occurs. The occurrence of one or the other of these situations depends upon the proximity of the supercritical fluid phase to its critical point.

Figure 6.3, taken from the work of Schmitz et al. [8], shows all of this clearly. What is particularly interesting about these data is the fact that *isomeric compounds* were

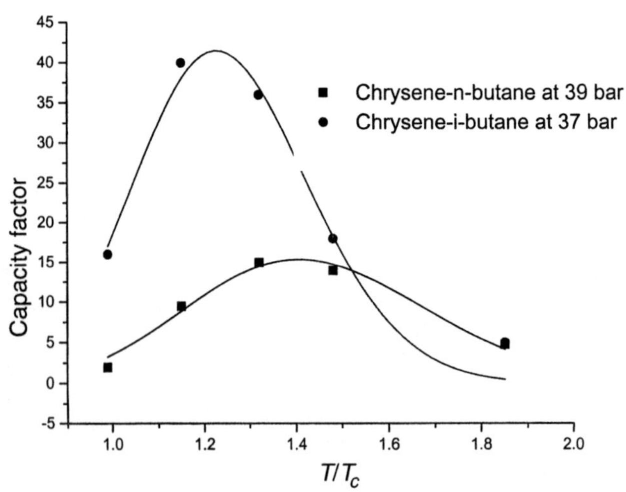

Figure 6.3 Capacity-factor data for isometric fluid species measured using supercritical fluid chromatography

used for the supercritical fluid solvent phase. Even with such closely related solvent molecules, the solute adsorption isobars are strikingly different at virtually the same reduced conditions in the fluid phase. Small molecular differences between solvent species translate into large differences in the adsorption behavior of the same solute species, showing the *amplification* effect instituted by the onset of critical-point divergence in the fluid phase. These effects in the near-critical region of the fluid phase are especially useful in *supercritical fluid chromatography* (SFC), where they have the effect of providing maximum intrinsic molecular resolution between species as the critical point of the supercritical mobile phase is approached.

The case for adsorption isotherms can be described using analogous equations to those discussed here, except that the pressure derivatives of the capacity factor provide the key relationships. These are considered in the additional exercises at the end of this chapter.

The Relationship between the Capacity Factor and the Density

The measurement of adsorption coefficients by chromatographic techniques is closely related to species retention times in the column as shown in earlier results. A very convenient relationship exists between these adsorption coefficients and the density of the supercritical mobile phase in SFC, which we now develop.

As in our previous analysis, we view the system physically as a compressible fluid mixture in equilibrium contact with an incompressible adsorbed (stationary) phase, itself a composite of solute, solvent, and stationary-phase materials. The definition of the solute's capacity factor k_i can be recast into the following equation:

$$k_i \equiv \frac{\theta_i}{y_i}\frac{v^{\mathrm{m}}}{v^{\mathrm{s}}} \cdot \frac{1}{K_{\mathrm{o}}} \tag{6.53}$$

where θ_i, and y_i are solute mole fractions in stationary and fluid phases, respectively, and $v^{\mathrm{m}}/v^{\mathrm{s}}$ is the ratio of the molar volumes of these phases. In terms of the respective fugacity coefficients of the species, this equation can be expressed as

$$k_i = \frac{\hat{f}_i^{\mathrm{s}}}{\hat{\phi}_i^{\mathrm{s}}}\frac{\hat{\phi}_i^{\mathrm{m}}}{\hat{f}_i^{\mathrm{m}}}\frac{1}{K_{\mathrm{o}}}\frac{v^{\mathrm{m}}}{v^{\mathrm{s}}} \tag{6.54}$$

The equilibrium condition requires that

$$\hat{f}_i^{\mathrm{s}} = \hat{f}_i^{\mathrm{m}} \tag{6.55}$$

in which case the capacity factor is given by

$$k_i = \frac{\hat{\phi}_i^{\mathrm{m}}}{\hat{\phi}_i^{\mathrm{s}}}\frac{1}{K_{\mathrm{o}}}\frac{v^{\mathrm{m}}}{v^{\mathrm{s}}} \tag{6.56}$$

Therefore, along the phase equilibrium envelope,

$$d\ln k_i = d\ln\hat{\phi}_i^{\mathrm{m}} - d\ln\hat{\phi}_i^{\mathrm{s}} + d\ln(1/\rho^{\mathrm{m}}) - d\ln(1/\rho^{\mathrm{s}}) \tag{6.57}$$

Since there are three intensive thermodynamic degrees of freedom in this system along the coexistence envelope, the expansion of $\ln k_i$ as a derivative quantity is given by the equation

$$d\ln k_i = \left(\frac{\partial \ln \hat{\phi}_i^{\mathrm{m}}}{\partial T}\right)_{y_i,\rho} dT + \left(\frac{\partial \ln \hat{\phi}_i^{\mathrm{m}}}{\partial \rho^{\mathrm{m}}}\right)_{T,y_i} d\rho^{\mathrm{m}} + \left(\frac{\partial \ln \hat{\phi}_i^{\mathrm{m}}}{\partial y_i}\right)_{T,\rho} dy_i - d\ln \rho^{\mathrm{m}}$$
$$+ d\ln \rho^{\mathrm{s}} + \text{similar adsorbed phase terms} - \left(\frac{\partial \ln \hat{\phi}_i^{\mathrm{s}}}{\partial T}\right)_{y_i,\rho} + \cdots \tag{6.58}$$

Using the thermodynamic identity

$$\left(\frac{\partial \ln \hat{\phi}_i^{\mathrm{m}}}{\partial \rho^{\mathrm{m}}}\right)_{T,y_i} = \left(\frac{\partial \ln \hat{\phi}_i^{\mathrm{m}}}{\partial P}\right)_{T,y_i} \frac{1}{\rho^{\mathrm{m}} K_T}$$

and its analogous form for the stationary phase, we find the following approximate equation describing the isothermal variation of the solute's capacity factor with fluid density in the critical region:

$$\left(\frac{\partial \ln k_i}{\partial \ln \rho^{\mathrm{m}}}\right)_{T,y_i} \approx \frac{\bar{V}_i^{\mathrm{m}}}{RTK_T} - 1 \tag{6.59}$$

The divergence of the fluid-phase compressibility K_T eliminates the stationary-phase terms in equation (6.58) from any further consideration.

We can use critical scaling results to examine the term $\bar{V}_i^{\mathrm{m}}/RTK_T - 1$ very near the solvent's critical point. In chapter 5 we showed that, at the limit of infinite dilution, the property $\bar{V}_i^{\infty}/K_T$ approaches a finite constant value both classically and nonclassically at the solvent's critical point. Therefore, data for the isothermal capacity factor taken near the fluid's critical temperature should be linear in a plot of $\log k_i$ against $\log \rho$, in a small density range around ρ_c, if the variation of $\bar{V}_i^{\mathrm{m}}/K_T$ with density is small in this region. Most available data shows this to be the case over wide temperature and density ranges around the critical values [9, 10]; a set of typical data is shown in figure 6.4 [2].

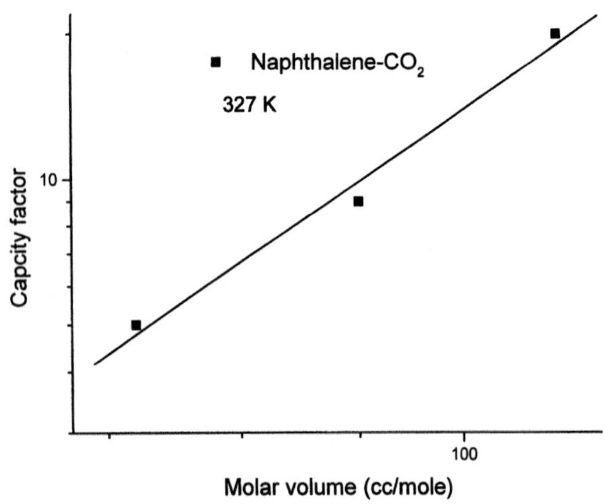

Figure 6.4 Capacity-factor–density data in naphthalene–carbon dioxide mixture

Another useful feature of equation (6.59) is that, in the linear regime, slopes of the data plot of $\log k_I$ against $\log \rho$ data allow $\bar{V}_i^{\mathrm{m}}$ for the solute to be established at various temperatures and densities if K_T for the supercritical solvent is known, as is frequently the case [11]. In addition, using (6.59), only two data points are, in principle, required to provide a relationship that predicts the solute adsorption coefficients for a given species over a wide range of density conditions. This is potentially a useful result for engineering applications where information of this sort is often required.

The Relationship between the Capacity Factor and Solubility Measurements

One useful consequence of the results developed in the previous section is their application to the rapid estimation of solubilities in supercritical solvents using chromatographic techniques [12]. Such measurements are done much more efficiently than those obtained with gravimetric techniques usually used for solubility measurements.

Assume that the solubility of a compound, designated j, is known at a given temperature and pressure designated as T and P_1, and that we want to find its solubility at T and P_2. Using an expression like (6.56), the ratio of the solute's fugacity coefficients at temperature T and the two pressures can be approximated as

$$\frac{\hat{\phi}_i^{\mathrm{m}}(T, P_1)}{\hat{\phi}_i^{\mathrm{m}}(T, P_2)} \approx \frac{k_i(T, P_1)}{k_i(T, P_2)} \frac{\hat{\phi}_i^{\mathrm{s}}(T, P_1)}{\hat{\phi}_i^{\mathrm{s}}(T, P_2)} \frac{\rho(T, P_1)}{\rho(T, P_2)} \tag{6.60}$$

The ratio of the solubilities of i at the two pressures is also given by the standard thermodynamic equation

$$\frac{y_i(P_1)}{y_i(P_2)} = \frac{\hat{\phi}_i^{\mathrm{m}}(T, P_2)}{\hat{\phi}_i^{\mathrm{m}}(T, P_1)} \frac{P_2}{P_1} \exp\left[\frac{v_i^{\mathrm{s}}(P_1 - P_2)}{RT}\right] \tag{6.61}$$

In the adsorbed phase, the effect of pressure on $\hat{\phi}_i^{\mathrm{s}}$ is very small, and the Poynting correction factor in equation (6.61) may also be neglected. Following this approach with (6.60) in (6.61), we arrive at the equation

$$\frac{y_i(P_1)}{y_i(P_2)} \approx \frac{k_i(T, P_2)}{k_i(T, P_1)} \frac{\rho(T, P_2)}{\rho(T, P_1)} \frac{P_2}{P_1} \tag{6.62}$$

Equation (6.62) shows how capacity-factor data which are relatively easy to find, can be used to estimate solubility data for the same solute in a given solvent. The densities of the fluid at the two pressures can be taken to be that of the pure supercritical fluid at these conditions because these solutions are very dilute. A similar analysis leads to useful equations for estimating a much wider variety of properties in these systems using only chromatographic measurements; a problem addressing this point is left for the additional exercises.

6.3 Optimal Separations Performance in Supercritical Fluid Chromatography

In the retrograde adsorption region where molecular selectivity is highest, solute species migration through the system is slowest; thus, favorable conditions from a viewpoint of molecular selectivity are often adverse from the perspective of processing time. The resolution of this problem with supercritical fluid chromatography (SFC) requires the use of *density programming*. The objective with this approach is to find the best dynamic density profile for the mobile phase, namely one that provides the required molecular resolution in a given processing time [1]. Our purpose here is to outline such a solution strategy for optimizing the performance of SFC, using optimal-control ideas applied to SFC. These methods are called gradient-programming SFC since gradients (dynamic changes) are made to process variables to achieve optimal outcomes.

The dynamic equation often used as the basic equation in modeling gradient-programming SFC is given for species i by

$$\frac{dx_i}{dt} = \frac{1}{1+k_i}u \tag{6.63}$$

where x_i is the position of the maximum of the solute peak and u is the superficial velocity of the supercritical mobile phase. The capacity factor for each species is usually taken to be a function of density in the form $k_i = a\rho^b$, as would be suggested by (6.59), with a and b constants for a particular solute–solvent system.

EXERCISE 6.9

Show that equation (6.63) can be expressed in the form:

$$\frac{dx}{dt} = \frac{1}{1+a\rho^{\mathrm{b}}}\,\frac{\dot{m}_0 + A(L-x)\dfrac{d\rho}{dt}}{A\rho} \tag{6.64}$$

where L is the column length and $\dot{m}_0$ is the input mass flow rate of supercritical solvent to the column.

We now define a *dynamic objective functional* for the separation, an example of which is given by Φ:

$$\Phi = \left\{ W_0[x_N(t_{\mathrm{f}}) - L]^2 + \int_{j=1}^{t_{\mathrm{f}}} \sum_{j=1}^{N-1} W_j[x_j(t) - x_{j+1}(t) - \Delta_j]^2 dt \right\} \tag{6.65}$$

The W_j are weighting factors used to express the relative importance of the various terms in this objective functional. The solute components are numbered from fast-eluting (= 1) to slow-eluting (= N) species, respectively. L stands for the length of the column and Δ_j is the distance by which two neighboring peak maxima are required to be separated. The variable t_{f} is the prespecified time within which the species' separation has to be completed. Qualitatively, the minimization of Φ by changing the density of the incoming fluid phase attempts to balance two process objectives. The first is represented by the term $W_0[x_N(t_f) - L]^2$ which forces the slowest-eluting component to be at the end of the column within the specified time t_f. The integral part of the objective function promotes the required resolution of peak maxima (between successive species) by requiring these peaks to be separated by a distance Δ_j at the end of the process. The purpose of the optimal control solution is to minimize Φ in (6.65) by finding the optimal functional trajectory for $\rho(t)$, subject to the dynamic equations, for each species, given by an equation similar to that given in (6.63). This is structurally a classic problem in optimal control theory applied to linear dynamic systems [13]. The general problem in optimal control is to choose the dynamic behavior of a set of time-dependent control variables so that the different state variables describing system dynamics, $x_i(t)$, are transformed from an initial state to a prescribed terminal point at the prespecified final time t_f. Mathematically, we would like to find the optimal control variable(s) $v(t)$ which yields a minimum value for the performance index I given by

$$I = G[\boldsymbol{x}(t_f)] + \int_{t_0}^{t_f} F[\boldsymbol{x}(t), \boldsymbol{v}(t)]dt \tag{6.66}$$

subject to the dynamic equations

$$\frac{d\boldsymbol{x}(t)}{dt} = f/[\boldsymbol{x}(t), \boldsymbol{v}(t)] \tag{6.67}$$

over the interval specified by

$$0 \leq t \leq t_f \tag{6.68}$$

where $\boldsymbol{x}(t)$ is an n-dimensional vector of state variables and $v(t)$ is an m-dimensional vector of control parameters. The initial state of the system is assumed to be specified

$$\boldsymbol{x}(0) = \boldsymbol{x}_0. \tag{6.69}$$

The function G, which is only dependent on the final state of the system at t_f, is responsible for ensuring satisfaction of the static part of the performance index. The function, F, which is integrated over time, depends on the values of the state and the control variables at the intermediate process times when $t \leq t_f$. The SFC problem just described easily fits this problem description. Here the control variable of interest is $\rho(t)$ and the dynamic equations analogous to (6.65) and (6.64) are given by equations (6.65) and (6.64), respectively. This problem has been solved in detail [1] for a real system, and a typical optimizing density gradient found is shown in figure 6.5.

6.4 Chapter Review

Adsorption phenomena are relevant in applications to chemical processing and materials engineering where fluid–solid interactions are important. These play an important role in many areas of chemical and materials engineering. Typical application

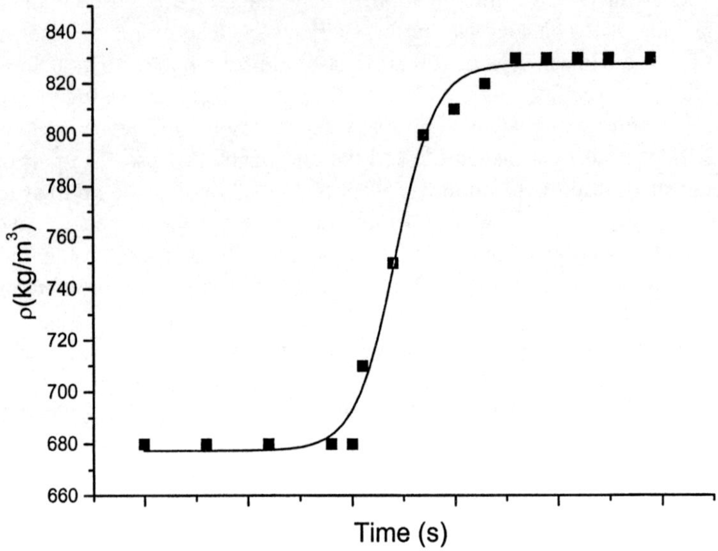

Figure 6.5 Density profile for the mobile phase in density-programmed supercritical fluid chromatography

areas include supercritical-fluid-based chemical separations, heterogeneous catalysis, porous-materials processing, and device fabrication in the microelectronics industry.

In this chapter, we focused upon analyzing the properties of adsorption systems where the mobile-fluid carrier phase is near its critical point. Mathematical models were developed for representing these adsorption processes showing how both thermodynamic and transport properties, particularly those in the fluid phase, are central to understanding the behavior of these systems. The key property in all of these models is the species' adsorption coefficient. In the critical region, thermodynamic results can be used to derive particularly simple equations for this property. These equations are especially useful in the mathematical modeling of SFC, and in their use for facilitating the measurement of thermodynamic and transportive properties in near-critical solvent mixtures.

From a standpoint of separation technology, the most important thermodynamic properties in these systems are again the adsorption coefficients of the various solute species. These show a maximum along a given adsorption isobar, with the maximum point being the boundary between endothermic and exothermic adsorption regimes. From a practical point of view, interspecies molecular resolution is highest in the region where these adsorption-coefficient maxima occur. However, species migration is also slowest in this region and, as a consequence, more sophisticated processing strategies are required to optimize separation performance using SFC. The approach taken in this regard requires the use of ideas from optimal control theory. This analysis leads to a method of determining the dynamic behavior of key process variables that enable one to achieve the required molecular resolution in the fastest possible processing times. A method for doing this was described using the supercritical fluid phase's density as the dynamic control variable.

6.5 Additional Exercises

Mathematical Dynamics in Packed-Bed Chromatographs

1. How would you measure the quantity $L/u = t_o$, the residence time of an unretained species in a chromatographic adsorption column?
2. A spherical particle is porous with radius R_p. The particle's porosity is ε_p, and the solute concentrations in the interstitial spaces and solid phase are $c_p(r_p)$ and $q_p(r_p)$, respectively. Show that the average concentration in the particle is given by

$$q_{avg} = \frac{3}{R_p^3} \int_0^{R_p} r_p^2 (\varepsilon_p c_p + (1 - \varepsilon_p) q_p) dr_p$$

3. Prove that the interfacial area per unit bed volume available for transport with spherical porous particles of radius R_p is given by $3(1 - \varepsilon)/R_p$, where ε is the overall bed porosity.
4. Show that the mass-balance differential equation for a solute diffusing into the interior of a porous spherical adsorbent particle is given by the equation

$$D_p \left(\frac{\partial^2 c_p}{\partial r_p^2} + \frac{2}{r_p} \frac{\partial c_p}{\partial r_p} \right) = (\varepsilon_p + (1 - \varepsilon_p) K) \frac{\partial c_p}{\partial t}$$

   if

$$q_p = K c_p$$

   What are the boundary conditions for solving this problem if the concentration on the particle's external surface is constant? Solve this system for the steady-state composition profile in the particle.
5. If we measure the dynamic response $c(t)$ of a chromatographic column to a delta function input of a tracer species of concentration c_0 at time $= 0$, we can establish a mean residence time for the species from a *moments analysis of the data*. If $c(t)$ is approximately Gaussian with variance $\sigma_{c(t)}$, show how $\sigma_{c(t)}$ can be found from the experimental data. [Hint: Show how the following integral quantity I can be related to the variance around the mean residence time $\bar{t}$ of the response function $c(t)$.]

$$I = \frac{\int_0^\infty (t - \bar{t})^2 c(t) dt}{\int_0^\infty c(t) dt}$$

6. Prove the following useful identities for Laplace transforms:

$$L_t \left[\frac{\partial c(t, z)}{\partial t} \right] = s\bar{c}(s) - c(0)$$

$$L_t \left[\frac{\partial c(t, z)}{\partial z} \right] = \frac{\partial \bar{c}(s, z)}{\partial z}$$

$$L_t \left[\frac{\partial^2 c(t, z)}{\partial z^2} \right] = \frac{\partial^2 \bar{c}(s, z)}{\partial z^2}$$

$$L_t \left[\frac{\partial^2 c(t, z)}{\partial t^2} \right] = s^2 \bar{c}(s) - sc(0) - c'(0)$$

7. The following convection–diffusion equation models solute flow through a hollow tube of length l:

$$\frac{\partial c}{\partial t} = D\frac{\partial^2 c}{\partial z^2} - u\frac{\partial c}{\partial z}$$

Using a moments analysis and the Laplace transform method, show that the mean residence time of a pulse input into this system is

$$\bar{t} = l/u$$

Use the boundary conditions

$$c(0, z) = 0, \qquad c(t, 0) = c_0\delta(t), \qquad c(t, z) \text{ finite for all } z \text{ and } t > 0$$

8. Solve the previous problem for $c(t, l)$ with the boundary conditions changed to accommodate a unit step input to the system.
9. The solution to the time-dependent diffusion equation in response to an input of solute at the origin of an infinite medium is given in the Laplace domain by the equation

$$\bar{c}(s, z) = \frac{\alpha}{D}\sqrt{\frac{D}{s}}e^{-\sqrt{s/D}z}$$

where α is a constant, D the diffusion constant, and z the isotropic spatial coordinate. Invert this equation and show that

$$c(t, z) = \beta\frac{1}{\sqrt{Dt}}e^{-1/4zDt}.$$

What is the value of the constant β?

Supercritical Adsorption

10. Derive the expression for the variation of $\ln k_i$ (k_i is the capacity factor for species i) versus pressure, at constant temperature, in a chromatographic column. Sketch the shape of $\ln k_i$ versus pressure for $T > T_c$
11. The enthalpy of transfer for solute b from the mobile to the stationary phase in a chromatograph is defined as

$$\Delta\bar{H}_b^{\text{trans}} \equiv \bar{H}_b^{\text{s}} - \bar{H}_b^{\text{m}}$$

What sign is $\Delta\bar{H}_b^{\text{trans}}$ in the endothermic adsorption regime? Determine also the sign of the corresponding term for entropy of transfer:

$$\Delta\bar{S}_b^{\text{trans}} \equiv \bar{S}_b^{\text{s}} - \bar{S}_b^{\text{m}}$$

How would one evaluate these terms for enthalpy and entropy of transfer from experimental chromatographic data in a system of dilute solute in supercritical solvent?

12. In a chromatographic system, the capacity factor and equilibrium relationships for solute species i are given by

$$k_i = \frac{\theta_i}{y_i} \cdot \frac{K_0}{\rho}, \qquad \hat{f}_i^{\mathrm{s}} = \theta_i H_i(T, P), \qquad \hat{f}_i^{\mathrm{m}} = y_i \hat{\phi}_i P$$

where $H_i(T, P)$ is the Henry's constant for the solute in the stationary phase (the other terms have been defined previously in this chapter). Assume that we take data for isothermal solubility and capacity factor for two species i and j at several identical pressures. Show that the ratio of solubilities to capacity factors, for both species, at each data point is given by

$$\frac{y_i}{y_j} = \alpha \frac{k_j}{k_i}$$

where α is a constant. What is α in terms of physical variables in the system? Note: this result can be readily adapted to rapidly measure solubilities in supercritical solvents using chromatographic measurements.

13. A supercritical fluid is used to extract a solute present as a collection of solid spheres in a packed bed. Show that the condition given by the maximum, on the profile of the capacity factor against T, is where the rate-determining step is most likely to be the solute desorption step from the sphere's surface to the flowing fluid phase. Develop an expression for the steady-state profile of the solute's concentration in the fluid phase as a function of bed length using the local equilibrium assumption to relate the solute concentration on the surface of the sphere with that in the fluid phase.

Langmuir Adsorption

14. For the Langmuir partition function given in this chapter, show that it can be expressed in matrix-vector form as follows:

$$\boldsymbol{G}\boldsymbol{q} = \boldsymbol{p}$$

$$\boldsymbol{G} = \begin{bmatrix} 1-\theta_1 & -\theta_1 & \cdots & -\theta_1 \\ -\theta_2 & 1-\theta_2 & \cdots & -\theta_2 \\ \vdots & \vdots & \ddots & \vdots \\ -\theta_N & -\theta_N & \cdots & 1-\theta_N \end{bmatrix}$$

and

$$\boldsymbol{q} = \begin{bmatrix} \lambda_1 q_1 \\ \lambda_2 q_2 \\ \vdots \\ \lambda_N q_N \end{bmatrix} \quad \text{and} \quad \boldsymbol{P} = \begin{bmatrix} \theta_1 \\ \theta_2 \\ \vdots \\ \theta_N \end{bmatrix}$$

where $\lambda_i \equiv e^{\beta\mu_i}$ with μ_i the chemical potential of species i. Show that the determinant of $\boldsymbol{G}$ is given by

$$|\boldsymbol{G}| = 1 - \sum_{j=1}^{N} \theta_j$$

and that the general solution for $\lambda_i q_i$ is

$$\lambda_i q_i = \frac{\theta_i}{1 - \sum_{k=1}^{N} \theta_k}$$

(Note: We often assume that the partition function for a single bound species is given by a form

$$q_i(T) = q_i(T_0)e^{-\beta U_i}$$

with U_i the bound configurational energy of the surface-bound species.)

15. Prove that, for the Langmuir model,

$$\mu_i - \mu_i^0 = -U_i$$

where μ_i^0 is defined to be the chemical potential in the case where $U_i = 0$ and $q_i(T) = q_i(T_0)e^{-\beta U_i}$.

16. Assume the following form for the individual-molecule partition function in the Langmuir model:

$$q_i(T) = q_i(T_0)e^{-\beta U_i}$$

where U_i is a constant parameter representing the potential well depth in which the adsorbed species vibrates. Formulate an expression for evaluating U_i from data on low-pressure adsorption modeled with this approach.

17. The canonical partition function for N ideal gas molecules in a volume V is given by the well-known statistical-mechanical expression

$$Q = \frac{1}{N!}\left(\frac{V}{\Lambda^3}\right)^N$$

where Λ is the thermal de Broglie wavelength. Prove that the chemical potential of this ideal gas is given by the equation

$$\mu_i^0(T, P) = -kT \ln\left(\frac{kT}{\Lambda^3}\right) + kT \ln P$$

18. The chemical potential of species i in a mixture $\hat{\mu}_i$ is related to its fugacity through the definition

$$\hat{\mu}_i - \mu_i^0 = RT \ln \frac{\hat{f}_i}{f_i^\circ}$$

where the superscript $^\circ$ designates a standard state condition—usually an ideal gas at the given conditions of T and P. The chemical potential applies to any species in a given phase.

If the Henry's law coefficient H_i in an adsorption system is furthermore defined by the equation

$$H_i = \lim_{\theta_i \to 0} \frac{\hat{f}_i}{\theta_i}$$

where θ_i is the fractional-coverage concentration of the adsorbed species i, show that H_i in a single-species adsorption system, modeled by the Langmuir partition function, is given by the simple expression

$$H_i = \frac{1}{\beta q_i \Lambda^3}$$

19. In an adsorption system, if the fluid phase is considered ideal, and the adsorbed phase is modeled with the Langmuir partition function, show that, in a pure system, the fractional coverage θ_i is given by the expression

$$\theta_i = \frac{PK_T}{1 + PK_T}, \quad K_T \text{ a function of } T \text{ only}$$

If we rewrite θ_i as

$$\theta_i = \frac{V}{V_{\mathrm{m}}}$$

where V is the volume of gas adsorbed and V_{m} the volume at complete monolayer coverage, show that the above equation can be rearranged to yield

$$\frac{P}{V} = \frac{P}{V_{\mathrm{m}}} + \frac{1}{K_T V_{\mathrm{m}}}$$

Describe how you would take data to find V_{m}. The *isosteric heat of adsorption* is the heat evolved during adsorption at a given value of coverage. How would you find the it from these equations?

20. Assume that we have a serial *catalytic chemical reaction* occurring on the surface of a nonporous catalyst for the hydrogenation of hexyne to hexene and eventually to hexane. These reactants are present only at dilute quantities, carried in a near-critical carrier solvent phase at a high (supercritical) pressure. If the kinetic constants for each of these surface-phase hydrogenation reactions are roughly equivalent and their adsorption constants follow the qualitative structure given in figure 6.2 (hexyne corresponding to naphthalene, hexene to o-xylene, and hexane to benzene), what conditions do you think will be optimal for the production of the alkene species?

Bibliography

[1] F. van Puyvelde, P. V. Rompay, and E. H. Chimowitz, "Optimal control of molecular resolution in supercritical fluid chromatography with density programming," *J. Supercrit. Fluids*, vol. 5, p. 227, 1992.

[2] F. D. Kelley, "Analysis and experimental study of near-critical phenomena and resolution in supercritical fluid chromatography," Ph.D thesis in Chemical Engineering. University of Rochester, NY, 1990.

[3] F. D. Kelley and E. H. Chimowitz, "Near-critical phenomena and resolution in supercritical fluid chromatography," *AIChE J.*, vol. 36, p. 1163, 1990.

[4] M. Suzuki and J. M. Smith, *Advances in Chromatography*, vol. 13. New York: Marcel Dekker, 1975.

[5] K. Kubin, *Czech. Chem. Commun.*, vol. 130, p. 1104, 1965.

[6] C. Erkey and A. Akgerman, "Chromatography theory: application to supercritical fluid extraction," *AIChE J.*, vol. 36, p. 1715, 1990.

[7] D. A. McQuarrie, *Statistical Mechanics*. New York: Harper & Row, 1976.

[8] F. P. Schmitz, D. Leyendecher, and E. Klesper, "Chromatography with mobile phases in the liquid and supercritical state," *Ber. Bunsen-Ges. Phys. Chem.*, vol. 88, p. 912, 1984.

[9] S. K. Kumar and K. P. Johnston, "Modeling the solubility of solids in supercritical fluids with density as the independent variable," *J. Supercrit. Fluids*, vol. 1, p. 15, 1988.

[10] E. H. Chimowitz and F. D. Kelley, "A new representation for retention time in supercritical fluid chromatography," *J. Supercrit. Fluids*, vol. 2, p. 106, 1989.

[11] M. P. Srinivasan and B. J. McCoy, "Partial molar volumes of ethyl acetate from supercritical carbon dioxide data," *J. Supercrit. Fluids*, vol. 4, p. 69, 1991.

[12] M. P. Ekart, K. L. Bennet, S. M. Ekart, G. S. Guirdial, C. L. Liotta, and C. A. Eckert, "Cosolvent interactions in supercritical fluid solutions," *AIChE J.*, vol. 39, p. 235, 1993.

[13] W. H. Ray, *Advanced Process Control*. New York: McGraw-Hill, 1980.

7 Mean-Field Theories

The most widely used analytic models for representing thermodynamic behavior in supercritical fluids are of the mean-field variety. In addition to the practical interest in studying this topic, this class of models is also the conceptual starting point for any microscopic discourse on critical phenomena. In this chapter we take up the basic ideas behind this approach, studying different physical models, showing how their mean-field approximations can be constructed as well as investigating their critical behavior.

A useful conceptual model for understanding mean-field ideas is the *Ising model* whose properties we consider in some detail, especially its mean-field approximation. The Ising model has the advantage of belonging to the same *critical universality class* as so-called *simple fluids*, defined as fluids with short-range intermolecular potentials. Most supercritical fluid solvent systems of practical interest fall within this class; hence results developed using the Ising model have important implications for understanding the critical behavior of this entire universality class. While we discuss universality and related ideas in more detail in subsequent chapters, suffice it to say here that the Ising system belongs to arguably the most important critical universality class from a process engineering standpoint.

7.1 The Ising Model

In its simplest form, the Ising model considers N spins arranged on a lattice structure (of 1, 2, or 3 dimensions) with each spin able to adopt one of two (*up* or *down*) orientations in its lattice position. A specific state of the system is determined by a given configuration of all the spins. The model can be made more complex by considering additional degrees of freedom to the spin orientations. For example, the *Heisenberg* model considers a 3-dimensional lattice with the spin orientation at each lattice site described by a 3-dimensional *vector* quantity. All that is required to facilitate the use of statistical mechanics with this model is the definition of the Hamiltonian (the system's energy function) associated with a particular lattice state v. This Hamiltonian usually consists of spin–spin interaction terms, as well as a term representing the presence of a magnetic field, which serves to orient the spins in its direction. For the spin *up–down* model, the usual form of the Hamiltonian is given by the function

$$E_v = -\sum_{i=1}^{N} H\mu_m S_i + \text{term due to spin interactions} \tag{7.1}$$

where S_i can take on the value $+1$ (up) or -1 (down), μ_m is the magnetic moment for spin i, and H is the *field strength*. The interaction term is usually given by a sum of nearest-neighbor terms of the form

$$-J\sum_{i,j}{}' S_i S_j \tag{7.2}$$

where J is the *spin–spin coupling constant* and the prime on the summation indicates that this sum is to be taken only between nearest neighbors. More complex Hamiltonians can include next-to-nearest-neighbor terms, and so on. In the simple Heisenberg model, the interaction term is a vector dot product of the form

$$-J\sum_{i,j}{}' \mathbf{s}_i \cdot \mathbf{s}_j \tag{7.3}$$

where the spin orientation directions are now given by the vector quantities $\mathbf{s}_i$ and $\mathbf{s}_j$. In the limit $T \to 0$, the system will be in its ground state with all spins oriented in the direction of the field; at the limit $T \to \infty$, we can imagine a random configuration of spins. In between these limits, there exists the possibility of a phase transition at a critical temperature, which we will designate T_c. At this temperature, there is a change from a disoriented to an oriented phase upon cooling. Characterizing the ordered phase is a property termed the *order parameter* which, in the case of the Ising model, is the average spin alignment defined by the property m as follows:

$$m \equiv \langle S_i \rangle = \frac{\langle M \rangle}{N\mu_m} \text{ with } M = \mu_m \sum_{i=1}^{N} S_i \tag{7.4}$$

where $\langle\ \rangle$ denotes the average, or expectation value, of the property, N represents the total number of spins in the lattice, and μ_m is the magnetic moment per spin. In a random disordered phase, $m = 0$, while $m = 1$ in the ground state. Often in this formulation the quantity μ_m is subsumed with the regular field term; that is, we define $\tilde{H} \equiv H\mu_m$.

Exact Solution of the Ising Model in One Dimension, with $\tilde{H} \neq 0$

One of the pedagogical advantages of the one-dimensional (1-d) Ising model is that it is one of the few statistical-mechanical models for which an exact solution is available for the partition function, and hence for the system's free energy. This is valuable for use in comparisons with the predictions given by other approximate theories, such as mean-field models. The mathematical-physics technique used in developing these results involves a matrix transfer method [1]. We follow this approach, which leads us to an analytical solution of the partition function in the 1-d Ising model with finite field. The energy of the lattice in state v is given by the equation

$$\begin{aligned} E_v &= -\sum_{i=1}^{N} \tilde{H} S_i + \text{ (energy due to interaction between spins)} \\ &= -\tilde{H} \sum_{i=1}^{N} S_i - J \sum_{i=1}^{N} S_i S_{i+1} \quad \text{with} \quad S_{N+1} = S_1 \end{aligned} \tag{7.5}$$

with $\langle S_1 \rangle = \langle S_2 \rangle = \cdots = \langle S_N \rangle$. The partition function is given by the equation

$$Q(N, h, T) = \sum_{S_1,\ldots,S_N} \exp\left[K \sum_{i=1}^{N} S_i S_{i+1} + h \sum_{i=1}^{N} S_i \right] \tag{7.6}$$

where

$$K \equiv \beta J = \frac{J}{k_B T} \tag{7.7}$$

with k_B the Boltzmann constant and

$$h \equiv \beta \tilde{H} \tag{7.8}$$

We now look for a factorization of the exponential terms in the partition function. If we define the function $v(s, s')$ as

$$v(s, s') \equiv \exp\left[Kss' + \frac{h}{2}(s + s') \right] \tag{7.9}$$

where

$$v(s, s') = v(s', s) \tag{7.10}$$

then we can rewrite $Q(N, h, T) \equiv Q_N$ as

$$Q_N = \sum_{s_i=\pm 1} v(s_1, s_2)v(s_2, s_3)v(s_3, s_4)\cdots v(s_{N-1}, s_N)v(s_N, s_1) \tag{7.11}$$

Examining the individual terms $v(s, s')$, we see that there are four possibilities for this term, which we can represent as the elements of a 2×2 matrix depending upon the respective values for s and s' (± 1). We define this (symmetric) matrix $\boldsymbol{V}$ as

$$\boldsymbol{V} \equiv \begin{bmatrix} v(1,1) & v(1,-1) \\ v(-1,1) & v(-1,-1) \end{bmatrix} = \begin{bmatrix} e^{K+h} & e^{-K} \\ e^{-K} & e^{K-h} \end{bmatrix} \tag{7.12}$$

and look at the squared matrix $\boldsymbol{V}^2$ expressed in terms of possible values of $v(s_1, s_2)$ and $v(s_2, s_3)$:

$$\begin{aligned} \boldsymbol{V}^2 = &\begin{bmatrix} v(s_1 = 1, s_2 = 1) & v(s_1 = 1, s_2 = -1) \\ v(s_1 = -1, s_2 = 1) & v(s_1 = -1, s_2 = -1) \end{bmatrix} \\ &\times \begin{bmatrix} v(s_2 = 1, s_3 = 1) & v(s_2 = 1, s_3 = -1) \\ v(s_2 = -1, s_3 = 1) & v(s_2 = -1, s_3 = -1) \end{bmatrix} \end{aligned} \tag{7.13}$$

which can be seen to comprise all the possible terms corresponding to the summation of $v(s_1, s_2)\, v(s_2, s_3)$ over the spin s_2. (The summation over a given spin is often referred to as the *trace* over that spin.) In the case $N = 2$, we have $s_3 = s_1$, and the relevant terms are in the diagonal of V^2; thus $Q_2 = \text{trace } V^2$. More generally,

$$Q_N = \text{trace } \boldsymbol{V}^N \tag{7.14}$$

Since $\boldsymbol{V}$ is symmetric and nonsingular, we can find orthogonal eigenvectors for it such that

$$\boldsymbol{V} = \boldsymbol{P} \begin{bmatrix} \lambda_1 & 0 \\ 0 & \lambda_2 \end{bmatrix} \boldsymbol{P}^{-1} \tag{7.15}$$

where $\boldsymbol{P}$ is the matrix of eigenvectors of $\boldsymbol{V}$ and λ_1 and λ_2 are its corresponding eigenvalues. It follows immediately that

$$Q_N = \text{trace} \begin{bmatrix} \lambda_1 & 0 \\ 0 & \lambda_2 \end{bmatrix}^N = \lambda_1^N + \lambda_2^N \tag{7.16}$$

If λ_1 is the larger of the two eigenvalues, then since

$$\frac{\ln Q_N}{N} = \ln \lambda_1 + \ln\left[1 + \left(\frac{\lambda_2}{\lambda_1}\right)^N\right] \tag{7.17}$$

and $|\lambda_2/\lambda_1| < 1$, we have that

$$\ln\left[1 + \left(\frac{\lambda_2}{\lambda_1}\right)^N\right] \to 0 \text{ as } N \to \infty \tag{7.18}$$

in which case the *free energy per site* becomes

$$f(h, T) \equiv -kT \lim_{N\to\infty} \frac{\ln Q_N}{N} \tag{7.19}$$

or

$$f(h, T) = -kT \ln \lambda_1 \tag{7.20}$$

where

$$\lambda_1 = e^K \cosh h + (e^{2K} \sinh^2 h + e^{-2K})^{1/2} \tag{7.21}$$

EXERCISE 7.1

Prove the value for λ_1 in (7.21).

The order parameter $m(h, T)$ for this model is given by the usual thermodynamic equation:

$$m(h, T) = -\beta \frac{\partial}{\partial h} f(h, T) \tag{7.22}$$

$$m(h, T) = \frac{e^K \sinh h}{e^{2K} \sinh h + e^{-2K}} \tag{7.23}$$

We see from this equation that $\partial f/\partial h \to 0$ as $h \to 0^+$, which implies that $m(h, T) \to 0$. Thus, at zero field, the exact solution for the 1-d Ising model predicts no spontaneous magnetization at any finite temperature.

EXERCISE 7.2

Show that at zero field, (7.6) yields a value for Q_N given by

$$Q_N = (2\cosh K)^N$$

in agreement with (7.17).

We now treat this system using a mean-field approach.

7.2 The Mean-Field Approximation to the Ising System

In order to understand the development of the mean-field model, it is instructive to follow the approach given by Plischke and Bergersen [1], who describe the energy of a particular spin in the Ising lattice, designated by the symbol o, by the form

$$E_o = -S_o \left(J \sum_j S_j + \tilde{H} \right) \tag{7.24}$$

where the summation over j in equation (7.24) refers to nearest neighbors of spin o. This equation can be rewritten as,

$$E_o = -S_o(zJm + \tilde{H}) - JS_o \sum_j (S_j - m) \tag{7.25}$$

where z, *the coordination number*, represents the number of nearest-neighbor spins to S_o on the lattice and depends upon its dimension. The last term in the above equation reflects *fluctuations* of the nearest neighbor spins S_j around the mean value of S_j, namely m. If, to first order, this term is ignored, we essentially have an independent spin S_o interacting with a field term given by $zJm + \tilde{H}$, where the fluctuating values of the nearest-neighbor spins S_j have been replaced by their *mean values* m. The mean value of $\langle S_o \rangle = m$ can now be found from the standard statistical-mechanical prescription for a single spin fluctuating in a *mean field* of strength $zJm + \tilde{H}$. The solution to this problem (see exercise 7.3) is given by the equation

$$m = \tanh \beta(zJm + \tilde{H}) \tag{7.26}$$

EXERCISE 7.3

Show that a single spin fluctuating in a mean field of strength $ZJm + \tilde{H}$ has a magnetization given by equation (7.26).

The solution given in equation (7.26) is the mean-field solution of the Ising model in any dimension. Symmetry requirements for the system are ensured by the result that

$$m(\tilde{H}, T) = -m(-\tilde{H}, T) \tag{7.27}$$

For zero field ($\tilde{H} = 0$) and $\beta zJ > 1$, there is always a solution to (7.26) at $m = 0$ as well as at two other conditions $\pm m_o$. In addition, for all values of $\tilde{H} \neq 0$, there is at least one solution for *m at any temperature*. We now consider these situations in more detail.

Mean-Field Solutions of the Ising Model: Magnetization at Zero Field ($\tilde{H} = 0$)

We now look at various solutions for the mean-field Ising model in order to assess its characteristics at both zero and finite field strengths. We define the function $f(m, \tilde{H})$ as follows:

$$f(m, \tilde{H}) \equiv \tanh \beta(zJm + \tilde{H})$$

or

$$f(m, \tilde{H}) = \frac{e^{am+b\tilde{H}} - e^{-(am+b\tilde{H})}}{e^{am+b\tilde{H}} + e^{-(am+b\tilde{H})}} \tag{7.28}$$

where

$$a \equiv \beta zJ > 0 \text{ and finite}, \qquad b \equiv \beta > 0 \text{ and finite} \tag{7.29}$$

We see that

$$f(m, \tilde{H}) < 1 \quad \text{for} \quad m,\ \tilde{H} \geq 0 \tag{7.30}$$

In addition, we have that

$$f(0,0) = 0, \qquad f(1,0) < 1 \tag{7.31}$$

The limiting value of a for which $f(m,0) = m$ is the value of a that satisfies the condition

$$\lim_{m \to 0} \frac{d \tanh am}{dm} = 1 \tag{7.32}$$

Since $d \tanh am/dm = a(1 - \tanh^2 am)$, it follows that this occurs when $a = 1$. Therefore, the first time at which finite real solutions to equation (7.26) at $\tilde{H} = 0$ appear is when

$$a > 1 \tag{7.33}$$

The equality condition we identify as the *critical point* for this mean-field model. This point represents the onset of an order–disorder phase transition in the system with the critical temperature given by the equation

$$\beta z J = 1, \quad \text{or} \quad T_{\mathrm{c}} = \frac{zJ}{k} \tag{7.34}$$

In figure 7.1, we show the function $f(m,0)$ versus m at various temperatures including the critical one. Note that we don't explicitly use stability arguments to get this result.

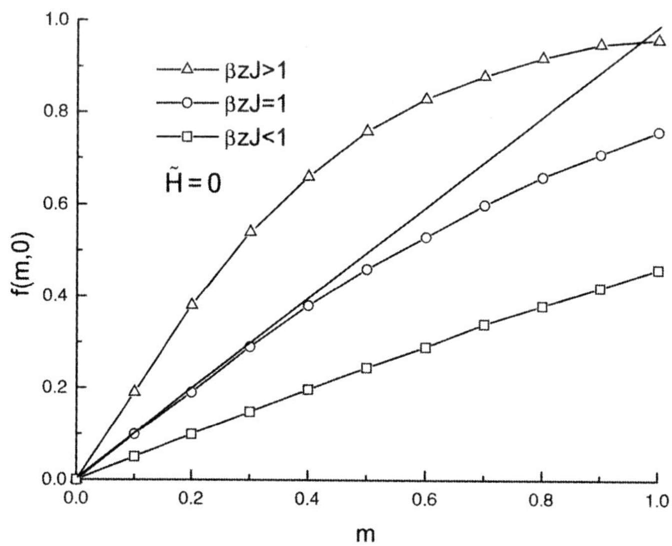

Figure 7.1 Magnetization in the mean-field Ising model at zero field ($\tilde{H} = 0$)

Finite-Field Solutions ($\tilde{H} \neq 0$)

Here we are interested in the character of the solutions to the equation

$$f(m, \tilde{H}) = m \tag{7.35}$$

We know that

$$0 < f(0, \tilde{H}) < 1 \tag{7.36}$$

and

$$0 < f(1, \tilde{H}) < 1 \tag{7.37}$$

Since

$$\frac{d \tanh(am + b\tilde{H})}{dm} > 0 \tag{7.38}$$

we conclude that

$$f(0, \tilde{H}) < f(1, \tilde{H}) < 1 \tag{7.39}$$

and that somewhere in the interval [0, 1] there must exist a solution $m = m^* < 1$ such that $f(m^*, \tilde{H}) = m^*$. In other words, at finite field strength, for finite temperatures, the mean-field model always has a solution m^* in the interval (0, 1) at each temperature—that is, it does not have a critical point. These situations are represented in figure 7.2, where we show the values of $f(m, \tilde{H})$ at various finite field conditions.

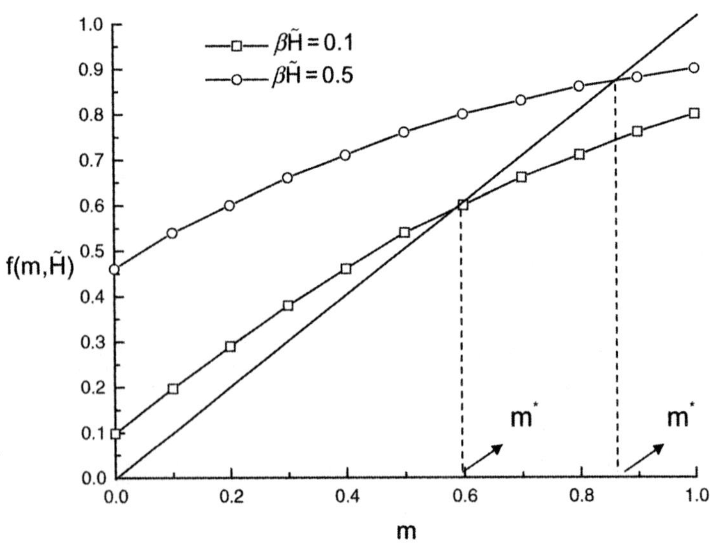

Figure 7.2 Magnetization in the mean-field Ising model at finite field ($\tilde{H} \neq 0$)

The Specific Heat

Another important thermodynamic property is the specific heat $C_{\tilde{H}}$ which is given by the definition:

$$C_{\tilde{H}} \equiv \left(\frac{\partial \langle E \rangle}{\partial T} \right)_{\tilde{H}} \tag{7.40}$$

This property can be found from the expression for the model's energy given by the equation

$$\langle E \rangle = -NJz \langle m \rangle^2 /2 - N\tilde{H} \langle m \rangle \tag{7.41}$$

from which it follows that

$$C_{\tilde{H}} = \frac{R(\beta zJ \langle m \rangle + \beta \tilde{H})^2 \operatorname{sech}^2(\beta zJ \langle m \rangle + \beta \tilde{H})}{1 - \beta zJ \operatorname{sech}^2(\beta zJ \langle m \rangle + \beta \tilde{H})} \tag{7.42}$$

In figure 7.3 we present results for the specific heat with this mean-field model, at finite field strength, over a wide temperature range.

EXERCISE 7.4

Prove equation (7.42) and show that, at zero field, it is indeterminate for $C_{\tilde{H}}$ at the critical point.

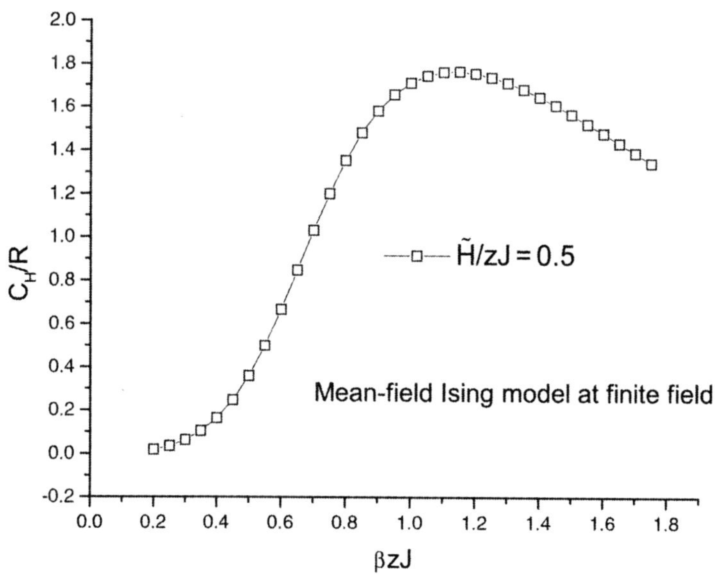

Figure 7.3 Specific heat in the mean-field Ising model at finite field ($\tilde{H} \neq 0$)

The Landau Model

In the evolution of the subject of critical phenomena, the *Landau* model was historically important and closely associated with mean-field ideas. In this model, the Gibbs free energy of the system is assumed to be an analytic (polynomial) function of the *order parameter m*. This function is specified so as to satisfy symmetry requirements of the order parameter around the coexistence line (a mean-field result). The stable equilibrium of the system occurs at the minimum of the free energy which, in the case of an Ising system above the critical temperature, would occur at the point $m = 0$. The simplest function that meets these requirements is of the form

$$G(m, T) = a(T) + b(T)m^2 + c(T)m^4 + d(T)m^6 + \cdots \tag{7.43}$$

If the temperature-dependent constants $a(T)$ and $b(T)$ are always positive, the model would not admit the possibility of a phase transition. To allow for the possibility of such a transition at a point of inflection of G, the constant $b(T)$ is assumed to take on the form

$$b(T) = b_0(T - T_c) \tag{7.44}$$

so that b is negative below the critical temperature. This completes the description for the free energy and allows for a function to display a critical point with the order parameter symmetrically distributed about the $m = 0$ axis. This is a mean-field model in that it assumes both that the free energy is analytic at the critical point and that the coexistence region is symmetric. In the additional exercises, we look at the scaling behavior of this model in the critical region. It is also useful conceptually in the next chapter when we study the effects of thermodynamic fluctuations on critical behavior.

7.3 First-Order Corrections to the Mean Field

The Bethe Model

Extensions to this mean-field model can be derived using larger groups of spins on which the mean field acts and *within which fluctuations are explicitly considered.* These larger groups of spins are often referred to as *clusters* and, in principle, a cluster of any reasonable size may be considered. In this section, we look at the *Bethe* model as discussed in [1] which considers a cluster consisting of a central spin surrounded only by its immediate nearest neighbors (see figure 7.4). The idea is to explicitly consider fluctuations *only* within the cluster. The spins on the rest of the lattice interact with those in the cluster through a mean-field term represented by the parameter $\tilde{H}'$. In this case, a Hamiltonian for the cluster can be written as

$$E_c = -JS_0 \sum_{j=1}^{z} S_j - \tilde{H} S_0 - \tilde{H}' \sum_{j=1}^{z} S_j \tag{7.45}$$

where z represents the number of nearest-neighbor spins surrounding a central spin designated as S_0 (in the 2-d Ising model, for example, $z = 4$, while in the 3-d model, $z = 6$).

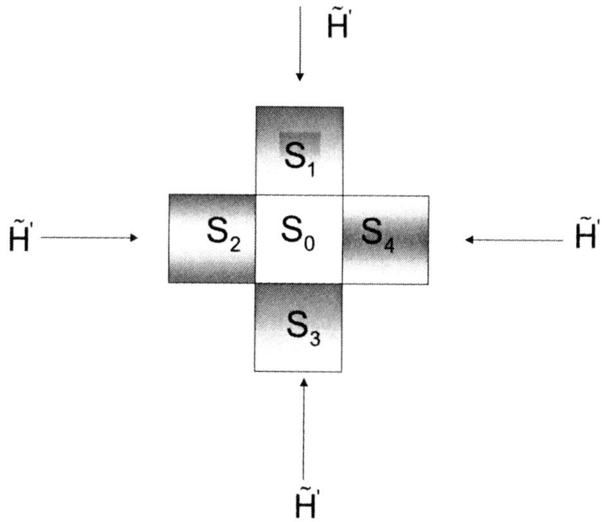

Figure 7.4 Schematic of spin cluster in the augmented Ising model

Finding Q_c explicitly for this model is reasonably straightforward. We begin by writing the partition function:

$$Q_c \equiv \sum_{\{S_i\}=\pm 1} e^{-\beta E_c} \tag{7.46}$$

then

$$Q_c = \sum_{\{S_j\}=\pm 1} \exp\left[\beta\left(JS_0 \sum_{j=1}^{z} S_j + \tilde{H} S_0 + \tilde{H}' \sum_{j=1}^{z} S_j\right)\right] \tag{7.47}$$

By summing over $S_0 = \pm 1$, we arrive at the following expression for $\langle S_0 \rangle$ (see steps in the derivation in the additional exercises):

$$\langle S_0 \rangle = \frac{1}{Q_c}[e^{\beta \tilde{H}}(2\cosh\beta(J + \tilde{H}'))^z - e^{-\beta \tilde{H}}(2\cosh\beta(J - \tilde{H}'))^z] \tag{7.48}$$

and that therefore $\langle S_0 \rangle \leq 1$.

EXERCISE 7.5

Find $\langle S_i \rangle$ for the Bethe model, where i is one of the spins surrounding the central one. Since the lattice must be translationally invariant, how does this lead to an equation for the critical temperature of this model. Using this result, show that the Bethe approximation for $\tilde{H} = 0$ leads to the following equation for the critical temperature of the lattice:

$$\coth \beta_c J = z - 1$$

which for the 2-d Ising model provides that $\beta_c J = 0.346$. Compare this with the mean-field prediction of 0.25, keeping in mind that the "exact" result available from computer simulation is $\beta_c J = 0.4407$.

The critical temperature given by the Bethe model is significantly more accurate than that given by the mean-field approximation given earlier. This shows the salutary effect of including fluctuations, even to first order, in the analysis. In table 7.1, we present both mean-field and "exact" reduced critical temperatures for the Ising model described here.

7.4 Continuum Fluid: Mean-Field Equations of State for Supercritical Fluids

The most common equations of state used to model thermodynamic behavior in supercritical fluids are continuum models. A *continuum* fluid is one where the individual molecular configurations are a function of the continuous coordinate vector variable $\boldsymbol{r}$. Here we show how a mean-field model can be constructed for these fluids using the canonical partition function as our starting point.

Imagine a fluid system consisting of N molecules in a volume V at temperature T. If we ignore rotational, vibrational, and electronic motions, the classical partition function of the system is given by the well-known result

$$Q_{\text{class.}} = \frac{Z_N}{N!\Lambda^{3N}} \tag{7.49}$$

where Λ, the thermal de Broglie wavelength, is defined as

$$\Lambda \equiv \left(\frac{h}{2\pi mkT}\right)^{1/2} \tag{7.50}$$

with h Planck's constant, m the mass of an individual molecule, and Z_N the system's *configuration integral* defined as

$$Z_N \equiv \int \cdots \int \exp(-\beta U_N) d\boldsymbol{r}_1 \cdots d\boldsymbol{r}_N \tag{7.51}$$

In (7.51), U_N is the potential energy of a given configuration of molecules in the system and $\boldsymbol{r}_i$ represents the position vector of molecule i. Evidently Z_N is a complicated quantity to calculate, given that the integral in (7.51) is over all possible configurations (*fluctuations*) of the N molecules in the system. At each given configuration, in

Table 7.1 Reduced critical temperatures for the exact Ising model and its mean-field approximation

Dimension, d	*"Exact" Ising Model $\beta_c J$*	*Mean-Field Ising (equation (7.34)) $\beta_c J$*
1	No finite critical temperature	0.5
2	0.4407, Onsager solution	0.25
3	0.2217, Computer simulation [2]	0.166

principle, U_N changes, so that the potential field in which the molecules find themselves is constantly fluctuating. The essential idea behind the mean-field approach here, as in the case of the previous Ising model, is to smooth out these fluctuations (conceptually speaking) so that a given molecule finds itself in a constant (mean) field environment.

To see how this might be mathematically rationalized, we borrow from a fundamental theorem of calculus that states that if a function f is continuous in an interval $[a, b]$, then its integral can be expressed algebraically as

$$\int_a^b f(x)dx = (b-a)f(\xi) \tag{7.52}$$

for some $\xi \in [a, b]$. If we apply this result to the multidimensional integral defining Z_N, we postulate that

$$Z_N = \exp(-\beta\xi_N)^N \cdot V^N \tag{7.53}$$

in accordance with the spirit of equation (7.52), where ξ_N is the mean-field potential experienced by a single molecule. The V^N term arises from the fact that each molecule's integration coordinate $d\boldsymbol{r}_i$, can range throughout the volume of the system. The use of equation (7.53) implicitly assumes that the molecules take up negligible volume, an assumption we can later relax, if need be.

Given this expression for Z_N, the system's equation of state immediately follows from the statistical-mechanical result which provides that

$$P = kT\left(\frac{\partial \ln Q_{\text{class.}}}{\partial V}\right)_{T,N} \tag{7.54}$$

with $Q_{\text{class.}}$ now given by the mean-field expression

$$Q_{\text{class.}}^{\text{MF}} = \frac{\exp(-\beta\xi_N)^N \cdot V^N}{N!\Lambda^{3N}} \tag{7.55}$$

In order to arrive at the *van der Waals equation of state* from this equation, for example, one makes the assumption that the volume available to a molecule is not the system volume V, but a free volume V_{f}, which is defined as V reduced by the volume taken up by the other molecules in the system [3]. In a simple approximation for V_{f}, the molecules are described as hard-core spheres of diameter σ, with only single nearest neighbors, in which case,

$$V_{\text{f}} \equiv V - \frac{2N\pi\sigma^3}{3} \tag{7.56}$$

The factor 2 implicit in this equation is to avoid double-counting the excluded volume for any pair of molecules. As such, this expression strictly only applies at very low densities. In addition, we also assume the mean-field term ξ_N to be linear in density, namely

$$\xi_N = -a\rho \tag{7.57}$$

where a is an empirical constant. Note that equation (7.57) is physically reasonable at low-to-moderate fluid densities since one would expect the molecules to experience a greater negative mean-field with increasing density at these conditions. We now define b as

$$b \equiv \frac{2\pi N_A \sigma^3}{3} \tag{7.58}$$

with N_A Avagadro's number. The net result is a mean-field thermodynamic model, better known as the *van der Waal's equation of state*, given by the equation:

$$P + \frac{a}{v^2} = \frac{RT}{v-b} \tag{7.59}$$

where $v = 1/\rho$. This equation of state exhibits a critical point with values for a and b at this condition given by

$$a = \frac{27R^2T_c^2}{64P_c}, \qquad b = \frac{RT_c}{8P_c} \tag{7.60}$$

EXERCISE 7.6

Prove the results given in equation (7.60), and show that the van der Waals equation of state can be expressed entirely in terms of *reduced thermodynamic quantities*, defined as follows:

$$P_r = P/P_c, \qquad V_r = V/V_c, \qquad T_r = T/T_c$$

The most common model continuum fluid is the *Lennard–Jones system*, and in table 7.2 we present its critical properties as established from computer simulation.

7.5 The Lattice Gas

Another conceptually popular statistical-mechanical model often studied in the mean-field limit is the *lattice gas*. This system usually consists of an ensemble of cells in three dimensions in which molecules may or may not reside. This occupancy condition defines

Table 7.2 Critical properties for the Lennard–Jones continuum fluid in reduced units $T_c^* = kT/\varepsilon$ and $\rho_c^* = \rho\sigma^3$

Full Potential	T_c^*	ρ_c^*
2-d [4]	0.515 ± 0.002	0.355 ± 0.003
3-d [5]	1.3120 ± 0.0007	0.316 ± 0.0006
Truncated and Shifted Potential		
2-d [5]	0.459 ± 0.001	0.36 ± 0.01
3-d [6]	1.085 ± 0.005	0.317 ± 0.006

the fundamental states available to each cell in the lattice. It should not be surprising that this two-state lattice-gas system can be made analogous to the two-spin-state Ising model. We now explore this correspondence in more detail.

Relationship between the Ising and Lattice-Gas Models

A Hamiltonian energy function can be defined for the lattice gas in which intermolecular interactions are considered between nearest-neighbor molecules, with the interaction constant designated by ε. Given these assumptions, we investigate the relationship between the Ising model and the lattice gas. To do this, we associate the two spin states $S_i = \pm 1$ in the Ising system with a variable n_i in the lattice gas, which can take on the values either 0 or 1. These values are identified with either the absence, or presence, of a molecule in a respective cell. The formal relationship between these variables is

$$S_i \equiv 2n_i - 1 \tag{7.61}$$

which can be substituted into the Ising partition function. In the case where only nearest-neighbor interactions are considered, the Ising partition function Q_{I} is given by

$$Q_{\mathrm{I}} = \sum_{\{S_i=\pm 1\}} \exp\left(\beta \tilde{H} \sum_{i=1}^{N} S_i + \beta J {\sum_{i,j}}' S_i S_j\right) \tag{7.62}$$

can be transformed into its lattice-gas equivalent as follows. The Ising *field term* becomes

$$\begin{aligned}\sum_{\{S_i=\pm 1\}} \exp\left(\beta \tilde{H} \sum_{i=1}^{N} S_i\right) &= \sum_{n_i=0,1} \exp\left(\beta \tilde{H} \sum_{i=1}^{N} (2n_i - 1)\right) \\ &= \exp(-\beta \tilde{H} N) \sum_{n_i=0,1} \exp\left(2\beta \tilde{H} \sum_{i=1}^{N} n_i\right)\end{aligned} \tag{7.63}$$

while the *nearest-neighbor interaction term* becomes

$$\begin{aligned}\sum_{\{S_i=\pm 1\}} \exp\left(\beta J {\sum_{i,j}}' S_i S_j\right) &= \sum_{n_1,\ldots,n_N=0,1} \exp\left(\beta J {\sum_{i,j}}' (2n_i - 1)(2n_j - 1)\right) \\ &= \sum_{n_1,\ldots,n_N=0,1} \exp\left(\beta J {\sum_{i,j}}' 4n_i n_j - 2z\beta J \sum_{i=1}^{N} n_i + \frac{N\beta J z}{2}\right)\end{aligned} \tag{7.64}$$

Thus, we see that

$$\begin{aligned}Q_{\mathrm{I}} &= \exp\left[\beta N \left(\frac{Jz}{2} - \tilde{H}\right)\right] \\ &\times \sum_{n_1,\ldots,n_N=0,1} \exp\left(4\beta J {\sum_{i,j}}' n_i n_j + 2\beta(\tilde{H} - zJ) \sum_{i=1}^{N} n_i\right)\end{aligned} \tag{7.65}$$

EXERCISE 7.7

To derive equation (7.64), one needs to prove that

$$-2\left(\sum_{i,j}{}' n_i + \sum_{i,j}{}' n_j\right) = -2z\sum_{i=1}^{N} n_i$$

Why is this the case?

The lattice gas is an open system, for which the grand canonical partition function Ξ_{LG} can be expressed as

$$\Xi_{\mathrm{LG}} = \sum_{n_1,\ldots,n_N=0,1} \exp\left(\beta\mu\sum_{i=1}^{N} n_i + \beta\varepsilon\sum_{i,j}{}' n_i n_j\right) \tag{7.66}$$

In (7.66), μ is the chemical potential of the lattice gas. Consider the following associations between the Ising model parameters J and $\tilde{H}$ with those of the grand canonical partition function of the lattice gas μ and ε:

$$\varepsilon = 4J \tag{7.67}$$

$$\mu = 2(\tilde{H} - Jz) \tag{7.68}$$

Then we can immediately write an equation linking the partition functions of each respective model:

$$Q_{\mathrm{I}} = \kappa\, \Xi_{\mathrm{LG}} \tag{7.69}$$

where κ is a constant defined as

$$\kappa \equiv \exp[\beta N(J - \tilde{H})] \tag{7.70}$$

Using (7.69), we see that a calculation of Q_{I} for the lattice gas yields a value for the grand canonical partition function in the corresponding lattice-gas system with parameters defined by (7.67) and (7.68).

On the face of it the analogy between these two systems is simple enough, but there are subtle distinctions between the two. One indication of this is given in equation (7.68). This shows that the coupling-constant analogy between these systems appears to be straightforward. However, the *field variable* μ in the lattice gas is a function of both $\tilde{H}$ and J in the Ising model. Furthermore, while $\tilde{H}$ can take on the value 0 in the Ising system, μ in the lattice gas must remain finite. Perhaps more importantly though, the critical scaling of the susceptibility χ_T in the Ising system is considered along the direction $\tilde{H} = 0$ which is a direction parallel to the T field. This is clearly not the case for the lattice gas since the coexistence curve is not parallel to either of the field variables μ or T (why?). These issues have important implications for scaling behavior in these systems, a topic we consider in greater detail in a subsequent chapter.

Thermodynamic and Statistical-Mechanical Comparisons between the Ising and LG Models

The fundamental thermodynamic equations describing each system are

$$dE = T\,dS - M\,dH + \mu\,dN \qquad \text{(Ising)} \tag{7.71}$$

$$dE = TdS - pdV + \mu\,dN \quad \text{(lattice gas)} \tag{7.72}$$

Legendre transforms of these equations lead to free energies $G(T, H, N)$ and $\Omega_{\text{LG}}(T, V, \mu)$ for each respective system such that

$$dG = -S\,dT - M\,dH + \mu\,dN \qquad \text{(Ising)} \tag{7.73}$$

$$d\Omega_{\text{LG}} = -S\,dT - p\,dV - N\,d\mu \quad \text{(lattice gas)} \tag{7.74}$$

where Ω_{LG} is the lattice-gas *grand potential.*

The statistical-mechanical partition functions for these two systems are given by $Q_{\text{I}}(\beta, H, N)$ and $\Xi_{\text{LG}}(\beta, V, \mu)$, respectively, from which their free energies naturally follow:

$$Q_{\text{I}} = e^{-\beta G(\beta,H,N)} \tag{7.75}$$

$$\Xi = e^{-\beta\Omega_{\text{LG}}(\beta,V,\mu)} \tag{7.76}$$

From these partition functions, standard statistical-mechanical results are easily derived; for example, the energies in both systems are given by the expressions

$$\langle E\rangle_{\text{Ising}} = -\left(\frac{\partial \ln Q_{\text{Ising}}}{\partial\beta}\right)_{N,H} \tag{7.77}$$

$$\langle E\rangle_{\text{LG}} = -\left(\frac{\partial \ln \Xi_{\text{LG}}}{\partial\beta}\right)_{\mu,V} \tag{7.78}$$

Fluctuations in extensive properties in these systems are also related to their susceptibilities χ_{I} and χ_{LG} through the following equations:

$$\left(\frac{-\partial^2 G}{\partial H^2}\right)_{T,N} = \left(\frac{\partial\langle M\rangle}{\partial H}\right)_T = \beta\langle(\delta M)^2\rangle \equiv \chi_{\text{I}} \tag{7.79}$$

$$\left(\frac{-\partial^2 \Omega_{\text{LG}}}{\partial\mu^2}\right)_{T,V} = \left(\frac{\partial\langle N\rangle}{\partial\mu}\right)_{T,V} = \beta\langle(\delta N)^2\rangle \equiv \chi_{\text{LG}} \tag{7.80}$$

where

$$\delta M \equiv M - \langle M\rangle \tag{7.81}$$

and

$$\delta N \equiv N - \langle N\rangle \tag{7.82}$$

We have more to say about the relationships shown in (7.79) and (7.80) in chapter 9, where we explicitly address the subject of thermodynamic fluctuations. A pressure-explicit equation of state for the lattice gas can be established using these thermodynamic identities. Since $\Omega_{\mathrm{LG}} = -PV$, we get that

$$\beta PV = \ln \Xi_{\mathrm{LG}} \tag{7.83}$$

EXERCISE 7.8

Starting out with the general form of the grand canonical partition function

$$\Xi = \sum_{v} \exp(-\beta E_v + \beta\mu N_v)$$

show that the average molecular number density in the lattice gas at constant T is given by the formula $\langle N \rangle = 1/\beta(\partial \ln \Xi_{\mathrm{LG}}/\partial\mu)_{T,V} = -\beta(\partial\Omega_{\mathrm{LG}}/\partial\beta\mu)_{T,V}$. Use these results to prove equation (7.80).

Order Parameter Equivalence between Ising and Lattice-Gas Models

The results of the previous sections readily lead to a mean-field model for a lattice-gas system based upon this isomorphism with the Ising model. To show this, consider the following relationship for finding the order parameter $m \equiv \langle S \rangle$ in the Ising system:

$$\langle S \rangle = \frac{\sum_{\{S_i=\pm 1\}} S_i \exp(\beta \tilde{H} \sum_{i=1}^{N} S_i + \beta J {\sum}'_{i,j} S_i S_j)}{Q_{\mathrm{I}}} \tag{7.84}$$

If we substitute the Ising–LG transformation given by equation (7.61) in this equation, it becomes equivalent to

$$2\langle n \rangle - 1 = \sum_{n_1,\ldots,n_N=0,1} (2n_i - 1) \exp\left(\beta\varepsilon {\sum_{i,j}}' n_i n_j + \beta\mu \sum_{i=1}^{N} n_i\right) \Big/ \Xi_{\mathrm{LG}} \tag{7.85}$$

with $\varepsilon = 4J$ and $\mu = 2(\tilde{H} - Jz)$. A mean-field approximation of this equation easily leads to an analytic equation of state for the lattice-gas system (see exercise 7.9).

EXERCISE 7.9

Derive equation (7.85) and develop the following mean-field equation of state for a lattice gas with partition function shown in equation (7.66):

$$\rho = \frac{1}{2}\left[1 + \left(\tanh \frac{\beta}{2}(\varepsilon z \rho + \mu)\right)\right] \tag{7.86}$$

How does one get the pressure change in the mean-field lattice gas from this equation of state?

This correspondence between these two models can also be conceived in the following terms:

- In a lattice-gas system with nearest-neighbor coupling parameter ε and at state conditions β and μ, establish the Ising-model analogue parameters, namely $J = \varepsilon/4$ and $\tilde{H} = \mu/2 + Jz$
- Solve the Ising equation (7.26) for m
- Calculate $\langle n \rangle$ in the LG from equation $\langle n \rangle = m + 1/2$
- Repeat at various values of β and/or μ.

Mean-Field Lattice-Gas Models from the Canonical Ensemble

Another very useful route for deriving mean-field models for the lattice gas makes use of the *canonical partition function* for the system, as opposed to the grand canonical partition function used in the preceding section. If we imagine a system with M_V lattice-gas cells distributed throughout a volume V, once again a physical picture emerges of a system with cells able to be filled or remain empty. If the molecules *do not* interact in this system, we can define a partition function for N cell-bound molecules, which can be written as

$$Q(N, M_V, T) = \frac{M_V!q^N}{N!(M_V - N)!} \tag{7.87}$$

where q is the partition function of an isolated cell-bound molecule. The degeneracy term $M_V!/N!(M_V - N)!$ accounts for the number of ways N molecules can be configured on M_V lattice sites. However, the more interesting case is when intermolecular interactions are allowed, as discussed in [7].

Let us assume that these are of the nearest-neighbor type, in which case the canonical partition function can be written as

$$Q(N, M_V, T) = \sum_{N_{11}} \Omega(N, M_V, N_{11})e^{\beta\varepsilon N_{11}/2} \tag{7.88}$$

where the summation sign $\sum_{N_{11}}$ represents a summation over all the possible states (consistent with N) containing N_{11} nearest-neighbor interactions on the lattice. The term $\Omega(N, M_V, N_{11})$ now represents the degeneracy associated with the given value of N_{11}. The partition function in (7.88) is difficult to sum exactly, since both N_{11} and $\Omega(N, M_V, N_{11})$ are complicated functions of N and difficult to calculate exactly; however, a simple approximation for it can be found by assuming that the N molecules are *randomly* distributed within the lattice structure. In this case we have that

$$\Omega(N, M_V, N_{11}) \approx \frac{M_V!}{N!(M_V - N)!} \tag{7.89}$$

and thus

$$Q(N, M_V, T) \approx \frac{M_V!q^N}{N!(M_V - N)!}e^{-\beta E_N} \tag{7.90}$$

where

$$E_N = -\bar{N}_{11}\varepsilon \tag{7.91}$$

with ε the average molecule–molecule interaction energy. There still remains the question of how to get $\bar{N}_{11}$, which represents the number of nearest-neighbor pair interactions in a random lattice with N cell-bound molecules. This is easily seen to be given by

$$\bar{N}_{11} = \frac{zN^2}{2M_V} \tag{7.92}$$

The factor 2 in (7.92) accounts for the fact that the interaction is shared between both species, and z is the lattice coordination number. These approximations immediately lead to the following mean-field partition function for this model:

$$Q(N, M_V, T) = \frac{M_V!q^N}{N!(M_V - N)!} \exp \frac{\beta z N^2}{2M_V} \tag{7.93}$$

The approximation in (7.93) is a famous result in this area of statistical mechanics often called the *Bragg–Williams* model [7].

EXERCISE 7.10

Derive the expression for $\bar{N}_{11}$ in equation (7.92).

Using various statistical-mechanical identities it is now straightforward to develop the equation of state and other important thermodynamic properties of this system (see exercise 7.11, for example). The equation of state is given by

$$\beta p = \frac{\beta z \varepsilon \theta^2}{2} - \ln(1 - \theta) \tag{7.94}$$

where

$$\theta = \frac{N}{M_V} \tag{7.95}$$

This model shows a phase transition and critical point in 1, 2, and 3 dimensions, with the characteristic van der Waals loops (see figure 7.5).

EXERCISE 7.11

Develop expressions for the chemical potential, energy, and entropy of the Bragg–Williams lattice-gas model for a pure 3-d fluid.

7.6 Mean-Field Model Behavior at the Critical Point

In the preceding sections, we have shown how mean-field models are constructed for some important physical systems. All of the mean-field models developed so far show

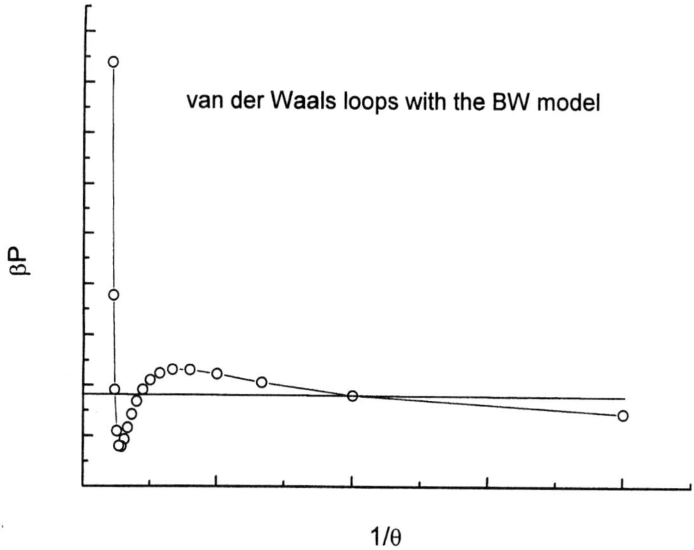

Figure 7.5 Phase behavior predictions in a lattice with the Bragg–Williams model

the existence of a phase transition and critical behavior in specific circumstances, for example, the zero-field limit of the Ising model as schematically illustrated in figure 7.6. While these models represent diverse physical systems, they share certain important characteristics in the critical region, and this grouping of physical models into broad classes describing their critical behavior is a statement of *critical-point universality*. Models within a given universality class are considered to display similar thermodynamic scaling behavior in the critical region. All of the models described in this chapter fall within the *Ising universality class* whose scaling behavior we now investigate within the mean-field approximation.

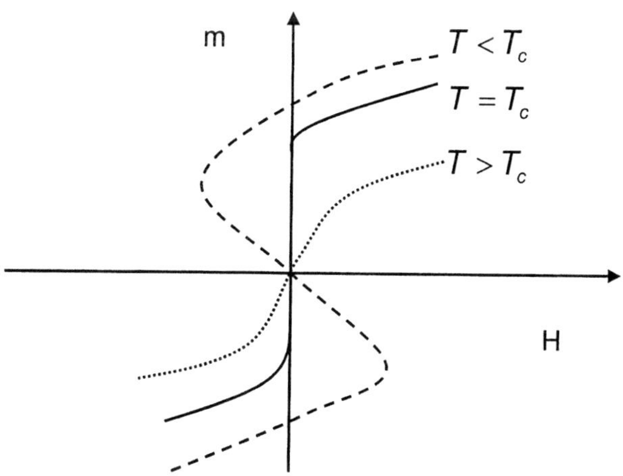

Figure 7.6 Schematic of isothermal magnetization in the Ising model

The Divergence of the Susceptibility with the Ising Model

An important property that diverges at the critical point in the Ising system, at zero field, is the susceptibility, which we define here (per spin) as

$$\chi_{\mathrm{m}}(\tilde{H}, t) \equiv \left(\frac{\partial m}{\partial \tilde{H}}\right)_t \tag{7.96}$$

The exponent of this divergence with temperature, along a path parallel to the coexistence curve, was defined as γ (see chapter 2). This particular path in the Ising model is along the $\tilde{H} = 0$ axis. We now investigate this property for the mean-field model given by (7.26) as $T \to T_c^+$. In order to do this, we make use of various trigonometric identities, stated for convenience at the end of this chapter; we also make use of the following calculus identity for implicit differentiation: If $y = f(y, x)$, it follows that

$$\frac{dy}{dx} = \left(\frac{\partial f}{\partial x}\right)_y + \left(\frac{\partial f}{\partial y}\right)_x \frac{dy}{dx} \tag{7.97}$$

that is,

$$\frac{dy}{dx}\left[1 - \left(\frac{\partial f}{\partial y}\right)_x\right] = \left(\frac{\partial f}{\partial x}\right)_y \tag{7.98}$$

Associating y and x with m and $\tilde{H}$, respectively, in (7.26) we have that

$$\left(\frac{\partial m}{\partial \tilde{H}}\right)_t = \left(\frac{\partial \tanh(A)}{\partial \tilde{H}}\right)_m \Bigg/ \left[1 - \left(\frac{\partial \tanh A}{\partial m}\right)_{\tilde{H}}\right] \tag{7.99}$$

where

$$A \equiv \beta(zJm + \tilde{H}) \tag{7.100}$$

$$\left(\frac{\partial m}{\partial \tilde{H}}\right)_t = \frac{\beta \operatorname{sech}^2 A}{1 - (\operatorname{sech}^2 A)\beta zJ} \tag{7.101}$$

In the limit that $\tilde{H} \to 0$, then,

$$\lim_{\tilde{H}\to 0} \left(\frac{\partial m}{\partial \tilde{H}}\right)_t = \frac{\beta \operatorname{sech}^2(\beta zJm)}{1 - \beta zJ \operatorname{sech}^2(\beta zJm)} \tag{7.102}$$

and since $m = 0$ for $T \to T_c^+$, we have

$$\lim_{\tilde{H}\to 0} \left(\frac{\partial m}{\partial \tilde{H}}\right)_t = \frac{\beta}{1 - \beta zJ} \tag{7.103}$$

Previously we showed with the mean-field model at $\tilde{H} = 0$ that

$$\beta_c zJ = 1$$

Therefore, using this result in (7.103), we have that,

$$\lim_{\tilde{H}\to 0}\left(\frac{\partial m}{\partial \tilde{H}}\right)_t = \frac{\beta}{1 - T_c/T} \tag{7.104}$$

or

$$\lim_{\tilde{H}\to 0}\left(\frac{\partial m}{\partial \tilde{H}}\right)_t = \frac{1}{k(T - T_c)} \tag{7.105}$$

which implies from (7.96) that

$$\chi_m(0, t) \approx t^{-1} \tag{7.106}$$

This equation shows that, with the mean-field Ising model, the susceptibility at zero field, $\chi_m(0, t)$, diverges with a value of the exponent $\gamma = 1$. This is exactly the same result as found with the van der Waals equation of state for fluids (see exercise 7.12).

EXERCISE 7.12

Show that the susceptibility with the continuum model using the van der Waals equation of state given in (7.59) also diverges with $\gamma = 1$ along a path parallel to the coexistence curve. What is the particular significance of choosing this path?

Specific Heat at the Critical Point ($\tilde{H} = 0$)

Once again we look at the mean-field Ising model. The specific heat is defined as

$$C_{\tilde{H}} \equiv \left(\frac{\partial \langle E\rangle}{\partial T}\right)_{\tilde{H}} \tag{7.107}$$

At zero field, we showed previously that $\langle E\rangle = -\frac{1}{2}NzJm^2$ for this model and, since m is zero for $T > T_c$, if follows that $C_H = 0$ when $T \to T_c^+$ However, for the case $T \to T_c^-$, we use an asymptotic expansion for the order parameter m to study this behavior. A low-order Taylor expansion of the mean-field model around the critical point [1] yields the result

$$m = \beta zJm - \tfrac{1}{3}(\beta zJ)^3 m^3 + \cdots \tag{7.108}$$

or

$$m^2 = 3\frac{\beta zJ - 1}{(\beta zJ)^3} \tag{7.109}$$

and since

$$T_c = \frac{zJ}{k}$$

we find that

$$m^2 = 3\left[\left(\frac{T_c}{T}\right) - 1\right] \Big/ \left(\frac{T_c}{T}\right)^3 \tag{7.110}$$

or

$$m = \pm\sqrt{3}\left(\frac{T_c}{T} - 1\right)^{1/2}\left(\frac{T}{T_c}\right)^{3/2} \tag{7.111}$$

Now, since

$$C_{\tilde{H}} = -\frac{N}{2}ZJ\left(\frac{\partial m^2}{\partial T}\right)_{\tilde{H}} \tag{7.112}$$

we find that

$$\left(\frac{\partial m^2}{\partial T}\right)_{\tilde{H}} = 3\left[\frac{2T_c^4}{T^5} - \frac{3T_c^3}{T^4}\right]\Big/\left(\frac{T_c}{T}\right)^6 \tag{7.113}$$

and so, as $T \rightarrow T_c^-$, we find that $C_{\tilde{H}}$ approaches the value $-\frac{1}{2}NJZ\,(\partial m^2/\partial T)_{\tilde{H}} = \frac{3}{2}Nk$. We immediately see that a *discontinuity* is predicted at T_c for $C_{\tilde{H}}$ by this mean-field theory. In fact, experimental data show that $C_{\tilde{H}}$ has a *diverging singularity* at T_c. This failure to predict the singularity of the specific heat at the critical point is a general feature of mean-field theories, and historically was important in spurring the development of modern theoretical approaches to the subject of critical phenomena.

All the other important scaling exponents can be developed using analogous reasoning to that just used (see the additional exercises at the end of this chapter). Table 7.3 shows the integer or rational values for the mean-field exponents of all systems within the *Ising universality class* which we expect based upon the results given in chapter 3.

7.7 Bounds on the Mean-Field Free Energy

Given these mean-field models, an important question arises concerning the nature of *the partition function* or *free energy* that is arrived at. Chandler [8] discusses an important bound that can be established for these properties in the mean-field approximation. Simply put, we seek to answer the question: does the mean-field approximation under- or overestimate the system's free energy? (The reader should try to answer this question using intuition before proceeding with the next paragraphs).

In any system, we can consider the mean-field energy to be an approximation to the real system energy. The difference between the two can be written as a *perturbation term* ΔE:

$$E - E_{\text{MF}} = \Delta E \tag{7.114}$$

Table 7.3 Critical exponents for mean-field models in the Ising universality class

Critical exponent:	γ	α	β	δ
Mean-field value:	1	0	$\frac{1}{2}$	3

from which the exact partition function can be expressed as

$$Q = \sum_{\{S_i\}} e^{-\beta[E_{\mathrm{MF}}+\Delta E]} \tag{7.115}$$

Equation (7.115) can also be written in the form

$$Q = \frac{Q_{\mathrm{MF}} \sum_{\{S_i\}} e^{-\beta E_{\mathrm{MF}}} e^{-\beta\,\Delta E}}{Q_{\mathrm{MF}}} \tag{7.116}$$

where Q_{MF} is the partition function provided by the mean-field model. Upon closer observation of (7.116) we see that this equation can be expressed as a (strange) ensemble average of the quantity $e^{-\beta\,\Delta E}$ which can be written as

$$\frac{Q}{Q_{\mathrm{MF}}} = \langle e^{-\beta\,\Delta E} \rangle_{\mathrm{MF}} \tag{7.117}$$

where $\langle\ \rangle_{\mathrm{MF}}$ means the average is taken in the mean-field ensemble. The relative magnitude of the exact partition function Q and its mean-field approximation Q_{MF} is thus equal to the quantity $\langle e^{-\beta\,\Delta E} \rangle_{\mathrm{MF}}$.

EXERCISE 7.13

What does it mean to take the average of a property within a mean-field ensemble? In other words, how would you establish the mean field with which to find the ensemble average $\langle e^{-\beta\,\Delta E} \rangle_{\mathrm{MF}}$?

Analysis of the Quantity $\langle e^{-\beta\,\Delta E} \rangle_{\mathrm{MF}}$

Using the mathematical identity

$$e^x = \sum_{i=0}^{\infty} \frac{x^i}{i!} \tag{7.118}$$

we expand the exponential and take the mean, as follows:

$$\langle e^{-\beta\,\Delta E} \rangle_{\mathrm{MF}} = \langle 1 - \beta\,\Delta E + \beta^2 \Delta E^2/2! + \cdots \rangle_{\mathrm{MF}} \tag{7.119}$$

If ΔE is small, as it should be for all perturbation-expansion approaches, and we use only the first-order terms of this expansion, then

$$\langle e^{-\beta\,\Delta E} \rangle_{\mathrm{MF}} \approx \langle 1 - \beta\,\Delta E \rangle_{\mathrm{MF}} = 1 - \langle \beta\,\Delta E \rangle_{\mathrm{MF}} \tag{7.120}$$

and so, to first order, we approximate the exact partition function Q as

$$Q \approx Q_{\mathrm{MF}}[1 - \langle \beta\,\Delta E \rangle_{\mathrm{MF}}] \tag{7.121}$$

Using (7.118)–(7.120), strictly speaking, we have the inequality

$$\frac{Q}{Q_{\mathrm{MF}}} \geq 1 - \beta\,\langle \Delta E \rangle_{\mathrm{MF}} \tag{7.122}$$

we see that

$$\frac{Q}{Q_{\text{MF}}} \geq e^{-\beta\langle\Delta E\rangle_{\text{MF}}} \geq 1 - \beta\langle\Delta E\rangle_{\text{MF}} \tag{7.123}$$

These inequalities are sometimes called the *Gibbs–Bogoliubov–Feynmann* (GBF) bounds. They have found use in developing error estimates in computer simulation studies [9].

EXERCISE 7.14

Prove equation (7.123) and is $\langle e^{-\beta\,\Delta E}\rangle_{\text{MF}} > 1$? What does the GBF inequality say about the magnitude of the free energy given by a mean-field approximation to any partition function of a given thermodynamic system?

7.8 Chapter Review

In this chapter, we described the fundamental ideas behind the development of mean-field models. These theories are useful for predicting the thermodynamic behavior of systems, and here we focused upon those belonging to the Ising universality class. These systems are characterized by short-range interaction potentials, characteristic of most supercritical solvents of practical interest. We spent a considerable amount of attention on the mean-field approach because it remains the standard by which more advanced theories are often judged. In particular, we developed results for three different physical system types: (1) the Ising model, (2) continuum fluids, and (3) the lattice-gas fluid. In each case, we showed how a mean-field model could be constructed by approximating the rigorously defined partition function for the system. The models thereby developed were used to predict a variety of important properties in these systems. Special consideration was given to the close correspondence between the Ising and lattice-gas models, including the presentation of equations showing how the partition functions of each are related when simple transformations are used to relate their respective order parameters.

Finally, the scaling behavior of this class of models in the critical region was studied using the Ising model paradigm. The critical exponents for this class of systems were extracted from this analysis and these are referred to as the mean-field or classical exponents. They assume importance in the study of critical phenomena as benchmarks against which to evaluate the results of more rigorous approaches.

Useful Trigonometric Identities

$$\tanh z \equiv \frac{e^z - e^{-z}}{e^z + e^{-z}} \tag{7.124}$$

$$\cosh z \equiv \tfrac{1}{2}(e^z + e^{-z}) \tag{7.125}$$

$$\sinh z \equiv \tfrac{1}{2}(e^{z} - e^{-z}) \tag{7.126}$$

$$\operatorname{sech} z \equiv \frac{1}{\cosh z} = \frac{2}{e^{z} + e^{-z}} \tag{7.127}$$

$$\frac{d \tanh z}{dz} = \operatorname{sech}^2 z = 1 - \tanh^2 z \tag{7.128}$$

$$\frac{d \cosh z}{dz} = \sinh z \tag{7.129}$$

$$\frac{d \sinh z}{dz} = \cosh z \tag{7.130}$$

$$\frac{d \operatorname{sech} z}{dz} = - \sinh z \cosh^2 z \tag{7.131}$$

7.9 Additional Exercises

Mean-Field Analysis

1. The mean-field model given in equation (7.34) shows that the critical temperature increases with system dimension. Explain this in terms of the effects of fluctuations in the system. Does this imply that mean-field models improve in accuracy in the critical region with increasing system dimensionality?
2. Show that, for an Ising model with Hamiltonian given by

$$H = -H_0 \sum_i S_i$$

   the free energy is given by

$$A = -NkT \ln(2 \cosh \beta H_0)$$

3. If the 2-d Ising model has an "exact" critical temperature calculated to be $\beta_c J = 0.4407$, explain how this result can be used to estimate the critical temperature of the 2-d lattice gas?
4. Show that the Ising mean-field model given by (7.26) is a corresponding-states model. What are the reducing parameters?
5. At a given temperature, does the presence of an external field h increase or decrease the specific heat of the Ising model? Try to justify your answer on physical grounds.
6. Does the *Bragg–Williams* (BW) model predict a phase transition in the one-dimensional case? How does this compare with the 1-d mean-field Ising model?
7. Prove equation (7.48) for the Bethe cluster model.

8. The *Langmuir model* can be found from the BW model if the intermolecular interaction energy ε is taken to be zero. Does the Langmuir model admit the possibility of a phase transition for 1-, 2-, and 3- dimensional systems, respectively?
9. Develop the BW model using next-to-nearest-neighbor intermolecular interactions for a pure fluid. What is the predicted critical temperature for this model in 2 dimensions? How does this compare to the Bethe approximation for the 2-d Ising model?
10. Assuming that a lattice-gas system has two species, of type 1 and 2, respectively, develop the BW partition function and find expressions for the chemical potentials of each species in the system.
11. Rationalize the basis for the mean-field–density relationship given in equation (7.57) for the van der Waals equation of state.
12. If the Hamiltonian for a lattice gas is given by the equation

$$-H_{\mathrm{LG}} = 4J \sum_{\langle i,j \rangle}{}' n_i n_j + \mu \sum_i n_i$$

show that, in the mean-field regime, we can derive the following equation of state

$$2\langle n \rangle - 1 = \tanh\left[\beta J \left(2z\langle n \rangle + \frac{\mu}{2J}\right)\right]$$

Prove that, at the critical point of this model,

$$\langle n \rangle_{\mathrm{c}} = \tfrac{1}{2}, \qquad \beta_{\mathrm{c}} J z = 1, \qquad \mu_{\mathrm{c}}/J = -2z$$

13. Derive the values for the critical exponents γ, β, δ for the model given in the previous question.
14. Prove that the lattice-gas mean-field equation of state derived earlier, given by

$$2\langle n \rangle - 1 = \tanh\left[\beta J \left(2z\langle n \rangle + \frac{\mu}{2J}\right)\right]$$

has a symmetric coexistence curve about the critical point for the situation where $\mu = \mu_{\mathrm{c}} = -2Jz$.
15. If we use the lattice-gas Hamiltonian

$$-H_{\mathrm{LG}} = 4J \sum_{\langle i,j \rangle}{}' n_i n_j + \mu \sum_i n_i$$

and utilize the approximation

$$\langle f(x) \rangle \approx f\langle x \rangle$$

prove that

$$\frac{-\langle H_{\mathrm{LG}} \rangle^*}{N} = \langle n \rangle \left(\frac{z\varepsilon\langle n \rangle}{2} + \mu\right)$$

Show that $\langle e^x \rangle \geq e^{\langle x \rangle}$. Use this identity to establish a relationship between the free energies of the system calculated with $-H_{\mathrm{LG}}$ and $-H^*_{\mathrm{LG}}$.

16. Using the Hamiltonian $-H^*_{\text{LG}}$ described in exercise 15, show that

$$\langle n \rangle = \frac{\sum_{n=0,1} n \exp[\beta n(\frac{z\varepsilon}{2}\langle n \rangle + \mu)]}{\sum_{n=0,1} \exp[\beta n(\frac{z\varepsilon}{2}\langle n \rangle + \mu)]}$$

which upon rearrangement leads to the result,

$$\langle n \rangle = \frac{e^{\beta(\frac{z\varepsilon}{2}\langle n \rangle + \mu)}}{1 + e^{\beta(\frac{z\varepsilon}{2}\langle n \rangle + \mu)}} \leq 1$$

Show that this approach yields the pressure equation of state:

$$p = kT\langle n \rangle^2 \ln \frac{\langle n \rangle}{1 - \langle n \rangle}$$

which implies $p > 0$ if and only if $½ < \langle n \rangle < 1$.

17. If $\Xi(T, V, \mu)$ is the grand canonical partition function in the lattice gas, given by

$$\Xi = e^{\beta PV}$$

we have that the grand potential $\Omega_G = -PV$, and so $\Xi = e^{-\beta\Omega_G}$. Prove that the relationship between $\Xi(T, V, \mu)$ and the canonical partition function $Q(N, V, T)$ is given by

$$\Xi(V, T, \mu) = \sum_{N=0}^{\infty} Q(N, V, T)e^{\beta\mu N}$$

From this result, derive the famous *Boltzmann* equation: $S = k \ln \Omega(N, V, \bar{E})$, where $\bar{E}$ is the system's average energy.

18. The *isothermal–isobaric ensemble* partition function $\Delta(T, P, N)$ is given by the summation

$$\Delta(T, P, N) = \sum_E \sum_V \Omega(N, V, E)e^{-\beta E}e^{-\beta PV}$$

Prove that, for any mean-field approximation, the isobaric–isothermal partition function in this regime designated Δ_{MF} obeys the relationship

$$\Delta_{\text{MF}} \leq \Delta_{\text{exact}}$$

19. If the mean-field Ising model is represented by a Hamiltonian of the form

$$E_{\text{MF}} = -(\tilde{H} + \Delta\tilde{H}) \sum_{i=1}^{N} S_i$$

where $\Delta\tilde{H}$ is the "surrounding spin mean-field contribution" to the system, show that:

$$Q_{\text{MF}} = (2\cosh[\beta(\tilde{H} + \Delta\tilde{H})])^N$$

How would you find $\Delta\tilde{H}$?

The Cluster Model

20. If we look at the case $h = 0$ in the Ising system and try and derive a spin cluster model, another plausible Hamiltonian for the cluster is given by

$$H_c = -JS_0 \sum_{j=1}^{z} S_j - h' \sum_{j=0}^{z} S_j$$

where h' is the mean field due to all the other spins outside of the cluster. Show that, in this case,

$$Q_c = e^{\beta h}(2\cosh\beta(J+h'))^z + e^{-\beta h'}(2\cosh\beta(J-h'))^z$$

and that the model with this Hamiltonian only yields a value of $\langle S_0 \rangle = 0$ in the 1-d Ising model when

$$h' = \frac{1}{2\beta}\ln\frac{\cosh^2\gamma}{\cosh^2\alpha}$$

where $\alpha \equiv \beta(J+h')$ and $\gamma \equiv \beta(h'-J)$. Is there a finite positive value for h' that satisfies this equation? How about the value $h' = 0$?

The Landau Model

21. The Landau model discussed earlier in this chapter for the free energy is given by an expansion in terms of the order parameter m of the form

$$G(m(T),T) = a(T) + b(T)m^2 + c(T)m^4 + d(T)m^6 + \cdots$$

If a, c, and d are positive, and we assume a temperature dependence for $b(T)$ of the form $b(T) = b_0 \cdot (T - T_c)$, show that this model yields: (1) a value for the order-parameter coexistence exponent $\beta = 1/2$ and (2) a discontinuity for the specific heat at the critical temperature. Why are there no odd powers of m in the above expansion?

22. In this chapter, we derived expressions for χ_m in the 1-d Ising model using a mean-field approach as well as an exact analysis. Compare the predictions of both of these equations for χ_m in this system at both zero-field and finite-field conditions. Which result gives the largest value for χ_m, and why is this the case? How does this comparison seem at higher lattice dimensions?

23. Consider the Ising model given by equations (7.1) and (7.2) If we denote the number of up-spins by N_1 and the number of down-spins by N_2, show that the Ising model partition function can be written as

$$Q_1(N,\tilde{H},T) = \sum_{1,2} \exp\left[\beta\tilde{H}(2N_1 - N) + \beta J\left(\frac{z}{2}N - 2zN_1 + 4N_{11}\right)\right]$$

where

$$N = N_1 + N_2$$

and N_{11} is the number of nearest-neighbor up-spins. Furthermore, derive an expression for $Q_1(N, \tilde{H}, T)$ in terms of summations over N_1 and N_{11} variables. Can you use these results to establish a relationship between $Q_1(N, \tilde{H}, T)$ and the partition function of a pure lattice-gas of the same lattice size? How would you find an average value of N_{11} at given N, $\tilde{H}$, and T?

24. Consider a *binary alloy* consisting of two components distributed throughout a lattice structure. We designate these components using the subscripts 1 and 2, respectively. Nearest-neighbor interactions in the lattice are represented by the Hamiltonian function

$$H = \varepsilon_{11}N_{11} + \varepsilon_{22}N_{22} + \varepsilon_{12}N_{12}$$

where the subscripts refer to the type of two-body interactions and N_{ij} refers to the number of ij pairs in the lattice. Use this Hamiltonian and the result of exercise 23 to show that the partition of the binary alloy system $Y(\beta, N, \mu_1, \mu_2)$ can be related to that of a pure lattice-gas of the same lattice dimension N by the equation

$$Y(\beta, N, \mu_1, \mu_2) = \left[\exp\left(\frac{\beta}{2}(\mu_1 - \mu_2) - \frac{1}{2}\beta z\varepsilon_{22}\right)\right]^N \Xi_{\mathrm{LG}}(\beta, N, \mu)$$

In these equations, μ_1, μ_2 are the chemical potential of components 1 and 2, respectively. How is μ in Ξ_{LG} in the above expression related to the binary alloy variables?

25. Solution for the Bethe model: equation (7.47) can be simplified into two terms:

$$\begin{aligned} Q_{\mathrm{c}} = & \sum_{S_1,\ldots,S_z=\pm 1} \exp\left[\beta\left(\sum_j S_j + \mu_{\mathrm{m}}H + \mu_{\mathrm{m}}H'\sum_j S_j\right)\right] \\ & + \sum_{S_1,\ldots,S_z=\pm 1} \exp\left[-\beta\left(\sum_j S_j + \mu_{\mathrm{m}}H - \mu_{\mathrm{m}}H'\sum_j S_j\right)\right] \end{aligned} \tag{7.132}$$

from which it follows that

$$\begin{aligned} Q_{\mathrm{c}} = & e^{\beta\mu_{\mathrm{m}}H} \sum_{S_1,\ldots,S_z} \exp\left[\beta(J + \mu_{\mathrm{m}}H')\sum_j S_j\right] \\ & + e^{-\beta\mu_{\mathrm{m}}H} \sum_{S_1,\ldots,S_z} \exp\left[-\beta(J - \mu_{\mathrm{m}}H')\sum_j S_j\right] \end{aligned} \tag{7.133}$$

The evaluation of the terms summed over $S_1, \ldots, S_z$ in this expression can be found easily, since

$$\begin{aligned} & \sum_{S_1,\ldots,S_z} \exp\left[\beta\left(J + \mu_{\mathrm{m}}H'\sum_j S_j\right)\right] \\ & = \sum_{S_1=\pm 1} e^{\beta(J+\mu_{\mathrm{m}}H')S_1} \cdots \sum_{S_z=\pm 1} e^{\beta(J+\mu_{\mathrm{m}}H')S_z} \end{aligned} \tag{7.134}$$

and

$$\sum_{S_i=\pm 1} e^{\beta(J+\mu_m H')S_i} = e^{\beta(J+\mu_m H')} + e^{-\beta(J+\mu_m H')}$$

$$= 2\cosh\beta(J + \mu_m H') \qquad (7.135)$$

Therefore we see that

$$Q_c = e^{\beta\mu_m H}(2\cosh\beta(J + \mu_m H'))^z + e^{-\beta\mu_m H}(2\cosh\beta(J - \mu_m H'))^z \qquad (7.136)$$

The value of $\langle S_0 \rangle$ can be found from a similar analysis using the statistical-mechanical equation

$$\langle S_0 \rangle = \frac{\sum_{\{S_i\}=\pm 1} S_0 e^{-\beta E_c}}{Q_c} \qquad (7.137)$$

and using (7.45) for E_c yields equation (7.48) easily.

Bibliography

Mean-field theories and their application to myriad situations are well described in references [1, 3, 7, 8], which were all sources for this chapter.

[1] M. Plischke and B. Bergersen, *Equilibrium Statistical Physics*. Singapore: World Scientific, 1994.

[2] A. M. Ferrenberg and D. P. Landau, "Critical behavior of the three-dimensional Ising model: A high-resolution Monte Carlo study," *Phys. Rev. B*, vol. 44, p. 5081, 1991.

[3] D. A. McQuarrie, *Statistical Mechanics*. New York: Harper & Row, 1976.

[4] B. Smit and D. Frenkel, "Vapor–liquid equilibria of the two-dimensional Lennard–Jones fluid," *J. Chem. Phys.*, vol. 94, p. 5664, 1991.

[5] J. J. Potoff and A. Z. Panagiotopoulos, "Critical point and phase behavior of a pure fluid and a Lennard–Jones mixture," *J. Chem. Phys.*, vol. 109, p. 10914, 1998.

[6] B. Smit, "Phase diagram of Lennard–Jones fluids," *J. Chem. Phys.*, vol. 96, p. 8639, 1992.

[7] T. L. Hill, *Introduction to Statistical Thermodynamics*. Reading, MA: Addison-Wesley, 1962.

[8] D. Chandler, *Introduction to Modern Statistical Mechanics*. New York: Oxford University Press, 1987.

[9] F. M. Munoz, T. W. Li, and E. H. Chimowitz, "Henry's law and synergism in dilute near-critical mixtures: theory and simulation results," *AIChE J.*, vol. 41, p. 3899, 1995.

8 Fluctuations and Critical Behavior

An important historical result with the Ising model was the achievement of an exact analytical solution by Onsager for the two-dimensional case. It is one of the few statistical-mechanical models for which an exact solution is available, and yields insight into the role played by thermodynamic fluctuations in determining critical behavior. We know that mean-field approaches, for example, which ignore the effects of fluctuations, do not predict the correct scaling exponents in the critical region. Also, we showed in the previous chapter that a mean-field analysis of the Ising model showed its specific heat C_H to have a discontinuity at T_c remaining finite on either side of T_c, namely for $T > T_c$ and $T < T_c$. The Onsager solution, however, gives a logarithmic singularity for C_H of the form

$$C_H(T) \sim \ln \left| 1 - \frac{T}{T_c} \right| \tag{8.1}$$

Nonclassical theories attempt to account for the role of fluctuations, an issue we focus upon in this chapter, beginning with an analysis in discrete spin systems, followed by consideration of the concepts in the context of continuum fluid models.

8.1 Thermodynamic Fluctuations

The idea of a fluctuation in a thermodynamic property is central to the subject of statistical mechanics. In a given macroscopic system, the microscopic states consistent with the macroscopic state are highly dynamic; at any given time, a mechanical property like energy can be in a microscopic state with energy different from its mean value. This deviation from the average is called a *fluctuation*, with the "size" of the fluctuation usually represented by a variance function. The variance of a property W can be defined as:

$$\sigma_W = \langle W^2 \rangle - \langle W \rangle^2 \tag{8.2}$$

Important relationships between functions of the type shown in (8.2), and thermodynamic stability coefficients can be derived (see exercise 8.1).

EXERCISE 8.1

Using the canonical partition function for a fluid system given by the definition

$$Q_N(N, V, T) = \sum_{\nu} e^{-\beta E_\nu}$$

show that the specific heat defined as $C_V \equiv \left(\frac{\partial E}{\partial T}\right)_{N,V}$ is related to the fluctuation in energy by the equation

$$C_V = \frac{1}{kT^2} \sigma_E$$

where

$$\sigma_E = \langle E^2 \rangle - \langle E \rangle^2$$

Recall that the probability of a microstate E_ν is given by $P_\nu = e^{-\beta E_\nu}/Q_N$

Fluctuations in the Ising System

We now analyze fluctuations in the Ising model; first though, we introduce the concept of the *pair correlation function*. This function designated by Γ_{ij} for spins

i and j, relates the correlation of spin states at points i and j in the lattice. It is defined as

$$\Gamma_{ij} \equiv \langle S_i S_j \rangle - \langle S_i \rangle \langle S_j \rangle \tag{8.3}$$

We immediately note that Γ_{ij} is zero within the mean-field approach. However, we seek an exact expression for Γ_{ij} within the Ising formulation when the fluctuations are explicitly accounted for. The approach of Plischke and Bergersen [1] is particularly clear in its formulation of the problem. Using the usual statistical-mechanical approach, they define lattice spin correlations in this system as

$$\langle S_i S_{i+j} \rangle \equiv \frac{1}{Q_N} \sum_{\{S_e\}} S_i S_{i+j} \exp\left\{\beta \sum_{e=1}^{N} J S_e S_{e+1}\right\} \tag{8.4}$$

Since we know that $S_i^2 = 1$, the first term in the above summation, $\sum S_i S_{i+j}$, can be written as follows:

$$S_i S_{i+j} = (S_i S_{i+1})(S_{i+1} S_{i+2}) \cdots (S_{i+j-1})(S_{i+j}) \tag{8.5}$$

in which case we can write

$$\langle S_i S_{i+j} \rangle = \frac{1}{Q_N} \sum_{\{S_e\}} (S_i S_{i+1})(S_{i+1} S_{i+2}) \cdots (S_{i+j-1} S_{i+j}) \exp\left\{\beta \sum_{e=1}^{N} J_e S_e S_{e+1}\right\} \tag{8.6}$$

where for generality the coupling constant J_e depends upon the specific nearest-neighbor interaction. In this case, Q_N is given by the equation

$$Q_N \equiv \sum_{S_1} \cdots \sum_{S_N} \exp\left(\beta \sum_{i=1}^{N} J_i S_i S_{i+1}\right) \tag{8.7}$$

Therefore

$$\langle S_i S_{i+j} \rangle = \frac{1}{Q_N \beta^j} \left(\frac{\partial^j Q_N}{\partial J_i \cdots \partial J_{i+j-1}} \right) \tag{8.8}$$

leading to the result that, in the particular limit for which $J_i \to J$ for each i, we have, using the results from exercise 7.2, that

$$\langle S_i S_{i+j} \rangle = (\tanh \beta J)^j \tag{8.9}$$

with $\tanh \beta J \leq 1$. Since $\partial \langle S_i S_{i+j} \rangle / \partial \beta > 0$ at finite temperatures, this shows that, at given i and j, higher temperatures always decrease the intensity of the interspin correlations, as would be expected. Furthermore, at a given temperature, the correlations between spins weaken as their separation distance represented by j increases.

EXERCISE 8.2

Using a similar approach to that used in developing equations (8.5)–(8.8) find an expression for the value of $\langle S_i S_j S_k \rangle$ in the 1-d Ising model at finite field $\tilde{H}$. The Hamiltonian (energy function) E for this model is given by

$$E = -\sum_{i=1}^{N} J_i S_i S_{i+1} - \tilde{H} \sum_{i=1}^{N} S_i \tag{8.10}$$

Is the quantity $\langle S_i S_j S_k \rangle$ larger at zero field than in the finite-field case? -Explain your answer physically.

8.2 Correlation Functions and Thermodynamic Stability Coefficients

The term $\langle S_i S_{i+j} \rangle$ can also be related to the *thermodynamic susceptibility* for the system. To see this, we first write the basic thermodynamic equation describing the lattice energy. This equation is

$$dE = T\,dS - M\,dH + \mu\,dN \tag{8.11}$$

where S is the entropy, M the magnetization, H the external field, and μ the chemical potential. The usual definition of M in terms of spin states is

$$M \equiv \mu_{\rm m} \sum_{i=1}^{N} S_i \tag{8.12}$$

where $\mu_{\rm m}$ is the magnetic moment. From thermodynamic stability theory, the lattice susceptibility is defined by

$$\chi_{\rm m} \equiv \frac{1}{\beta}\left(\frac{\partial M}{\partial H}\right)_{\beta} \tag{8.13}$$

with the property that $\lim_{H\to 0} \chi_{\rm m} \to \infty$ at the critical point. The magnetization per spin m is also sometimes used, and is defined by

$$m \equiv \frac{M}{N\mu_{\rm m}}$$

in which case,

$$\chi_{\rm m} \equiv \left(\frac{\partial m}{\partial \beta H}\right)_{\beta} = \frac{1}{N\mu_{\rm m}}\left(\frac{\partial M}{\partial \beta H}\right)_{\beta,N} \to \infty \tag{8.14}$$

at the critical point.

From a *fluctuation analysis* of the Ising lattice variables, similar to that required in exercise 8.1, using the partition function $Q(N, T, H)$, we can show that

$$\chi_{\mathrm{m}} = \frac{1}{N}\langle(\delta M)^2\rangle = \frac{1}{N}[\langle M^2\rangle - \langle M\rangle^2] \tag{8.15}$$

where

$$\delta M \equiv M - \langle M\rangle \tag{8.16}$$

from which it follows that

$$\chi_{\mathrm{m}} = \frac{\mu_{\mathrm{m}}^2}{N}\left\langle\left(\sum_{i=1}^{N} S_i - \sum_{i=1}^{N}\langle S_i\rangle\right)^2\right\rangle \tag{8.17}$$

or

$$\chi_{\mathrm{m}} = \frac{\mu_{\mathrm{m}}^2}{N}\left[\left\langle\sum_{i=1}^{N} S_i \sum_{j=1}^{N} S_j\right\rangle - \sum_{i=1}^{N}\langle S_i\rangle\sum_{j=1}^{N}\langle S_j\rangle\right] \tag{8.18}$$

EXERCISE 8.3

Prove equation (8.15) and show how it leads to (8.18).

EXERCISE 8.4

From the general Ising partition function, given by

$$Q(T, H, N) = \sum_{\nu} e^{-\beta E_{\nu}}$$

and, with the help of routine statistical mechanical manipulations, prove the following results:

$$E = kT^2\left(\frac{\partial \ln Q}{\partial T}\right)_{N,H}, \quad S = k\ln Q + kT\left(\frac{\partial \ln Q}{\partial T}\right)_{N,H}$$

Show that the probability of observing a state with average energy $\langle E\rangle$in the system is given by $e^{S/k}$.

Equation (8.18) is a general result that applies to Ising lattices of any dimension and geometry. Using equation (8.18), we can now begin to solidify the relationship between χ_{m} and the various spin correlation functions Γ_{ij}. We note that terms of the form Γ_{ij} are prevalent in (8.18) and that any divergence in χ_{m} at a critical point must necessarily show up in these terms. Before discussing this further, however, we first introduce the concept of the correlation length.

The Correlation Length ξ

We now discuss the important idea of the *correlation length*, which describes the length scale over which correlations persist in the lattice. We rewrite equation (8.18) as

$$\chi_{\mathrm{m}} = \frac{\mu_{\mathrm{m}}^2}{N} \sum_{i,j=1}^{N} \Gamma_{ij} \tag{8.19}$$

or

$$\chi_{\mathrm{m}} = \mu_{\mathrm{m}}^2 \sum_{j=1}^{N} \Gamma_{1j} \tag{8.20}$$

Equation (8.20) follows from (8.19) since all sites in the lattice are equivalent, and thus $\sum_{i,j=1}^{N} \Gamma_{ij} = N \sum_{j=1}^{N} \Gamma_{1j}$, where 1 represents a particular spin.

It is clear that, for χ_{m} to diverge at the critical point, so must the number of spins correlated to spin 1—that is, Γ_{1j} must be *finite* all the way out to $j = N$, as $N \to \infty$. In other words, the length scale over which these correlations persist must diverge at the critical point. This idea of a *diverging correlation length scale* permeates modern theories of critical phenomena and is one of the most fundamental properties of systems at their critical points.

We express this concept formally by defining the *correlation function* in terms of the distance between two particular spins (in lattice units) and a generalized length representing the distance over which correlations couple. We designate this length *the correlation length* ξ, a useful definition for which can be given by

$$\langle S_i S_{i+j} \rangle \equiv e^{-j/\xi} \tag{8.21}$$

This definition obeys important boundary conditions for ξ. First, at finite ξ, it shows that correlations decay (exponentially here) with distance. It also provides that, at very large j (i.e., $j \to \infty$), for the quantity $\langle S_i S_{i+j} \rangle$ to stay finite, we must have that $\xi \to \infty$; this divergence in ξ we identify with the critical point. For the 1-d Ising model, analyzed earlier, ξ can be expressly related to the model parameters through equation (8.9). In this case,

$$\xi = -\frac{1}{\ln \tanh \beta J} \tag{8.22}$$

and since

$$\tanh \beta J = \frac{1 - e^{-2\beta J}}{1 + e^{-2\beta J}} < 1 \tag{8.23}$$

we have that $\xi > 0$. We note that ξ cannot diverge in this system, consistent with there not being a critical point in the 1-d Ising model at finite temperatures.

EXERCISE 8.5

At the critical point the susceptibility χ_{m} found from equation (8.14) diverges. However, in a finite lattice, the quantity χ_{m} must of necessity be finite since there are only a finite number of spins for which correlations can be enumerated. Can you think of potential ideas for resolving this conundrum at the critical point, keeping in mind that in computer simulation, for example, χ_m must be found using finite system sizes.

8.3 Higher-Order Correlation Functions

Specific Heat in the 1-d Ising Model

It is of interest to show how the method of the last section can be used to find properties involving *higher-order correlations*. From fluctuation results (see, for example, exercise 8.1), the specific heat of the Ising model can be found from the equation

$$kT^2 C_H = \langle (\delta E)^2 \rangle \tag{8.24}$$

where

$$\delta E \equiv E - \langle E \rangle \tag{8.25}$$

At zero field and for the 1-d case, we have that

$$E = -\sum_{i=1}^{N} J_i S_i S_{i+1} \tag{8.26}$$

from which it follows that

$$kT^2 C_H \equiv \sum_i \sum_j J_i J_j (\langle S_i S_{i+1} S_j S_{j+1} \rangle - \langle S_i S_{i+1} \rangle \langle S_j S_{j+1} \rangle) \tag{8.27}$$

The quantities $\langle S_i S_{i+1} \rangle$ can be found as shown previously in the 1-d case. However, the properties $\langle S_i S_{i+1} S_j S_{j+1} \rangle$ are 4-spin correlation functions which, for a uniform coupling parameter ($J_i = J$, for all i), can be easily shown to be given by the equation (see exercise 8.6)

$$\langle S_i S_{i+1} S_j S_{j+1} \rangle = \begin{Bmatrix} 1 & \text{if } i = j, \\ \tanh^2 \beta J & \text{if } i \neq j \end{Bmatrix} \tag{8.28}$$

Substituting equation (8.28) into (8.27), we find that [2]

$$kT^2 C_H = J^2 \sum_i \sum_j \delta_{ij} (1 - \tanh^2 \beta J) \tag{8.29}$$

where $\delta_{ij} = 1$ if $i = j$ and $\delta_{ij} = 0$ if $i \neq j$, from which we have that

$$C_H = k\beta^2 J^2 N \,\text{sech}^2 \beta J \tag{8.30}$$

We thus see that C_H for the 1-d Ising model at zero field is always finite and positive at finite temperatures, so that there is no phase transition associated with the divergence of this property.

EXERCISE 8.6

Prove equations (8.28)–(8.30) and compare the exact solution for C_H in (8.30) to its mean-field prediction at zero field, available from results in the previous chapter. How would you expect this comparison to look like at finite-field conditions?

8.4 Fluctuation Properties in Fluids

Fluctuations in the Grand Canonical Ensemble

In the case of continuum fluids, a particularly appropriate independent set of thermodynamic variables is that of the grand canonical ensemble, namely β, μ, and V, respectively. The natural thermodynamic potential corresponding to these coordinates is the grand potential Ω_G defined by the equation

$$d\Omega_G = -S\,dT - P\,dV - N\,d\mu \tag{8.31}$$

The Gibbs–Duhem equation for a pure fluid provides that

$$S\,dT - V\,dP + N\,d\mu = 0 \tag{8.32}$$

which implies that

$$d\Omega_G = -P\,dV - V\,dP = -d(PV)$$

or

$$\Omega_G = -PV \tag{8.33}$$

The thermodynamic identity linking the system compressibility K_T to fluctuations in Ncan be found from an analysis of fluctuations in the grand canonical ensemble (see exercise 8.7). This result states that

$$\frac{1}{\beta}\left(\frac{\partial\langle N\rangle}{\partial\mu}\right)_{\beta,V} = \langle(\delta N)^2\rangle \tag{8.34}$$

which in turn can be related to the isothermal compressibility K_T, a susceptibility, through the equation

$$\frac{1}{\beta}\rho K_T = \frac{\langle(\delta N)^2\rangle}{\langle N\rangle} \tag{8.35}$$

The property $\langle(\delta N)^2\rangle$ plays an important role in fluids—analogous to the quantity $\langle(\delta M)^2\rangle$ in the Ising model.

EXERCISE 8.7

In the grand canonical ensemble, the partition function emerges in the following form:

$$\Xi(V,T,\mu) = \sum_{v}\exp(-\beta E_v + \beta\mu N_v) \tag{8.36}$$

where ν denotes the state with N_υ particles and energy E_υ. If the fluid compressibility is defined as $K_T \equiv -V^{-1}(\partial V/\partial P)_{T,N}$, show that it is related to fluctuations in N by equation (8.35) using this ensemble. Note that $(\partial^2\Omega_G/\partial\mu^2)_{T,V} = -\langle N\rangle\rho K_T/\beta$.

The Density–Density Correlation Function in Fluids

We now describe the development of the correlation function in fluids similar to those for the spin system described previously. By definition,

$$\begin{aligned}\langle(\delta N)^2\rangle &\equiv \langle(N-\langle N\rangle)^2\rangle \\ &= \langle N^2\rangle - \langle N\rangle^2 \end{aligned} \tag{8.37}$$

In order to relate this property to a spatially varying quantity in the fluid, we define *a number density* at a point $\boldsymbol{r}$ in the system $n(\boldsymbol{r})$ as,

$$n(\boldsymbol{r}) \equiv \sum_{i=1}^{N} \delta(\boldsymbol{r}-\boldsymbol{r}_i) \tag{8.38}$$

where $\boldsymbol{r}_i$ is the spatial coordinate of the ith molecule. In a spatially uniform (*isotropic*) system of volume V, we expect $\langle n(\boldsymbol{r})\rangle$ to be independent of position, and so

$$\langle n(\boldsymbol{r})\rangle \equiv \frac{\int n(\boldsymbol{r})dV}{V} = \left\langle \frac{N}{V}\right\rangle = \rho \tag{8.39}$$

For a *continuum fluid* this average can obviously be found in the grand canonical ensemble where N varies. Equation (8.37) can also be reexpressed in terms of the following integrals:

$$\langle(N-\langle N\rangle)^2\rangle = \left\langle \int_V d\boldsymbol{r}[n(\boldsymbol{r}) - \langle n(\boldsymbol{r})\rangle] \int_V d\boldsymbol{r}'[n(\boldsymbol{r}') - \langle n(\boldsymbol{r}')\rangle]\right\rangle \tag{8.40}$$

where $n(\boldsymbol{r})$ is the number density for a particle at position $\boldsymbol{r}$, with $n(\boldsymbol{r}')$ at $\boldsymbol{r}'$, and both $\boldsymbol{r}$ and $\boldsymbol{r}'$ are allowed to vary throughout the entire system volume V.

We now define a function $G(\boldsymbol{r},\boldsymbol{r}')$, which measures the correlations of fluctuations of the number density from the average value at points $\boldsymbol{r}$ and $\boldsymbol{r}'$, respectively. This function, called *the density–density correlation function*, is defined by

$$G(\boldsymbol{r},\boldsymbol{r}') \equiv \langle[n(\boldsymbol{r}) - \langle n(\boldsymbol{r})\rangle][n(\boldsymbol{r}') - \langle n(\boldsymbol{r}')\rangle]\rangle \tag{8.41}$$

In a spatially uniform (isotropic) system we insist that $G(\boldsymbol{r},\boldsymbol{r}')$ must depend only upon the scalar quantity $r \equiv |\boldsymbol{r}-\boldsymbol{r}'|$, that is, $G(\boldsymbol{r},\boldsymbol{r}')$ can be written as $G(r)$. With these considerations in mind, we rewrite (8.40) as

$$\begin{aligned}\langle(N-\langle N\rangle)^2\rangle &= \int d\boldsymbol{r}\int d\boldsymbol{r}' G(\boldsymbol{r}-\boldsymbol{r}') \\ &= V\int_V G(r)d\boldsymbol{r}' \end{aligned} \tag{8.42}$$

Using (8.35) in (8.42), we get

$$\frac{\langle N\rangle \rho K_T}{\beta} = V \int_V G(r) d\boldsymbol{r}' \tag{8.43}$$

or

$$K_T = \frac{\beta}{\rho^2} \int_V G(r) d\boldsymbol{r}' \tag{8.44}$$

In equation (8.44), the fluid analogy to the Ising lattice equation (8.20) is clear. A direct correspondence exists between the lattice's spin–spin correlation function Γ_{ij} and the fluid's density–density correlation function $G(r)$. Both are related to susceptibilities in their respective systems, which diverge at the critical point. At this point, $K_T \to \infty$, which implies that the integral in (8.44) must diverge there; that is, in the integral quantity $\int_V G(r) d\boldsymbol{r}'$, the integrand $G(r)$ remains finite, even as $r \to \infty$. In the situation where $G(r)$ vanishes,

$$G(\boldsymbol{r}, \boldsymbol{r}') \equiv \langle n(\boldsymbol{r}) n(\boldsymbol{r}')\rangle - \rho^2 = 0 \tag{8.45}$$

and so

$$\langle n(\boldsymbol{r}) n(\boldsymbol{r}')\rangle = \langle n(\boldsymbol{r})\rangle \langle n(\boldsymbol{r}')\rangle = \rho^2 \tag{8.46}$$

Equation (8.46) would apply to a random, uncorrelated, fluid. The more common way to represent $G(r)$ though is in terms of the pair correlation function $g(r)$.

The Fluid Pair Correlation Function $g(r)$

The *pair correlation function* $g(\boldsymbol{r}, \boldsymbol{r}')$ is related to $G(\boldsymbol{r}, \boldsymbol{r}')$ by the definition

$$G(\boldsymbol{r}, \boldsymbol{r}') \equiv \rho^2 [g(\boldsymbol{r}, \boldsymbol{r}') - 1] \tag{8.47}$$

which implies that

$$\frac{\langle n(\boldsymbol{r}) n(\mathbf{r}')\rangle}{\rho^2} = g(\boldsymbol{r}, \boldsymbol{r}') \tag{8.48}$$

According to this definition, the function $g(\boldsymbol{r}, \boldsymbol{r}')$ represents a measure of the departure of the density–density correlations in the system from those of a purely random configuration of molecules. In a random system, one would expect the number-density correlation property to obey equation (8.46) in which case,

$$g(\boldsymbol{r}, \boldsymbol{r}') = 1 \tag{8.49}$$

A value greater than 1 signifies an increased probability of finding two molecules at $\boldsymbol{r}$ and $\boldsymbol{r}'$, respectively—the converse being true when $g(\boldsymbol{r}, \boldsymbol{r}')$ is less than $1 - g(\boldsymbol{r}, \boldsymbol{r}')$ is thus a measure of the correlation between any two molecules in the system separated by a distance $r = |\boldsymbol{r} - \boldsymbol{r}'|$. A graph showing $g(\boldsymbol{r}, \boldsymbol{r}')$ in the critical region is shown in figure 8.1; the effects of long-range correlations are evident in the "tail" region.

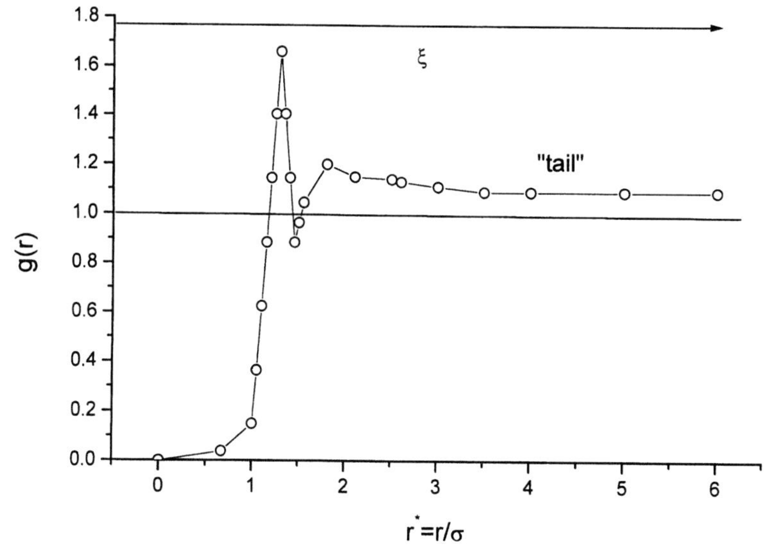

Figure 8.1 Schematic of the pair correlation function in a fluid in its critical region

Another description of $g(\boldsymbol{r}, \boldsymbol{r}')$ draws upon ideas similar to those that led to the 2-spin correlation function given in (8.20). The property $\langle n(\boldsymbol{r})n(\boldsymbol{r}')\rangle$ is the expectation of finding any two molecules at $\boldsymbol{r}$ and $\boldsymbol{r}'$, respectively. This average must be found by considering all possible configurations of the remaining $3, \ldots, N$ molecules, through what amounts to a standard *Boltzmann sampling* process. The positions of these remaining molecules are represented by the continuous vector variables $\boldsymbol{r}_i \quad (i = 3, \ldots, N)$ which can range over the entire volume V. Expressed this way, we can evaluate $\langle n(\boldsymbol{r})n(\boldsymbol{r}')\rangle$ as

$$\langle n(\boldsymbol{r})n(\boldsymbol{r}')\rangle = \frac{N!}{(N-2)!}\frac{\iint e^{-\beta U_N} d\boldsymbol{r}_3 \cdots d\boldsymbol{r}_N}{Z_N} \tag{8.50}$$

where Z_N is the *configuration integral* and U_N is the potential function associated with all the N molecules given two molecules fixed at $\boldsymbol{r}$ and $\boldsymbol{r}'$, respectively.

EXERCISE 8.8

Prove equation (8.50) and use it to derive an expression for $g(\boldsymbol{r}, \boldsymbol{r}')$ in terms of the integrals given in that equation.

In the literature, functions of the kind $\langle n(\boldsymbol{r})n(\boldsymbol{r}')\rangle$ are often referred to as *probability distribution functions* denoted by $\rho^{(n)}(\boldsymbol{r}_1, \ldots, \boldsymbol{r}_n)$, where n refers to the order of the distribution function, namely the number of molecules whose correlated positions are of interest. For example,

$$\rho^{(2)} = \langle n(\boldsymbol{r})n(\boldsymbol{r}')\rangle \tag{8.51}$$

which implies from (8.48) that

$$\rho^{(2)} = \rho^2 g \tag{8.52}$$

Thermodynamic Properties in Terms of $g(\boldsymbol{r}, \boldsymbol{r}')$

The fact that the potential energy of an N-molecule system is usually found by pairwise additivity of the respective 2-body pair potentials confers a special status on $g(\boldsymbol{r}, \boldsymbol{r}')$ in the theory of fluids. To illustrate this, we derive an equation for the system energy E in terms of $g(\boldsymbol{r}, \boldsymbol{r}')$.

From standard thermodynamic identities [3], we can easily show that

$$E = \tfrac{3}{2}NkT + \bar{U} \tag{8.53}$$

where $\bar{U}$ is the average configurational energy, given by

$$\bar{U} = \frac{\int\int U_N e^{-\beta U_N} d\boldsymbol{r}_1 \cdots d\boldsymbol{r}_N}{Z_N} \tag{8.54}$$

Since $U_N = \sum_{i<j} u(r_{ij})$, where $u(r_{ij})$ is a typical (i, j)-pair intermolecular potential, the right-hand side of equation (8.54) can be factorized as

$$\bar{U} = \frac{N(N-1)}{2} \frac{\int\int u(r_{12})[\int\int e^{-\beta U_N} d\boldsymbol{r}_3 \ldots d\boldsymbol{r}_N] d\boldsymbol{r}_1 d\boldsymbol{r}_2}{Z_N} \tag{8.55}$$

where we have focused upon a specific 1–2 pair in factorizing the above integral. The quantity in square brackets in (8.55) is related to $\rho^{(2)}$ and therefore $\rho^2 g$ (see (8.50)–(8.52)). Using these equations for an isotropic system, it is easy to show that

$$\bar{U} = \frac{N\rho}{2} \int_0^{\infty} u(r)g(r)4\pi r^2 dr \tag{8.56}$$

where $u(r)$ is the pair potential at a distance r separating two molecules. Thus, the *total energy* of the system is given by the equation

$$E = \frac{3NkT}{2} + \frac{\rho N \int_0^{\infty} u(r)g(r)4\pi r^2 dr}{2} \tag{8.57}$$

This is a rigorous thermodynamic result for the system's energy, dependent only upon the assumption of pairwise additivity in finding U_N. Fluctuations in the system are implicitly incorporated into this result through $g(r)$, which itself takes into account all possible configurations of the other molecules $3, \ldots, N$, at each value of r separating the given 1–2 pair.

Similar results can be developed for other thermodynamic properties of interest in the fluid, where once again the centrality of $g(r)$ becomes apparent. Particularly useful ones are the equation of state for the fluid—the so-called *pressure equation*—and the *Kirkwood equation* for the chemical potential, which are routinely derived in books on statistical mechanics. McQuarrie [3] provides the following results for these two quantities:

$$\frac{p}{kT} = \rho - \frac{\rho^2}{6kT} \int_0^{\infty} r u'(r) g(r) 4\pi r^2 dr \tag{8.58}$$

$$\frac{\mu}{kT} = \ln \rho \Lambda^3 + \frac{\rho}{kT} \int_0^1 \int_0^{\infty} u(r) g(r, \zeta) 4\pi r^2 dr\, d\zeta \tag{8.59}$$

The Behavior of $G(r)$ and $g(r)$ in the Critical Region

Equation (8.44) suggests that the long-range behavior of both $G(r)$ and $g(r)$ at the critical point is important, given that the compressibility diverges there. This behavior is also of relevance in determining the system pressure and chemical potential near the critical point using equations (8.58) and (8.59). Note, however, that the integrals in these equations depend upon the product of $g(r)$ with the intermolecular potential $u(r)$ and its derivative.

In this section, we establish some fundamental results in this area and introduce two new scaling exponents. In order to do this, we investigate a 3-d fluid system with a spatially dependent order parameter $\rho(\boldsymbol{r})$ and ask the question: how does the Helmholtz free energy, in a control volume V, change in response to a fluctuation in density within the control volume?

To answer this question, we need to evaluate the integral

$$\Delta A = \int_{\mathrm{V}} (a(\boldsymbol{r}) - \langle a \rangle) d\boldsymbol{r} \tag{8.60}$$

where a is the free energy per unit volume. We expand the integrand functional in (8.60), in terms of the variable $\Delta\rho(\boldsymbol{r}) \equiv \rho(\boldsymbol{r}) - \langle\rho\rangle$, as follows [4]:

$$\Delta a(\boldsymbol{r}) = \mu\,\Delta\rho(\boldsymbol{r}) + \frac{1}{2\langle\rho\rangle}\left[\frac{\partial P}{\partial \rho}\right]_T \Delta\rho(\boldsymbol{r})^2 + h[\nabla_{\boldsymbol{r}}\Delta\rho(\boldsymbol{r})]^2 + \cdots \tag{8.61}$$

This expansion is very similar to the *Landau model* for the free energy given in the previous chapter, except that now we are dealing with a spatially dependent order parameter. In equation (8.61), $h > 0$, since this term is the leading contribution to the free energy of the system which is displaced from an equilibrium configuration.

The relationship between $\Delta\rho(\boldsymbol{r})$ and its Fourier transform is given by

$$\Delta\rho(\boldsymbol{r}) = \sum_{\mathbf{q}} \Delta\rho(\boldsymbol{q}) e^{i\boldsymbol{q}\cdot\boldsymbol{r}} \tag{8.62}$$

If we use this result in (8.61) and do the integral in (8.60), we get the following equation [4]:

$$\Delta A = \frac{V}{2}\sum_{q}\left[\frac{1}{\langle\rho\rangle}\left(\frac{\partial P}{\partial \rho}\right)_T + 2hq^2\right]|\Delta(\rho(\boldsymbol{q}))|^2 \tag{8.63}$$

or

$$\langle|\Delta(\rho(\boldsymbol{q}))|^2\rangle \approx \frac{kT}{V\sum_q [1/\langle\rho\rangle(\partial P/\partial\rho)_T + 2hq^2]} \tag{8.64}$$

where $\langle\Delta A\rangle \propto kT$ for small fluctuations. (These equations are the discrete form of the Fourier transform.) The Fourier transform of $G(r)$ in equation (8.41) is given by

$$G(r) = \sum_{q} \langle|\Delta\rho(\boldsymbol{q})|^2\rangle e^{i\boldsymbol{q}\cdot\boldsymbol{r}} \tag{8.65}$$

If we substitute (8.65) into (8.64) and Fourier-invert the result we end up with the following equation for $G(r)$:

$$\beta G(r) = \frac{1}{8\pi hr} \exp \frac{-r}{\xi} \tag{8.66}$$

where

$$\xi = \left[2h\langle\rho\rangle^{-1} \left(\frac{\partial \rho}{\partial P} \right)_T \right]^{1/2} \tag{8.67}$$

here we identify ξ as the characteristic length scale over which fluctuations are correlated. There is something quite novel about (8.67); for the first time, in this text, the divergence of the correlation length is given an explicit form. Along the coexistence direction, we have that $\xi \propto t^{-\gamma/2}$, with $\gamma = 1$ classically. However, in general we describe the scaling of ξ as

$$\xi \propto t^{-\nu} \tag{8.68}$$

where the exponent ν, newly introduced, characterizes the scaling of ξ with temperature along the coexistence direction; as just stated, it is equal to $1/2$ classically (why is the development leading to (8.67) classical?). From this analysis, it also follows that

$$G(r) \propto g(r) \propto r^{-1} \quad \text{(classically)} \tag{8.69}$$

and again we postulate that

$$G(r) \propto g(r) \propto r^{-(1+\eta)} \quad \text{(nonclassically)} \tag{8.70}$$

where η, another new nonclassical scaling exponent, describes the scaling of the correlation function in the critical region.

Fluctuations and System Dimensionality: The Landau–Ginzburg Criterion

The historically important Landau model described in the previous chapter is of the mean-field type. Using the Landau approach, it is possible to develop another interesting result called *the Landau–Ginzburg criterion* (LGC), which allows one to estimate more precisely the relative significance of fluctuations in the system. For example, it seems reasonable to assume that, as we move away from the critical region, the effects of fluctuations should diminish, with the consequence that mean-field theories should become increasingly accurate there. The LGC sheds important light on this issue.

Imagine a system with a spatially dependent order parameter like that considered in the previous section. Consider that the system is perturbed by an external field whose effects propagate throughout the medium. We would like to answer the question: how are order-parameter fluctuation correlations important to this propagation process?

The correlation function between the order parameter, designated by $m(\boldsymbol{r})$, at the origin and position $\boldsymbol{r}$ is defined by the usual equation:

$$\Gamma(r) \equiv \langle m(\boldsymbol{r})m(\boldsymbol{0})\rangle - \langle m(\boldsymbol{r})\rangle\langle m(\boldsymbol{0})\rangle \tag{8.71}$$

where $r \equiv |\boldsymbol{r}|$. Furthermore, in (8.66) we showed that, classically at least, the correlation function $\Gamma(r)$ is related to r by the scaling relationship

$$\Gamma(r) \sim \frac{e^{-r/\xi}}{r} \tag{8.72}$$

where the correlation length, $\xi(T) \sim |T - T_c|^{-\nu}$. We note that, at higher dimensions, the classical scaling of $\Gamma(r)$ can be shown to be given by the scaling relationship [2]

$$\Gamma(r) \sim \frac{\exp[-r/\xi]}{r^{d-2}} \tag{8.73}$$

We now express the effects of order-parameter fluctuation correlations in the medium, compared to the situation in an uncorrelated system, through the ratio quantity [1]

$$R_{LG} \equiv \frac{\int_V [\langle m(\boldsymbol{r})m(\boldsymbol{0})\rangle - \langle m(\boldsymbol{r})\rangle\langle m(\boldsymbol{0})\rangle]dV}{\int_V m_o^2 dV} \tag{8.74}$$

where m_o is the mean-field value for the order parameter, prior to the perturbation. Assuming that the system is close to criticality,

$$m_o^2 \sim |T - T_c|^{\beta} \tag{8.75}$$

and

$$\Gamma(r) \sim \frac{\exp[-r/\xi]}{r} \tag{8.76}$$

We now seek to find the conditions for which $R_{LG} \ll 1$ which would indicate that fluctuation effects are insignificant relative to mean-field behavior. If we use (8.75) and (8.76) in (8.74), do the integration over a three-dimensional spherical volume, and impose the condition $R_{LG} \ll 1$, we arrive at the following result [1]. It spells out the conditions when the effects of order parameter fluctuation correlations are small:

$$\frac{3\int_0^{\xi} r\exp[-r/\xi]dr}{\xi^3|T - T_c|^{2\beta}} \ll 1 \tag{8.77}$$

EXERCISE 8.9

Prove equation (8.77)

In (8.77), we have integrated out to a value of $r = \xi$. This equation is one manifestation of the Landau–Ginzburg criterion. By employing the scaling relationship $\xi \propto t^{-\upsilon}$ in (8.77), we can then estimate the range of temperatures, relative to T_c, for which fluctuations are significant. It is also possible to evaluate the ratio in (8.74) using

(8.75) and (8.76) in a spherical hyperspace of dimension $d > 3$. This analysis [1] leads to the interesting result that the inequality $R_{LG} \ll 1$ is satisfied as long as

$$d > 2 + \frac{2\beta}{\upsilon} \tag{8.78}$$

For the Landau, and other mean-field theories with $\beta = \upsilon = \frac{1}{2}$, this inequality holds for $d \geq 4$. The dimension $d = 4$ is referred to as the *upper critical dimensionality*, and can be thought of as representing the system dimension above which fluctuation corrections to mean-field predictions become small.

8.5 Integral-Equation Theory

The results to this point in this chapter show how statistical-mechanical ideas can be used to develop broad conclusions about the connection between order-parameter fluctuations and critical phenomena in both discrete and continuous systems. In the final part of this chapter, we describe a particular statistical-mechanical theory (the Kirkwood–Buff theory) that makes use of many of these concepts. It has been probably been the most widely used fundamental theory used for calculating the behavior of model supercritical fluid mixtures.

The Kirkwood–Buff Theory

Consider an open system of volume V that contains $N_1, \ldots, N_\nu$ molecules of ν species in a multicomponent system. Kirkwood and Buff [5] defined various *densities* of species analogous to those defined earlier for pure fluids. The *singlet density* for species i at a point $\boldsymbol{r}_1$ in the volume V is defined by

$$\nu_i^{(1)}(\boldsymbol{r}_1) \equiv \sum_{m=1}^{N_i} \delta(\boldsymbol{r}_{i_m} - \boldsymbol{r}_1) \tag{8.79}$$

where δ is the three-dimensional Dirac delta function. The density of *ordered pairs* of molecules (i, j) with species i at $\boldsymbol{r}_1$ and species j at point $\boldsymbol{r}_2$ is defined as

$$\nu_{ij}^{(2)}(\boldsymbol{r}_i, \boldsymbol{r}_j) = \sum_{m=1}^{N_i} \sum_{n=1}^{N_j} (1 - \delta_{ij}\delta_{mn})\delta(\boldsymbol{r}_{i_m} - \boldsymbol{r}_1)\delta(\boldsymbol{r}_{j_n} - \boldsymbol{r}_2) \tag{8.80}$$

where δ_{ij} is the Kronecker delta. We look at the term $(1 - \delta_{ij}\delta_{mn})\delta(\boldsymbol{r}_{i_m} - \boldsymbol{r}_1)\delta(\boldsymbol{r}_{j_n} - \boldsymbol{r}_2)$ in (8.80). When a molecule of species i is at $\boldsymbol{r}_1$ and j is at $\boldsymbol{r}_2$, then $\delta(\boldsymbol{r}_{i_m} - \boldsymbol{r}_1)\delta(\boldsymbol{r}_{j_n} - \boldsymbol{r}_2) = 1$. In this case, both δ_{ij} and δ_{mn} are equal to zero, and only then is there a contribution to the sum shown in (8.80) of value 1.

These respective densities possess the following integrals, which follow directly from their definitions:

$$\int_V \nu_i^{(1)}(\boldsymbol{r}_1) d\boldsymbol{r}_1 = N_i \tag{8.81}$$

$$\int_V \nu_{ij}^{(2)}(\boldsymbol{r}_1, \boldsymbol{r}_2) d\boldsymbol{r}_1 d\boldsymbol{r}_2 = N_i N_j - N_i \delta_{ij} \tag{8.82}$$

The average number densities of the singlet and pair functions are

$$\rho_i^{(1)}(\boldsymbol{r}_1) = \langle v_i^{(1)} \rangle \tag{8.83}$$

$$\rho_{ij}^{(2)}(\boldsymbol{r}_1, \boldsymbol{r}_2) = \langle v_{ij}^{(2)}(\boldsymbol{r}_1, \boldsymbol{r}_2) \rangle \tag{8.84}$$

where these averages are taken over the complete number of molecules $N_1, \ldots, N_v$ using the probability distribution function of the grand canonical ensemble. Integrating these equations, we get

$$\int_V \rho_i^{(1)}(\boldsymbol{r}_1) d\boldsymbol{r}_1 = \langle N_i \rangle \tag{8.85}$$

and

$$\int_V \int_V \rho_{ij}^{(2)}(\boldsymbol{r}_1, \boldsymbol{r}_2) d\boldsymbol{r}_1 d\boldsymbol{r}_2 = \langle N_i N_j \rangle - \delta_{ij} \langle N_i \rangle \tag{8.86}$$

therefore

$$\int_V \int_V [\rho_{ij}^{(2)}(\boldsymbol{r}_1, \boldsymbol{r}_2) - \rho_i^{(1)} \rho_j^{(2)}] d\boldsymbol{r}_1 d\boldsymbol{r}_2 = [\langle N_i N_j \rangle - \langle N_i \rangle \langle N_j \rangle] - \delta_{ij} \langle N_i \rangle \tag{8.87}$$

In fluid systems, these mean densities have the more recognizable forms:

$$\rho_i^{(1)}(\boldsymbol{r}_1) \equiv \frac{\langle N_i \rangle}{V} = \rho_i \tag{8.88}$$

$$\rho_{ij}^{(2)}(\boldsymbol{r}_1, \boldsymbol{r}_2) \equiv \rho_i \rho_j g_{ij}(r) \tag{8.89}$$

where

$$r \equiv |\boldsymbol{r}_1 - \boldsymbol{r}_2| \tag{8.90}$$

with $g_{ij}(r)$ the radial distribution function of species i and j, dependent (in the isotropic case) upon the scalar distance between the i–j pair of points. These equations lead to important results linking radial distribution functions and density fluctuations in the system. For example, equation (8.87) becomes

$$\rho_i \rho_j V \int_V [g_{ij}(\boldsymbol{r}) - 1] dV = \langle N_i N_j \rangle - \langle N_i \rangle \langle N_j \rangle - \delta_{ij} \langle N_i \rangle \tag{8.91}$$

If we define a quantity G_{ij}, called the *total correlation fluctuation integral* (TCFI), by the equation

$$G_{ij} \equiv \int_V [g_{ij}(\boldsymbol{r}) - 1] dV \tag{8.92}$$

then (8.91) becomes

$$G_{ij} = \frac{V[\langle N_i N_j \rangle - \langle N_i \rangle \langle N_j \rangle]}{\langle N_i \rangle \langle N_j \rangle} - \frac{\delta_{ij}}{\rho_j} \tag{8.93}$$

We can now relate these results to fluctuation identities available from the grand canonical ensemble to further develop the theory. The grand canonical partition function Ξ in a multicomponent system is given by the equation

$$\Xi(V, T, \mu_1, \ldots, \mu_v) = \sum_{N_1,\ldots,N_v}^{\infty} \prod e^{\beta\mu_\alpha N_\alpha} Q(N_1, \ldots, N_v, T, V) = e^{\Omega} \tag{8.94}$$

where $\Omega = \beta PV$ and $Q(N_1, \ldots, N_v, T, V)$ is the canonical ensemble partition function. (Π here means the product of all the terms of the form $e^{\beta\mu_\alpha N_\alpha}$.) Using this result we can show that (see additional exercises)

$$\frac{1}{\beta}\left(\frac{\partial \langle N_i\rangle}{\partial \mu_j}\right)_{T,V,\mu_k} = \langle N_i N_j\rangle - \langle N_i\rangle\langle N_j\rangle \tag{8.95}$$

and substituting equation (8.93) in (8.95) leads to the following equation in the multicomponent system:

$$kT\left(\frac{\partial \rho_i}{\partial \mu_j}\right)_{T,V,\mu_k} = \rho_i\delta_{ij} + \rho_i\rho_j G_{ij} \tag{8.96}$$

This equation, for species i, is the essential result that emerges from the Kirkwood–Buff theory. It relates thermodynamic stability coefficients to the pair correlation functions. We point out that the G_{ij} quantity in equation (8.96) is not to be confused with the density–density correlation defined earlier (8.47), but is an integral of the pair correlation function. (We have used this particular nomenclature to be consistent with common literature usage.)

In a pure fluid, equation (8.96) reduces to the result

$$\beta\left(\frac{\partial \mu}{\partial \rho}\right)_T = -\frac{\beta}{\rho K_T} = \frac{1}{\rho(1+\rho G)} \tag{8.97}$$

where G is the total correlation fluctuation integral for the pure fluid:

$$G \equiv \int_V (g(r) - 1)dV \tag{8.98}$$

which must diverge at the critical point consistent with the scaling relation shown in equation (8.70).

EXERCISE 8.10

If we define the i–j elements of the matrices $\boldsymbol{\Omega}$ and $\boldsymbol{A}$ by

$$\Omega_{ij} \equiv kT\left(\frac{\partial N_i}{\partial \mu_j}\right)_{T,V,\mu_k} \tag{8.99}$$

and

$$A_{ij} = \frac{1}{kT}\left(\frac{\partial \mu_i}{\partial N_j}\right)_{T,V,N_k} \tag{8.100}$$

show that these are thermodynamic stability matrices with the following relationship between them:

$$\boldsymbol{\Omega} = \boldsymbol{A}^{-1} \tag{8.101}$$

We are now in a position to employ the KB theory to derive thermodynamic properties that are useful in the study of near-critical mixtures.

Compressibility in Multicomponent Mixtures

Both the isothermal compressibility K_T and partial molar volumes are important thermodynamic properties of solutions, particularly in the critical region. From a standard thermodynamic identity, we have in multicomponent mixtures that

$$K_T^{-1} = \rho \left(\frac{\partial P}{\partial \rho} \right)_{T,x} = \sum_i \rho_i \left(\frac{\partial P}{\partial \rho_i} \right)_{T,\rho_j} \tag{8.102}$$

We now use the KB equations developed earlier in (8.102). First, we define a matrix $\boldsymbol{B}$ with (i, j) element B_{ij} as follows:

$$B_{ij} \equiv \left(\frac{\partial \rho_i}{\partial \beta \mu_j} \right)_{T,V,\mu_k} \tag{8.103}$$

in which case we have that,

$$VB_{ij} = \langle N_i N_j \rangle - \langle N_i \rangle \langle N_j \rangle \tag{8.104}$$

Using these definitions expressed in (8.102) we find that:

$$K_T = \frac{|\boldsymbol{B}|}{kT \sum_i \sum_j \rho_i \rho_j |\boldsymbol{B}_{ij}|} \tag{8.105}$$

where $|\boldsymbol{B}|$ is the determinant of $\boldsymbol{B}$ while $|\boldsymbol{B}_{ij}|$ is the cofactor of the (i, j) matrix element of $\boldsymbol{B}$ (see the definition for matrix cofactors given in chapter 1).

EXERCISE 8.11

Prove equation (8.105) and show that, in a binary mixture, the KB theory yields the following results:

$$B_{11} = \rho_1 + \rho_1^2 G_{11}, \quad B_{12} = B_{21} = \rho_1 \rho_2 G_{21}, \quad B_{22} = \rho_2 + \rho_2^2 G_{22}$$

$$|\boldsymbol{B}| = B_{11} B_{22} - B_{12}^2, \quad |\boldsymbol{B}_{11}| = B_{22}, \quad |\boldsymbol{B}_{22}| = B_{11}$$

$$|\boldsymbol{B}_{12}| = -\rho_1 \rho_2 G_{12} \quad \text{(etc.)}$$

Use these results in (8.105) to develop the KB result for K_T at a binary mixture's critical point. Is it finite there? How does this result square with those given in chapter 4?

Partial Molar Volumes in Multicomponent Mixtures

Since the partial molar property $\bar{V}_k$ is given by the definition

$$\bar{V}_k \equiv \frac{(\partial P/\partial \rho_k)_{T,\rho_j}}{\rho(\partial P/\partial \rho)_{T,x}} \tag{8.106}$$

the KB theory provides that

$$\bar{V}_k = \frac{\sum_i \rho_i |\boldsymbol{B}_{ik}|}{\sum_i \sum_j \rho_i \rho_j |\boldsymbol{B}_{ij}|} \tag{8.107}$$

Using the results of equations (8.105) and (8.107) for binary mixtures, we can show that

$$\rho k T K_T = \frac{1 + \rho_1 G_{11} + \rho_2 G_{22} + \rho_1 \rho_2 (G_{11} G_{22} - G_{12}^2)}{1 + x_1 x_2 \rho (G_{11} + G_{22} - 2G_{12})} \tag{8.108}$$

with

$$\rho \bar{V}_1 = \frac{1 + \rho_2 (G_{22} - G_{12})}{1 + x_1 x_2 \rho (G_{11} + G_{22} - 2G_{12})} \tag{8.109}$$

EXERCISE 8.12

Using the KB results show that the osmotic susceptibility in a binary solution is given by the equation

$$\frac{kT}{x_1} \left(\frac{\partial x_1}{\partial \mu_1} \right)_{T,P} = 1 + x_1 x_2 \rho (G_{11} + G_{22} - 2G_{12}) \tag{8.110}$$

What is the value of the term $G_{11} + G_{22} - 2G_{12}$ at finite compositions along the mixture's critical locus? How about at the limit $x_1 \to 0$ along the critical locus? Use these results to determine the values of K_T and $\bar{V}_1$ along the binary mixture's critical line, predicted by the KB theory, at both conditions (with finite composition and in the limit $x_1 \to 0$).

In order to use the KB theory, one needs to be able to calculate the various G_{ij}, which in turn requires finding all of the solution's pair correlation functions $g_{ij}(r)$. These are usually found from numerical solutions of the *Ornstein–Zernike* and related integral equations. This has been done for dilute supercritical mixtures [6]; see a schematic representation of such results in figure 8.2. There is also an interesting tie that can be established between the KB and Ornstein–Zernike theory, the development of which we leave for an exercise at the end of this chapter (see the additional exercises). We now investigate the relationship of these fluctuation integrals to experimental data in dilute supercritical binary systems.

Representing Solubility Data in Terms of the Total Correlation Fluctuation Integrals

We now use the TCFI to describe solvation behavior in a near-critical fluid focusing upon a two-phase system of a dilute solute (1) in a fluid solvent (2) in equilibrium with

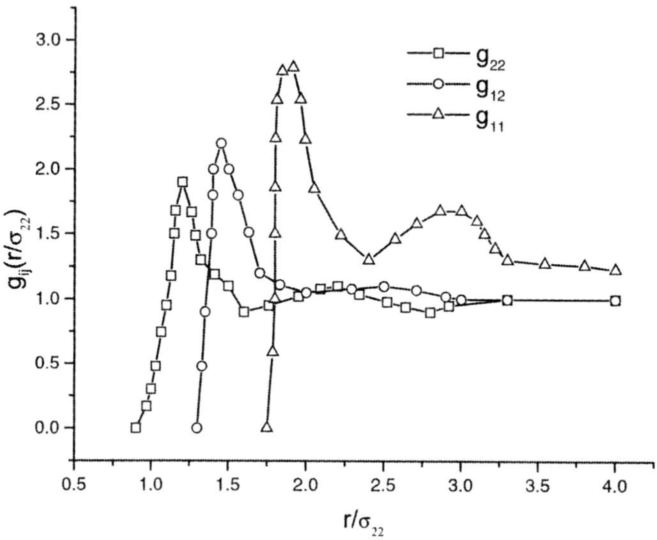

Figure 8.2 Mixture pair correlation functions in a binary Lennard–Jones fluid at the solute's infinite-dilution limit

a pure solute phase. The variation of solubility of the solute with solvent density, along the phase envelope, can be approximated by the thermodynamic equation

$$\left(\frac{\partial \ln \rho_1}{\partial \ln \rho_2}\right)_{T,\sigma} \approx \left(\frac{\partial \ln \rho_1}{\partial \ln \rho_2}\right)_{T,\mu_1} \tag{8.111}$$

where the σ denotes a derivative property evaluated along the phase coexistence envelope. From equation (8.96), we have that

$$\left(\frac{\partial \rho_1}{\partial \mu_2}\right)_{T,V,\mu_1} = \beta\rho_1\rho_2 G_{12} \tag{8.112}$$

and

$$\left(\frac{\partial \rho_2}{\partial \mu_2}\right)_{T,V,\mu_1} = \beta\rho_2(1 + \rho_2 G_{22}) \tag{8.113}$$

which means that

$$\left(\frac{\partial \ln \rho_1}{\partial \ln \rho_2}\right)_{T,\mu_1} = \frac{\rho_2 G_{12}}{1 + \rho_2 G_{22}} \tag{8.114}$$

Using this result, one can imagine three distinct solvation regimes at a near-critical temperature depending upon the TCFIs on the right-hand side of (8.114):

1. Low solvent density (pressure) where small ρ_2 and G_{12} implies that $\rho_2 G_{12}/(1 + \rho_2 G_{12}) \approx \rho_2 G_{12}$
2. Near-critical regime of most interest here, where the long-range effects in the G_{ij} given by (8.70), imply that

$$\frac{\rho_2 G_{12}}{1 + \rho_2 G_{22}} \approx \frac{G_{12}}{G_{22}} \tag{8.115}$$

In attractive solutions, we would expect G_{12}/G_{22} to be positive and insensitive to density (why?).

3. A high-density region where one might anticipate that $G_{12} < 0$ and $G_{22} > 0$.

The Near-Critical Region

We focus our discussion upon the near-critical region where

$$\left(\frac{\partial \ln \rho_1}{\partial \ln \rho_2}\right)_{T,\sigma} \approx \frac{G_{12}}{G_{22}} = \gamma \text{ (a constant)} \tag{8.116}$$

Integrating this equation leads to the result that, in dilute binary mixtures, the solute mole fractions at two different solvent densities in the critical region are related by the following power law

$$\left(\frac{x_1''}{x_1'}\right) = \left(\frac{\rho''}{\rho'}\right)^{\gamma-1} \tag{8.117}$$

The superscripts $'$ and $''$ in (8.117) denote two isothermal solubility conditions at different solvent densities. This result predicts that solubility-density data in the solvent's critical region will be linearly related when represented in log–log coordinates. This prediction appears to be consistent with most available experimental data, as observed in chapter 5. Here we see that the slopes of such data plots are related to the theoretical TCFI in the KB theory.

Equation (8.117) also illustrates why solvent power is easily "tuned" in the critical region. Since the isothermal compressibility is large here, small changes in pressure lead to large density changes in the solvent which in turn exponentially amplify solubility differences; the "amplification index" $\gamma - 1$ given by this analysis is the quantity $(G_{12} - G_{22})/G_{22}$.

EXERCISE 8.13

Establish the circumstances under which the amplification index given in (8.117) is positive. In which types of mixture is this likely to be the situation?

8.6 Chapter Review

In this chapter, we studied the role that thermodynamic fluctuations play in determining system behavior, especially in the critical region. Both discrete (Ising-like) and continuum models were analyzed, since they belong to the same critical universality class. In each case, general relationships linking fluctuations to thermodynamic stability coefficients were provided. These results illustrate how thermodynamic divergences in properties such as the susceptibility at the critical point can be related to the divergence in the correlation length ξ, a length scale that characterizes the distance over which order-parameter fluctuations are correlated in the system. The correlation length assumes special significance in nonclassical theories of critical phenomena such as the *renormalization group* method, which we discuss in a subsequent chapter.

Two new (nonclassical) critical exponents were introduced in the context of this analysis of fluctuations. The first is ν, the exponent characterizing the divergence of the correlation length ξ with temperature as the critical temperature is approached. The second new exponent η characterizes the correlation of order-parameter fluctuations in the critical region. Both of these exponents are not present in mean-field theories, which explicitly ignore the effects of fluctuations and their correlations in describing thermodynamic behavior.

Finally, we described the Kirkwood–Buff statistical-mechanical theory for describing the thermodynamic behavior of multicomponent mixtures. This theory explicitly includes fluctuation properties in its conception, and has been widely used to study the properties of modelled supercritical mixtures with well-defined intermolecular potentials.

8.7 Additional Exercises

Fluctuation Analysis: Ising Systems

1. Rationalize the following fundamental thermodynamic equation for the energy of an Ising (magnet) system:

$$dE = TdS - MdH + \mu dN$$

Develop all the Legendre transforms and show that the susceptibility

$$\chi_{\mathrm{M}} = \left(\frac{\partial M}{\partial H}\right)_T$$

diverges at a stability limit. What other conditions apply at the magnet's critical point besides the divergence of χ_{M}? Do these pertain at the critical point of the mean-field model given in chapter 7?

2. Prove the following identity in the magnet:

$$C_H = C_{\mathrm{M}} + \frac{T}{\chi_{\mathrm{M}}}\left(\frac{\partial M}{\partial T}\right)_H^2$$

How does this identity couple the various critical exponents in the magnet?

3. Prove that the susceptibility in a magnet is given by the following *fluctuation formula*:

$$-\frac{1}{\beta}\left(\frac{\partial \langle M\rangle}{\partial H}\right)_\beta = \langle M^2\rangle - \langle M\rangle^2$$

4. Using the Ising lattice partition function, derive a fluctuation formula for the specific heat of the model that is defined by

$$C_H = \left(\frac{\partial E}{\partial T}\right)_{H,N} = -T\left(\frac{\partial^2 A}{\partial T^2}\right)_{H,N}$$

where the free energy in this equation is a function of N, T, H. Given these results and those of the previous question, derive a formula for the *cross-correlations* $\langle(\delta M)^2(\delta E)^2\rangle$.

5. In the Ising system, assuming that we express the partition function $Q(T, H, N)$ in the form

$$Q(T, H, N) = \sum_{M} \sum_{E_{\upsilon M}} e^{-\beta E_{\upsilon M}}$$

where $E_{\upsilon M}$ is the energy microstate consistent with a value of the order parameter M, what is the probability $P(M)$ of observing this state? Below the critical temperature, sketch the function $P(M)$ against M.
6. Derive an equation for the 3-spin function $\langle S_i S_{i+j} S_{i+k} \rangle$ in the 1-d Ising model. In addition, compare the 2-spin correlation function Γ_{ij} derived in this chapter with the 3-spin correlation function $\Gamma_{ijk} \equiv \langle S_i S_{i+j} S_{i+k} \rangle - \langle S_i \rangle \langle S_{i+j} \rangle \langle S_{i+k} \rangle$.
7. Chandler [7] provides a description of *Onsager's solution* to the 2-d Ising model at zero field as follows:

$$Q(T, N, 0) = [2(\cosh \beta J)e^{I}]^{N}$$

$$I = \frac{1}{2\pi} \int_0^{\pi} \ln \left\{ \frac{1}{2}[1 + (1 - \theta^2 \sin^2 \phi)^{1/2}] \right\} d\phi$$

$$\theta = 2\frac{\sinh 2\beta J}{\cosh^2 2\beta J}$$

Why does this solution imply that the free energy at the critical point is nonanalytic. How would you find the specific-heat scaling exponent α from these equations?
8. The Gibbs formula for the entropy is given by

$$S = -k \sum_{\upsilon} P_{\upsilon} \ln P_{\upsilon}$$

where P_{υ} is the probability of observing state υ. Show that the Boltzmann result for S given by the equation

$$S = k \ln \Omega(N, V, E)$$

where $\Omega(N, V, E)$ is the degeneracy associated with this state, is not in conflict with the Gibbs result.
9. If a spontaneous fluctuation occurs in an isolated system taking it from states 1 to 2, show that the probability of this occurring is given by:

$$e^{\Delta S/k}$$

where $\Delta S = S_2 - S_1$.

Analysis in Fluid Systems

10. Prove the identity

$$\left(\frac{\partial \mu}{\partial \rho}\right)_T = \frac{1}{\rho^2 K_T}$$

Express the left-hand side of this equation in terms of the pair correlation function and discuss its behavior in these terms as the critical point is approached.

11. Derive the following Gaussian approximation to $P(N)$, the probability of observing a state with N molecules ($T > T_c$):

$$P(N) = \frac{1}{\rho\sqrt{2\pi kTK_T V}} \exp\left[\frac{-(N-\bar{N})^2}{2kT\rho^2 K_T V}\right]$$

where $\rho = \bar{N}/V$. In the two phase region how would you expect this probability distribution to look?

12. The *dielectric constant* of a medium ε is related to its density by the Clausius–Mossotti equation:

$$\frac{\varepsilon - 1}{\varepsilon + 2} = A\rho, \quad A \text{ constant}$$

Find σ_ε^2 in terms of σ_ρ^2 from this equation and discuss what happens to ε at the critical point (see figure 8.3 for a schematic of water's dielectric constant in the near-critical region). How would you use this analysis to measure the scaling exponent γ in a fluid from dielectric measurements?

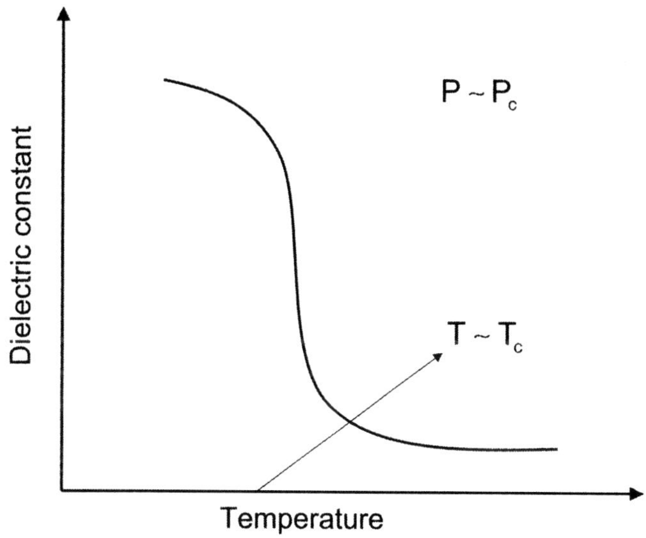

Figure 8.3 Schematic of dielectric constant of water in its critical region

13. Various ensemble averages play an important role in *statistical-mechanical perturbation theory*. In this approach, the actual system's configurational energy is expanded, rather like a fluctuation term, around that of a reference system as follows:

$$U_N = U_N^\circ + U_N'$$

where U_N° is the reference system's potential and U_N' the perturbation term. Show that the configuration integral Z_N for the actual system can be written as

$$Z_N = Z_N^\circ \langle \exp(-\beta U_N') \rangle_\circ$$

where $\langle\,\rangle_\circ$ designates an average taken in the *reference system's ensemble*. Find the free energy A of the system in terms of A° and a perturbation term A' which itself depends upon $\langle U'_N\rangle_\circ$. Find $\langle U'_N\rangle_\circ$ in terms of the reference system's pair correlation function $g_\circ(r_1, r_2)$. Derive the pressure-explicit equation of state for this system.

14. If we make the assumption that the three-body correlation function in a fluid is given by [3]

$$g^{(3)}(1,2,3) = g^{(2)}(1,2)g^{(2)}(1,3)g^{(2)}(2,3)$$

derive the thermodynamic equation for the system energy following the approach given earlier in this chapter that led to (8.57). In this case, the system energy is given by

$$U_N = \sum_{i<j} u_{ij} + \sum_{i<j<k} u_{ijk}$$

where u_{ijk} is the 3-body potential that depends upon the relative positions of the three molecules i, j, k, respectively.

Fourier Transforms

15. The Fourier transform (FT) of $f(x)$, designated $F_x[f(x)] \equiv g(k)$, is defined as

$$g(k) = \frac{1}{\sqrt{2\pi}} \int_{-\infty}^{\infty} f(x) e^{-ikx} dx$$

If follows that the inverse transform (IFT) is given by

$$f(x) = \frac{1}{\sqrt{2\pi}} \int_{-\infty}^{\infty} g(k) e^{ikx} dk$$

This definition and result can easily be extended to many dimensions, in particular, in 3 dimensions:

$$g(\mathbf{k}) = \frac{1}{(2\pi)^{3/2}} \int_{-\infty}^{\infty} f(\mathbf{x}) e^{-i\mathbf{k}\cdot\mathbf{x}} d\mathbf{x}$$

$$f(\mathbf{x}) = \frac{1}{(2\pi)^{3/2}} \int_{-\infty}^{\infty} g(\mathbf{k}) e^{i\mathbf{k}\cdot\mathbf{x}} d\mathbf{k}$$

The prefactors, powers of $(2\pi)^{-1/2}$, are often left out for convenience, when defining Fourier transforms, making it easier notationally to carry out algebraic manipulations. We adopt both conventions in this text. Write the above in polar variables.

16. Show that the FT of the function $f(x) = e^{-a|x|}$ $(a > 0)$ is given by

$$\left(\frac{2}{\pi}\right)^{\frac{1}{2}} \frac{a}{k^2 + a^2}$$

17. Find the FT of the *Dirac delta function* $\delta(x)$ defined as follows:

$$\int_{-\infty}^{\infty} \delta(x - x_0) f(x) dx = f(x_0)$$

for all continuous functions f. Note that setting $f(x) = 1$ and $x_0 = 0$ gives

$$\int_{-\infty}^{\infty} \delta(x) dx = 1$$

18. Prove equations (8.62)–(8.65).
19. If $f(t, z)$ is a function of both t and z, show that the following identities hold:

$$F_t \left[\frac{\partial^2 f(t, z)}{\partial z^2} \right] = \frac{\partial^2 \bar{f}(k, z)}{\partial z^2}, \qquad F_t \left[\frac{\partial^2 f(t, z)}{\partial t^2} \right] = -k^2 \bar{f}(k, z)$$

where

$$\bar{f}(k, z) = \frac{1}{\sqrt{2\pi}} \int_{-\infty}^{\infty} f(t, z) e^{-ikt} dt$$

20. Find the Fourier transform of the function e^{-ax^2}, where $a > 0$.
21. Find the Fourier transform of the function x^2.
22. Show that the Fourier coefficients $g(k)$ of an odd function are purely imaginary. An odd function is one for which

$$f(x) = -f(-x)$$

23. If the discrete form of the FT of a real-valued function $f(\boldsymbol{r})$ is given by

$$f(\boldsymbol{r}) = \sum_{\mathbf{k}} \bar{f}(k) e^{i\boldsymbol{k}\cdot\boldsymbol{r}}$$

show that

$$f(\boldsymbol{r})^2 = \left[\sum_{\mathbf{k}} \bar{f}(\mathbf{k}) \right]^2$$

24. Cylindrical coordinates $\boldsymbol{r} \equiv (r, \theta, z)$ are related to Cartesian ones as follows:

$$x = r\cos\theta, \qquad y = r\sin\theta, \qquad z = z$$

Prove that the gradient operator $\nabla_{\boldsymbol{r}}$ in cylindrical coordinates is given by

$$\nabla_{\boldsymbol{r}} = \boldsymbol{\delta}_r \frac{\partial}{\partial r} + \boldsymbol{\delta}_\theta \frac{1}{r} \frac{\partial}{\partial \theta} + \boldsymbol{\delta}_z \frac{\partial}{\partial z}$$

where $\boldsymbol{\delta}_r$, $\boldsymbol{\delta}_\theta$, $\boldsymbol{\delta}_z$ are unit vectors associated with the respective coordinates.

25. The Ornstein–Zernike equation is a famous result in thermodynamic theory for pure fluids. It partitions the particle–particle correlations into indirect $h(\boldsymbol{r},\boldsymbol{r}')$ and direct $c(\boldsymbol{r},\boldsymbol{r}')$ contributions through the integral equation

$$h(\boldsymbol{r},\boldsymbol{r}') = c(\boldsymbol{r},\boldsymbol{r}') + \rho \int c(\boldsymbol{r},\boldsymbol{r}'')h(\boldsymbol{r}'',\boldsymbol{r}')d\boldsymbol{r}''$$

We can define the Fourier transform of a vector function like $c(\boldsymbol{r},\boldsymbol{r}')$, whose value depends only on $\boldsymbol{r}-\boldsymbol{r}'$, as follows:

$$\bar{c}(\boldsymbol{k}) \equiv \iint e^{-i\boldsymbol{k}\cdot(\boldsymbol{r}_1-\boldsymbol{r}_2)}c(\boldsymbol{r}_1-\boldsymbol{r}_2)d\boldsymbol{r}_1 d\boldsymbol{r}_2$$

where we write $c(\boldsymbol{r}_1-\boldsymbol{r}_2) \equiv c(\boldsymbol{r}_1,\boldsymbol{r}_2)$ (where each integral variable corresponds to the 3-d position of a molecule). In this case, show that the FT of the Ornstein–Zernike equation can be written as

$$1+\rho\bar{h}(\boldsymbol{k}) = \frac{1}{1-\rho\bar{c}(\boldsymbol{k})}$$

Stanley [2] discusses the role played by $G(\boldsymbol{r},\boldsymbol{r}')$ in determining system behavior in the theory of quasi-elastic light scattering as the critical point is approached. The following Fourier integral defines the system's *structure factor* $\bar{S}(\boldsymbol{k})$:

$$\bar{S}(\boldsymbol{k}) \equiv \int e^{-i\boldsymbol{k}\cdot\boldsymbol{r}}G(\boldsymbol{r})d\boldsymbol{r}$$

and the scattering intensity $I(\boldsymbol{k})$ of the incident beam of light is related to the Fourier transform of $G(\boldsymbol{r},\boldsymbol{r}')$ by

$$\bar{R}(\boldsymbol{k}) \equiv \frac{\bar{I}(\boldsymbol{k})}{\bar{I}^{\circ}(\boldsymbol{k})} = \frac{1}{\rho}\int e^{-i\boldsymbol{k}\cdot\boldsymbol{r}}G(\boldsymbol{r})d\boldsymbol{r}$$

In this equation $\bar{I}^{\circ}(\boldsymbol{k})$ is the scattering intensity in an ideal (uncorrelated particles) system. Show that we can write

$$\frac{\bar{S}(\boldsymbol{k})}{\rho} = 1+\rho\bar{h}(\boldsymbol{k}) = \frac{1}{1-\rho\bar{c}(\boldsymbol{k})}$$

Rewrite the Fourier $\bar{c}(\boldsymbol{k})$ in terms of a unit vector $\boldsymbol{w}$ and magnitude k as follows:

$$\bar{c}(k) \equiv \int e^{-ik\boldsymbol{w}\cdot\boldsymbol{r}}c(\boldsymbol{r})d\boldsymbol{r}$$

in which case a Taylor series expansion gives

$$\bar{c}(k) = \bar{c}(0) + \bar{c}_1(\rho,T)k^2 + \cdots$$

Why is the first-order term equal to zero in this equation? Given this, following the result given in [2], show that

$$\rho^{-1}\bar{S}(\boldsymbol{k}) = \frac{\phi^{-2}}{\alpha^2+k^2}$$

with $\alpha^2\phi^2 = 1 - \rho\bar{c}(0)$ and the quantity ϕ a constant at given density and temperature. By taking the Fourier inverse of this equation (using the result proved in question 16) show that in the asymptotic limit:

$$G(r) \propto e^{-\alpha r}/r$$

What is the natural definition of the correlation length given by this approach?

26. If, classically in 1-d, $G(r) = re^{-\alpha r}$ with r the distance separating two species, show that the Fourier coefficient at the origin is given by

$$\bar{G}(k=0) = \left(\frac{2}{\pi}\right)^{\frac{1}{2}} \frac{1}{\alpha^2}$$

Since we just showed that $\alpha \propto 1/\xi$, this result illustrates the importance of the long-wavelength Fourier modes as the critical point is approached. Develop the 2-d classical expression for $\bar{G}(k=1)$.

27. Show that, for a randomly structured fluid (like an ideal gas), the structure factor is given by

$$\bar{S}(\boldsymbol{k}) = \rho(1 + \rho(2\pi)^3\delta(\boldsymbol{k}))$$

28. Show, for the singlet density function

$$n(\boldsymbol{x}) = \sum_{\alpha} \delta(\boldsymbol{x} - \boldsymbol{x}_\alpha)$$

that

$$\bar{n}(\boldsymbol{k}) = \sum_{\alpha} e^{-i\boldsymbol{k}\cdot\boldsymbol{x}_\alpha}$$

Where α represents a summation over all α species in the system. Note that the overbar denotes a Fourier variable.

29. Prove that, for the pair correlation function defined as

$$g(\boldsymbol{x}_1, \boldsymbol{x}_2) \equiv \langle n(\boldsymbol{x}_1) n(\boldsymbol{x}_2)\rangle$$

its FT is given by

$$\langle \bar{n}(\boldsymbol{k}) \bar{n}(-\boldsymbol{k})\rangle$$

Kirkwood–Buff Relationships

30. Prove the following identity from a fluctuation analysis in the grand canonical ensemble for multicomponent systems:

$$\langle N_i N_j \rangle = \frac{1}{\beta}\left(\frac{\partial \langle N_i\rangle}{\partial \mu_j}\right)_{T,V,\mu_k} + \langle N_i\rangle\langle N_j\rangle$$

This thermodynamic identity parallels the result of the KB theory. What specific *additional* information does the KB theory provide in this regard?

31. In a pure fluid, the probability of observing n molecules at positions $\boldsymbol{r}_1, \ldots, \boldsymbol{r}_n$ irrespective of the total number N is given by:

$$\rho^{(n)} = \sum_{N \geq n} \rho_N^{(n)} P_N$$

where P_N is the probability density taken in the grand canonical ensemble (see (8.94)). Show that

$$\int \cdots \int \rho^{(n)}(1, \ldots, n) d\boldsymbol{r}_1 \cdots d\boldsymbol{r}_n = \left\langle \frac{N!}{(N-n)!} \right\rangle$$

Using $n = 2$ in this equation, derive the KB equation given by (8.95).

32. Show that the probability $P(E, N)$ of observing a state with energy E and N fluid particles is given by the expression:

$$P(E, N) = \frac{\Omega(E, V, N) e^{-\beta E} e^{\beta \mu N}}{\Xi(E, V, N)}$$

where $\Omega(E, V, N)$ is the microcanonical ensemble partition function. How do we find $\Xi(E, V, N)$ from histograms developed during grand canonical simulations of the fluid at a given temperature? Given $\Xi(E, V, N)$, how would you find the fluid pressure at these conditions? Below the critical temperature why would the bimodal probability function have equal peak areas at phase equilibrium? Now generalize this analysis so that it applies to a binary fluid.

Bibliography

Fluctuations and their significance are described in most statistical-mechanics textbooks, good examples of which are references [1, 2, 3, 7]. The original Kirkwood–Buff article is reference [5], while its application to the Lennard–Jones system is given in reference [6]. An excellent review of fluctuation theory and supercritical solutions is given in reference [8].

[1] M. Plischke and B. Bergersen, *Equilibrium Statistical Physics*. Singapore: World Scientific, 1994.

[2] H. E. Stanley, *Introduction to Phase Transitions and Critical Phenomena*. Oxford: Clarendon Press, 1971.

[3] D. A. McQuarrie, *Statistical Mechanics*. New York: Harper & Row, 1976.

[4] P. G. Debenedetti and S. K. Kumar, "Infinite dilution fugacity coefficients and the general behavior of dilute binary systems," *AIChE J.*, vol. 32, p. 1253, 1986.

[5] J. G. Kirkwood and F. P. Buff, "The statistical mechanical theory of solutions. 1," *J. Chem. Phys.*, vol. 19, p. 774, 1951.

[6] F. Munoz and E. H. Chimowitz, "Integral equation calculations of the solute chemical potential in a near-critical fluid environment," *Fluid Phase Equilib.*, vol. 71, p. 237, 1992.

[7] D. Chandler, *Introduction to Modern Statistical Mechanics*. New York: Oxford University Press, 1987.

[8] L. L. Lee, P. G. Debenedett, and H. D. Cochran, "Fluctuation theory of supercritical solutions," in *Supercritical Fluid Technology: Reviews in Modern Theory and Applications*, T. J. Bruno and J. F. Ely (eds). Boca Raton, FL: CRC Press, 1991, p. 194.

9 Scaling Theory and Computer Simulation

In this chapter, we investigate the concept of scaling, in particular the *scaling hypothesis* that has proved so important in the development of the modern theory of critical phenomena. Scaling refers to the way in which properties such as the susceptibility behave

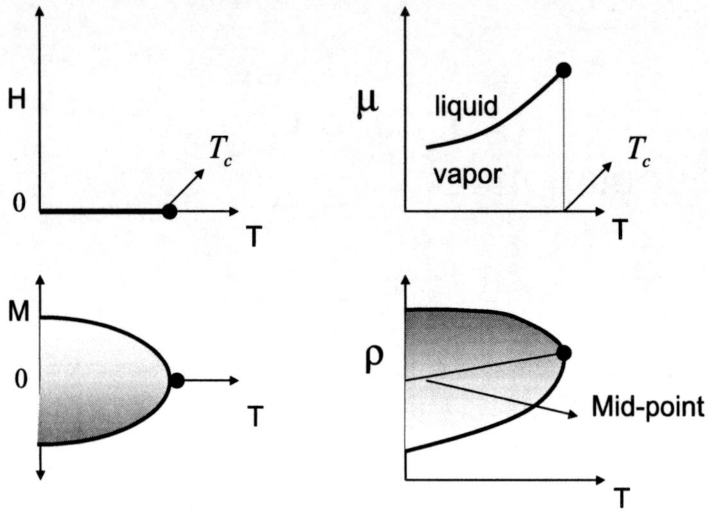

Figure 9.1 Co-existence-curve schematics in fluids and magnetic systems

in the critical region as they approach infinity (or zero). In this discussion, it is important to be specific about the thermodynamic field variables used to represent the system as well as the direction in which the critical point is approached. A useful schematic in this regard is shown in figure 9.1, which shows phase coexistence diagrams for both an Ising-type magnet and a typical fluid. The fields for the magnet are (H, T) while for the fluid they are (μ, T). We note immediately from this figure that the magnet's coexistence curve is coincident with the $H = 0$ axis and symmetric about the $M = 0$ axis. However, these features do not reflect in the corresponding diagrams for the fluid, with important implications for a scaling analysis in this system, which we take up later in this chapter.

9.1 Scaling Laws and Critical Exponents

We illustrate the basic critical scaling laws using the Ising (magnet) system; however, the fact that similar scaling laws also apply to fluids is an important consequence of *critical-point universality*, a topic we have often alluded to in this text and which is discussed in more detail in chapter 10 dealing with the renormalization-group method.

From our earlier definition of susceptibility in the Ising system $\chi_{\mathrm{M}}(H, T)$, we have that

$$\begin{aligned}\chi_{\mathrm{M}}(0, T) &= \frac{1}{\beta}\left(\frac{\partial M}{\partial H}\right)_{\beta} \\ &= \left(\frac{N\mu_{\mathrm{m}}}{\beta}\right)\lim_{H\to 0}\left(\frac{\partial m}{\partial H}\right)_{\beta}\end{aligned} \tag{9.1}$$

with $m = M/N\mu_{\mathrm{m}}$. As the critical point is approached, the divergent part of $\chi_{\mathrm{M}}(0, T)$ is given by the $\lim_{H\to 0}(\partial m/\partial H)_{\beta}$. Implicit to (9.1) is the notion that the divergence of

$\chi_M(H, T)$ is considered in a direction *parallel to the coexistence curve* at the critical point (why?). In this case, following the rationale given in chapter 3, we postulate the following scaling relationship for $\chi_M(H, T)$ as the critical temperature is approached:

$$\chi_M \approx \lim_{H \to 0} \left(\frac{\partial m}{\partial H} \right)_\beta \sim |T - T_c|^{-\gamma} \tag{9.2}$$

In addition to this scaling form for $\chi_M(H, T)$ we also propose the following scaling *power laws* for other quantities where each property is tied to the particular scaling exponent shown.

$$C_H(0, T) \sim |T - T_c|^{-\alpha} \tag{9.3}$$

$$m(0, T) \sim |T - T_c|^{\beta} \tag{9.4}$$

$$m(H, T_c) \sim \text{sgn}(H)|H|^{1/\delta} \tag{9.5}$$

In a fluid, the properties corresponding to (9.2)–(9.5) are K_T, C_V, and $\rho_l - \rho_v$, respectively. We emphasize that each of these scaling laws, and its respective exponent, requires that the divergent property be analyzed in a very specific direction, as discussed in chapter 3. For both χ_M and C_H this is a direction parallel to the coexistence curve's linear extension at the critical point. If any one of these properties were to be studied using another direction of approach to the critical point not parallel to the coexistence curve, then the scaling exponents would not necessarily be those given above. For β, the path is along the coexistence curve with $T < T_c$, while for δ it is perpendicular to the T axis, the coexistence direction. In fluids, for example, where the linear extension to the coexistence direction is not parallel to one of the field variables, a scaling analysis is more complicated. It requires subtleties to be addressed, as we shall see later on in this chapter.

EXERCISE 9.1

The notion of a scaling exponent being dependent upon the direction taken toward the function's singularity can easily be demonstrated by looking at the following function:

$$f(x, y) = \frac{1}{x^2 + y^{1/2}}, \quad y \geq 0$$

Show that the scaling of $f(x, y)$ depends upon the path of approach toward its singularity. What is the x exponent along the line $y = 2x$ and the parabolic path $y = x^2$?

Nonclassical values for these critical exponents have been found, either from exact analytical solutions (e.g., for the 2-dimensional Ising model), computer simulation results or experimental measurements. Classical exponents on the other hand can be directly obtained from mean-field models as discussed in chapter 7. Currently accepted values for these exponents [1] are given in table 9.1.

The differences between the classical and nonclassical exponents in table 9.1 were a major consideration that spurred the development of modern theories of critical phenomena.

Table 9.1 "Exact" critical exponents for the Ising universality class and the mean-field counterpart

	Mean Field	*Nonclassical ("Exact") 2-Dimensional Ising*	*Nonclassical ("Exact") 3-Dimensional Ising*
γ	1	$\frac{7}{4}$	1.239
α	0	Logarithmic	$\frac{1}{8}$
β	$\frac{1}{2}$	$\frac{1}{8}$	0.326
δ	3	5	4.8
η	—	$\frac{1}{4}$	0.04
ν	—	1	0.63

Relationships between the Critical Exponents

The relationships shown in equations (9.1)–(9.5) are known as *scaling laws*, and an immediate question that comes to mind is whether or not thermodynamics provides any independent relationships between these various exponents; the answer to this question is in the affirmative, and here we provide an example showing how one of these exponent relationships may be derived.

An important thermodynamic identity in the Ising magnet is

$$C_H - C_M = T\left(\frac{\partial H}{\partial \langle M \rangle}\right)_{T,N}\left[\left(\frac{\partial \langle M \rangle}{\partial T}\right)_{H,N}\right]^2 \tag{9.6}$$

EXERCISE 9.2

Prove the thermodynamic identity given by equation (9.6).

Since the magnet susceptibility is defined as

$$\chi_M = \frac{1}{\beta}\left(\frac{\partial \langle M \rangle}{\partial H}\right)_{\beta,N} \tag{9.7}$$

we can use a result from thermodynamic stability theory which states that, since χ_M, C_H, and C_M are all greater than or equal to zero, it follows from (9.6) that

$$C_H > \beta^{-1}\chi_M^{-1} T\left(\frac{\partial \langle M \rangle}{\partial T}\right)^2_{H,N} \tag{9.8}$$

At zero field and in the scaling region, which we define as the region $T \approx T_c$, we can use the previous scaling relationships (equations (9.1)–(9.5)) to show that

$$(T_c - T)^{-\alpha} > (T_c - T)^{\gamma}(T_c - T)^{2(\beta-1)} \tag{9.9}$$

which leads to a famous result called the *Rushbrooke inequality*; it relates these various critical exponents as follows:

$$\alpha + 2\beta + \gamma \geq 2 \tag{9.10}$$

We show in a subsequent section that (9.10) holds as an equality. Other important equations linking critical exponents are given by [1]

$$\gamma = \nu(2 - \eta) \quad \text{Fisher's law} \tag{9.11}$$

$$\nu d = 2 - \alpha \quad \text{Josephson's law} \tag{9.12}$$

$$2\beta\delta - \gamma = 2 - \alpha \quad \text{Griffiths' law} \tag{9.13}$$

where d is the system's dimensionality.

The Role of the Free Energy in Scaling

As we showed in an earlier chapter, classical results follow from the assumption that the free energy is analytic at the critical point. We thus surmise that nonclassical models are not analytic there, with a singular component present in the free energy. We show how the origin of this singularity can be seen quite simply by analyzing the free energy of the system, in the context of the scaling laws just presented.

From the fundamental thermodynamic equation for the Ising magnet given earlier, we can use Legendre transformations to develop all the other thermodynamic potentials, two important ones of which are

$$dA(T, M, N) = -S\,dT + H\,dM + \mu\,dN \tag{9.14}$$

$$dG(T, H, N) = -S\,dT - M\,dH + \mu\,dN \tag{9.15}$$

It is convenient to rewrite these equations in terms of the following deviation variables:

$$t \equiv T - T_c \tag{9.16}$$

$$h \equiv H - H_c \tag{9.17}$$

in which case $dG = -S\,dt - M\,dh + \mu\,dN$, from which it follows that

$$\left(\frac{\partial G}{\partial h}\right)_{t,N} = -M(t, h) \tag{9.18}$$

with

$$\lim_{h\to 0} M(t, h) \sim |t|^{\beta} \quad \text{along the coexistence curve}$$

and

$$\left(\frac{\partial^2 G}{\partial h^2}\right)_{t,N} = -\left(\frac{\partial M(t, h)}{\partial h}\right)_{t,N} \tag{9.19}$$

with

$$\lim_{h\to 0}\left[-\left(\frac{\partial M(t, h)}{\partial h}\right)_{t,N}\right] \sim |t|^{-\gamma} \tag{9.20}$$

Also we have that

$$C_h(t, h) = -T\left(\frac{\partial^2 G}{\partial t^2}\right)_{h,N} \tag{9.21}$$

with

$$\lim_{h \to 0} \left[-T \left(\frac{\partial^2 G}{\partial t^2} \right)_{h,N} \right] \sim |t|^{-\alpha} \tag{9.22}$$

We thus see that the singularities in the free energy are explicitly represented in equations (9.20) and (9.22). This is a clear indication of the problem associated with the assumption of analyticity in $G(t, h, N)$ or $A(t, m, N)$ at the critical point, and led initially to the adoption of *phenomenological scaling theory* as a remedy.

9.2 Phenomenological Scaling Theory

Intuitive ideas that were historically important in the development of the theory of critical phenomena are represented by phenomenological scaling theory. An important idea here is that the correlation length $\xi(t, h)$ diverges at the critical point, something we saw in chapter 8. This divergence is accorded *a special scaling exponent* ν and scales according to the power law

$$\xi(t, 0) \sim |t|^{-\nu} \tag{9.23}$$

We now show how these simple ideas can be used to construct a phenomenological theory in which length scale is a key parameter. Dimension is not normally a physical variable associated with describing thermodynamic phenomena, so this is conceptually quite a radical notion at this point.

Kadanoff Block Systems

In order to demonstrate the effect of length scale on the system behavior, Kadanoff [1] invoked a view of the system in terms of block spins (see figure 9.2) with individual spins grouped into square blocks of length L, where

$$L = S_d n_d \tag{9.24}$$

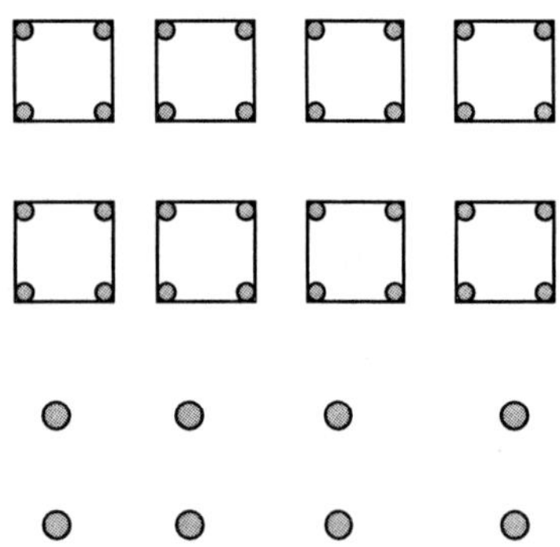

Figure 9.2 Schematic of rescaling in a square lattice

The *natural dimension* of the block spin system is L, while that of the site spin system is S_d and n_d represents the number of site spins per block. A long length in terms of the scale of the site spin system is manifestly shorter in the block system scale. Therefore, at a given absolute value of t and h *in both systems*, proximity to criticality, which depends upon the length scale of correlations, is likely to be different in the block system from the site spin system. Long-range correlations for the site system may be relatively short in scale for the block system. Hence, at a given absolute temperature, one may view the block and spin systems as being at different reduced temperatures. By *reduced* we mean a temperature normalized in some fashion relative to the critical one. Working out this "relative temperature relationship" between such so-called *rescaled* systems is one of the central objectives of the renormalization-group theory which brings to the fore the idea of geometric rescaling as a key concept in understanding critical behavior [2].

We now consider the intuitive notion of how to connect these respective systems as given in [1]. Consider a transformation between the site system's t and h and the analogous variables $\bar{t}$ and $\bar{h}$ in the rescaled block system. Clearly, as suggested, L defines a basic difference in both system descriptions, and it appears intuitively reasonable to use L in any transformation from t to $\bar{t}$ and h to $\bar{h}$. Furthermore, certain boundary conditions should be satisfied by this transformation. These are that, when $-h$ replaces h, then correspondingly $-\bar{h}$ replaces $\bar{h}$ (symmetry), and that, if $\bar{t} = \bar{h} = 0$, then $t = h = 0$. In addition, there is no *a priori* reason to assume that the singular part of the free energy for both systems should have different analytical forms; these assumptions clearly view the spin system as a microcosm of the block one. Relationships that are consistent with these conditions are

$$\bar{h} = hL^x \tag{9.25}$$

$$\bar{t} = tL^y \tag{9.26}$$

where x and y are unknown, but *positive* exponents (why?). Without loss of generality, we can assume that $S_d = 1$, in which case there are L^d sites per block. In this case, since the free energy is an extensive thermodynamic property, we should find the following relationship between this property in the site and block system, respectively:

$$G(t, h) = \frac{G(tL^y, hL^x)}{L^d} \tag{9.27}$$

The reader familiar with the concept of fractal structures can see that the free energy defined in equation (9.27) is a fractal object [3]. Since length scales in the block system have to be reduced by a factor L to accommodate scale considerations in the site spin system, one would also expect that the correlation length ξ is coupled in both systems by the equation

$$\xi(t, h) = L\xi(\bar{t}, \bar{h}) \tag{9.28}$$

Nothing we have said so far limits our choice of L, so in the particular case where we choose $L = |t|^{-1/y}$, equation (9.27) becomes (for $h = 0$)

$$G(t, 0) = \left[|t|^{-1/y}\right]^{-d} G(\pm 1, 0) \tag{9.29}$$

or

$$G(t, 0) = |t|^{d/y} G(\pm 1, 0) \tag{9.30}$$

and, from (9.28),

$$\xi(t, 0) = |t|^{-1/y} \xi(\pm 1, 0) \tag{9.31}$$

The rather remarkable result of these disarmingly simple ideas is that they lead to a form of the free energy that accommodates scaling at the critical point *without* resorting to analytic restrictions on the free energy at this condition. The relationships (9.29)–(9.31) enable us to relate y to certain critical exponents. For example, from the specific heat relationship given earlier (equation (9.21)), we find from equation (9.29) that

$$-T \left[\left(\frac{d}{y}\right) \left(\frac{d}{y} - 1\right) |t|^{d/y-2} \right] G(\pm 1, 0) \sim |t|^{-\alpha} \tag{9.32}$$

or, relating exponents in this equation,

$$\frac{d}{y} = 2 - \alpha \tag{9.33}$$

and since, from equations (9.23) and (9.31),

$$\nu = 1/y \tag{9.34}$$

we have

$$d\nu = 2 - \alpha \tag{9.35}$$

Equation (9.35) is a fundamental result in the theory of critical phenomena presented without proof earlier in equation (9.12). It is termed the *hyperscaling equation* or *Josephson's law*. It relates various critical exponents to the *dimensionality* of the system, a result that is totally absent from a classical analysis. One interpretation of the choice of L as $|t|^{-1/y}$ is that, as criticality occurs at t and h in the site spin system, one must, at the very least, force $L \to \infty$ (the natural length scale of the block system) in order for the correlation length also to diverge in terms of that length scale.

9.3 Widom's Scaling Hypothesis

The free-energy relationship coupling the site and block systems just derived can be seen as a particular case of a more general scaling relationship for G at the critical point, often referred to as *Widom's scaling hypothesis*. A general equation expressing this result provides that the free energy G is given by a *general homogeneous function* of the form

$$G(t, h) = \lambda G(\lambda^P t, \lambda^q h) \tag{9.36}$$

where one can identify λ with L^{-d} in equation (9.27). Given this general form for G, it is a straightforward matter to relate all the various critical exponents thus far presented to the indices p and q in this free-energy scaling function. For example, it is easy to show (see exercise 9.3) from (9.20) that the susceptibility will scale as

$$\chi(t, h) = \lambda^{2q+1}\chi(\lambda^p t, \lambda^q h) \tag{9.37}$$

EXERCISE 9.3

Prove equation (9.37).

Once again, one is free to choose values for the scale parameter λ. At zero field and with the choice of $\lambda = |t|^{-1/p}$, we find that

$$\chi(t, 0) = |t|^{-(2q+1)/p}\chi(\pm 1, 0) \tag{9.38}$$

which immediately implies that

$$-\frac{2q+1}{p} = \gamma \tag{9.39}$$

One can repeat this process for all the other properties that scale with particular exponents (e.g., $C_H(t, 0), m(t, 0), m(0, h)$) and find relationships linking q and p with all the critical exponents α, β, γ, and δ. What this analysis shows is that the critical exponents are not independent since they may all be found from p and q alone. Using this approach, it is straightforward to show that the following equality relations hold between the various critical exponents:

$$\alpha + 2\beta + \gamma = 2 \tag{9.40}$$

$$\beta(\delta - 1) = \gamma \tag{9.41}$$

Equation (9.40) was previously presented as a (Rushbrooke) inequality, while (9.41) is found from combining Griffiths' law and the Rushbrooke equation.

EXERCISE 9.4

Prove the scaling exponent relationships given by equations (9.40) and (9.41). What is the significance of the hyperscaling equation (9.35) vis-à-vis mean-field theories?

EXERCISE 9.5

If the free energy f is a homogeneous function of the form

$$f(t, h) = \lambda f(\lambda^y t, \lambda^x h)$$

show that

$$\lambda f(\lambda^y t, \lambda^x h) \sim t^{-1/y}\phi\left(\frac{h}{t^{x/y}}\right)$$

where ϕ is a function of the scaling quotient $h/t^{x/y}$ *and* $x/y \equiv \Delta = \gamma + \beta$

EXERCISE 9.6

Widom's scaling hypothesis posits that the free energy f scales as

$$f(t,h) = \lambda f(t\lambda^p, h\lambda^q)$$

The Kadanoff relationship posits that

$$f(t,h) = L^{-d} f(tL^y, hL^x).$$

Show that using $\lambda = L^{-d}$ in the Kadanoff equation leads to the result that

$$f(t,h) \sim t^{2-\alpha} f(1, h/t^{x/y})$$

with

$$tL^{1/\nu} = Lt^{\nu} = 1$$

9.4 The Role of Finite-Size Effects in Determining Critical Properties

We have repeatedly pointed out, in this and the previous chapter, that the divergence of properties such as the susceptibility depend upon the existence of a correlation length (and system size) that go to ∞ at the critical point. An immediate, potentially adverse, consequence of this result is found when attempting to do computer simulations of near-critical systems. Obviously simulations involve the use of finite particle numbers, a requirement that, on the face of it, appears to be at odds with being able to produce divergent (singular) properties. Clearly, we must find a way to "extrapolate" information from simulation data taken with finite system sizes, to provide more accurate values of the property that would occur if infinite system sizes were able to be used in the calculations.

In attempting to link the system size L to the scaling of a particular property, a quantity of obvious significance is the ratio $L/\xi(t)$. In the situation where $\xi(t) > L$ at a given temperature, finite-size effects are bound to adversely affect simulation results. We show this for the case of a fluid's compressibility K_T given by the equation

$$K_T = \beta \int_0^\infty [g(r) - 1] 4\pi r^2 dr \tag{9.42}$$

with the scaling of $g(r)$ given by

$$g(r) \sim \frac{e^{-r/\xi}}{r^{d-2+\eta}} \tag{9.43}$$

When $\xi \to \infty$, these equations show that $g(r)$ remains finite out to large r, and that contributions to $g(r)$ for $r > L$ are certainly not obtained in the finite system L. The role of finite-size scaling theory is to propose scaling functions for these types of divergent properties that preserve their correct scaling behavior in the finite system, even as the critical temperature of the corresponding infinite system is approached.

Finite-Size Scaling of the Specific Heat

We illustrate this issue by focusing on the specific heat C_H in the Ising system which scales in the infinite system according to the scaling law given earlier by

$$C_H \sim |t|^{-\alpha} \tag{9.44}$$

Here $t \equiv T - T_c(\infty)$, where $T_c(\infty)$ is the critical temperature in the infinite system. In the finite system however, C_H will be a function of both L and t. In the case where $\xi(t) > L$, one way [1] to preserve the scaling shown in equation (9.44) is to postulate that

$$C_H(L,t) = |t|^{-\alpha} g\left(\frac{L}{\xi(t)}\right) \tag{9.45}$$

with boundary conditions defined on the *scaling function* $g(L/\xi(t))$ that force the correct scaling behavior of $C_H(L,t)$. We now discuss these boundary conditions and their consequences.

If L is large enough to encompass correlations (i.e., $L \gg \xi(t)$), then the scaling of the specific heat should follow $C_H \sim |t|^{-\alpha}$. This will be true if $g(L/\xi)$ becomes a finite (unknown) constant in this limit in (9.45). At the other extreme when $L \ll \xi(t)$, more apposite to this discussion, we would expect C_H to remain finite and independent of t as $t \to 0$ since the effects of correlations beyond L are not accounted for as $t \to 0$. In order for this to occur, the scaling function $g(L/\xi(t))$ must effectively cancel out the $|t|^{-\alpha}$ term in equation (9.45). This can be accomplished if and only if $g(L/\xi(t)) \to (L/\xi(t))^{\alpha/\nu}$. In this situation, we will have

$$C_H(t,L) = |t|^{-\alpha}\left(\frac{L}{\xi(t)}\right)^{\alpha/\nu} \tag{9.46}$$

and substituting the scaling equation for $\xi(t)$, namely

$$\xi(t) = \xi_0 |t|^{-\nu} \tag{9.47}$$

into equation (9.46) leads to the result

$$C_H(t,L) = \frac{|t|^{-\alpha} L^{\alpha/\nu}}{[\xi_0 |t|^{-\nu}]^{\alpha/\nu}} \tag{9.48}$$

or

$$C_H(t,L) = \left(\frac{L}{\xi_0}\right)^{\alpha/\nu} \tag{9.49}$$

This latter result shows that, in a system of size L,

$$C_H(t,L) \sim L^{\alpha/\nu} \tag{9.50}$$

This *finite-size equation* says that $C_H(t,L)$ becomes finite as $t \to 0$ in the limit where $\xi(t) \geq L$, as required. Most importantly it shows that the value of C_H found near

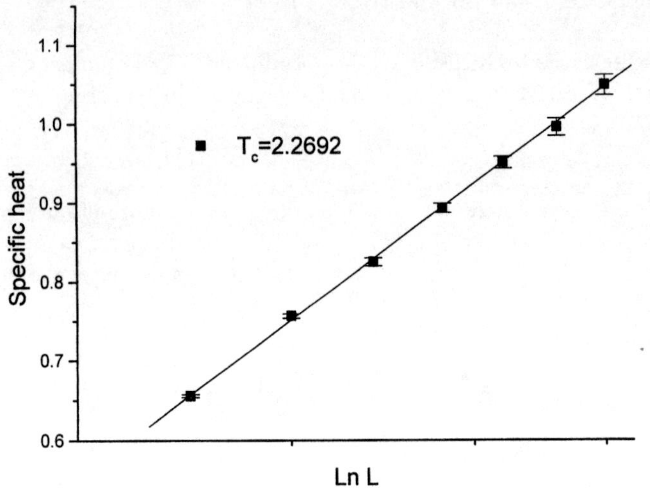

Figure 9.3 Simulation data for the specific-heat scaling with system size L in a 2d Ising system

the infinite-system's critical temperature will scale with the length of the system size to the power α/ν, a result that can, in principle, be checked by computer simulation. Figure 9.3 shows simulation results for various 2-d ($l \times l$) Ising lattices [4]. In this case, though, $\alpha = 0$ (see table 9.1) and $C_H \propto -\ln t$, so one expects the finite-size scaling to be of the form $C_H \propto -\ln L$. The results in figure 9.3 support this conclusion. Using this approach, the validity of the scaling hypothesis, and equation (9.50) in particular, is thus open to investigation through a series of simulations of this sort (see exercise 9.7).

EXERCISE 9.7

How would you calculate $C_H(t, L)$ from simulations in an $L \times L$ Ising lattice; that is, what specific ensemble and fluctuation properties would you calculate? If a plot of $\ln C_H(t, L)$ against $\ln L$ from your calculations did not yield a slope of value α/ν, what conclusions would you draw from this?

Extraction of Critical Properties Using Finite-Size Scaling

One of the major objectives of finite-size scaling is to be able to extract critical properties from computer simulation data. To investigate this, we look at a thermodynamic quantity $A(T, L)$, derivable from the free energy, that obeys a generic finite-size scaling relation of the form

$$A(T, L) \sim L^a \Phi(tL^y) \tag{9.51}$$

From values for $A(T, L)$ at two different system sizes L and L', we form the ratio

$$R_A(T, L, L') \equiv \frac{A(T, L)}{A(T, L')} = \left(\frac{L}{L'}\right)^a \frac{\Phi(tL^y)}{\Phi(t[L']^y)} \tag{9.52}$$

Choose the *length rescaling factor* b such that

$$\frac{L^{**}}{L^{*}} = \frac{L'}{L} = b \tag{9.53}$$

We see then that, if we plot simulation results for the quantities $R_A(T, L, L' = bL)$ and $R_A^*(T, L^*, L^{**} = bL^*)$ with temperature, they can only intersect at $t = 0$. At this point, both $R_A(T, L, L')$ and $R_A^*(T, L^*, L^{**})$ will have a common value equal to b^a. Hence, in principle, one can establish estimates of both T_c and the scaling exponent a from a series of calculations at different temperatures and system sizes with a given rescaling factor b. We now outline such a simulation algorithm for *finite-size scaling*.

The Barber–Selke Finite-Size Scaling Method [5]

1. Stipulate the scaling-function form for the property $A(T, L) \sim L^a \Phi(tL^y)$.
2. Establish the rescaling factor b and system sizes L and L^*.
3. Specify temperature T.
4. Calculate $R_A(T, L, bL)$ and $R_A^*(T, L^*, bL^{**})$—usually from computer simulations.
5. Change T and repeat step 4 until several temperatures have been used, after which proceed to step 6.
6. From the intersection of curves R_A and R_A^*, estimate T_c and the scaling exponent a.

The thermodynamic property $A(T, L)$ is usually found in the simulations from fluctuation formulae. In the case of χ_M in the Ising system, for example, this is given by the result

$$\chi_M = \frac{L^d}{T}(\langle M^2\rangle - \langle M\rangle^2) \tag{9.54}$$

EXERCISE 9.8

Show, with the Barber–Selke finite-size scaling method applied to χ_M, that the scaling exponent found is equal to γ/ν, while applied to C_H it yields α/ν.

This method works well provided that L is large enough to be in the *scaling region* for the property of interest. This scaling regime is the thermodynamic region where the regular (nondiverging) part of the property is considered negligible relative to its diverging counterpart and where the scaling laws are considered to hold. For properties where the singular part of the free energy diverges fast enough with L, this is likely to be straightforward. Thus χ_M is a good candidate for use with this approach while a more weakly divergent property like C_H needs special attention, because the scaling region is nominally harder to define as a result of the weaker divergence of this property [6].

The Special Role of $L^{1/\nu}|t|$: Locating the Scaling Region

Another way to investigate the finite-size scaling hypothesis discussed by Plischke and Bergersen [1] is to recognize that the term $L^{1/\nu}|t|$ plays a special role in this theory.

Since $\xi \sim t^{-\nu}$, and if $L \sim t^{-\nu}$, then $L/\xi = 1$, and we are in the scaling regime where we anticipate that system properties will be sensitive to L. This means that, if a property $X(|t|, L)$ scales as $|t|^y Y(L^{1/\nu}|t|)$, then

$$L^{y/\nu} X(|t|, L) = (L^{1/\nu}|t|)^y Y(L^{1/\nu}|t|) \tag{9.55}$$

Note that this equation preserves the scaling behavior of X, but the functionality on the right-hand side suggests that the property $L^{y/\nu}X$ should be a unique function of the term $L^{1/\nu}|t|$. In log–log coordinates, if we define

$$P \equiv L^{y/\nu} X \tag{9.56}$$

we should find

$$\ln P = y \ln z + \ln Y(z) \tag{9.57}$$

where

$$z \equiv L^{1/\nu}|t| \tag{9.58}$$

Calculations of P at various values of z can be used to check the validity of equations (9.56)–(9.58). In assuming a value for z, there are clearly many combinations of L and t that could and should be used for each z. If the predicted linearity between $\ln P$ and $\ln z$ is seen in the data from these calculations, this presumably shows that the given finite-scaling hypothesis is reasonable. From such results the scaling's critical exponent y can be estimated as well as the quantity $\lim_{z \to 0} Y(z)$.

EXERCISE 9.9

Sketch the data you would expect to find from simulation of a 3-dimensional Ising model using equation (9.55) for investigating C_H in the lattice under the hypothesis of finite-size scaling. How could you use this approach to estimate the scaling region in this system?

Histogram Methods for Determining T_c

Particularly powerful techniques for calculating critical properties using computer simulation, in conjunction with finite-size scaling, are referred to as *histogram methods.* These methods, pioneered by Binder [10] rely upon knowing the probability density function for the order parameter which must usually be estimated by simulation. This function $p_L(m)$, for a system of size L, follows a Gaussian distribution:

$$p_L(m) = \sqrt{\frac{\beta}{2\pi\chi(L, T)}} \exp \frac{-\beta m^2}{2\chi(L, T)} \tag{9.59}$$

where $\chi(L, T)$ is the susceptibility at temperature T in a system of size L. We now seek a finite-size scaling representation for the function $p_L(m)$. We know that

$$\langle m \rangle = \frac{\int m p_L(m) dm}{\int p_L(m) dm} \tag{9.60}$$

and that near the critical point,

$$\langle m\rangle = at^{\beta} \text{ with } a \text{ some constant} \tag{9.61}$$

Since, in the scaling region, $\xi \propto t^{-\nu}$ and $L \propto t^{-\nu}$, we have that $t \propto L^{-1/\nu}$. Therefore

$$\langle m\rangle \propto L^{-\beta/\nu} \quad \text{or} \quad mL^{\beta/\nu} \propto 1 \tag{9.62}$$

Therefore, if in (9.60) we try a form for $p_L(m)$ given by

$$p_L(m) = L^{\beta/\nu}\,\phi(mL^{\beta/\nu}, L/\xi(T)) \tag{9.63}$$

we get

$$\langle m\rangle = \frac{a\int L^{-\beta/\nu}L^{\beta/\nu}\phi\,dm}{\int L^{\beta/\nu}\phi\,dm} \tag{9.64}$$

or

$$\langle m\rangle = -aL^{-\beta/\nu} \tag{9.65}$$

which is the required result. We are now in a position to take various means of m and its higher powers—sometimes referred to as its *cumulants*. For example, if we define $z \equiv mL^{\beta/\nu}$, then

$$\langle m^2\rangle = L^{-2\nu/\beta}\int_{-\infty}^{\infty} z^2\bar{\phi}(z, L/\xi(T))dz \tag{9.66}$$

We can thus evaluate any of the powers of m, and note that, at the critical temperature, where $\xi(T_c) \to \infty$, certain functions of these cumulants become independent of L. For example, the function $R(T, L/\xi(T))$, defined as

$$R(T, L/\xi(T)) \equiv \frac{\langle m^4\rangle}{\langle m^2\rangle^2} \tag{9.67}$$

is independent of L as $T \to T_c$. Therefore, graphs of R versus T calculated for different system sizes in the scaling regime will cross at a temperature equal to T_c. A typical set of results in the 2-d Ising model [4] is shown in figure 9.4.

9.5 Scaling in Homogeneous Bulk Fluids

The scaling described previously was oriented toward Ising-type systems, where the relationship between the field variables and scaling directions are particularly straightforward to discern; this, however, is not the case with fluids, because of the inherent asymmetry in the phase coexistence boundary. In the Ising system, the direction parallel to the coexistence line at the critical point implies that we must approach the critical point along the temperature axis with $H = 0$. This is *not* the case with fluids, and since the direction parallel to the coexistence curve is fundamental to defining scaling exponents such as α and γ, we need to define a new coordinate system to analyze scaling in fluids.

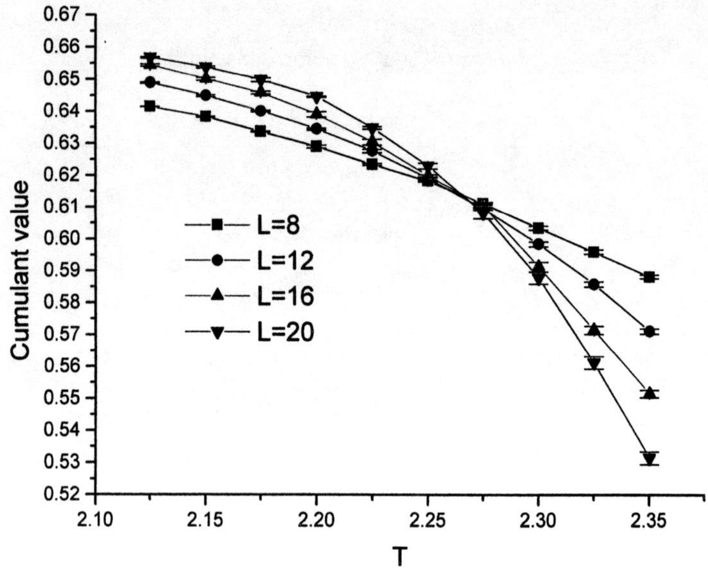

Figure 9.4 Simulation data for cumulants in 2-d Ising systems at various systems sizes L

In this case, we define new scaling fields (coordinates) to align themselves parallel (and orthogonal) to the *linear extension of the phase coexistence line* at the critical point. These new fields called g_t *and* g_ψ are linear combinations of the original field variables. One set of definitions for them is given by

$$g_t = a_{11}t + a_{12}\psi \tag{9.68}$$

$$g_\psi = a_{21}t + a_{22}\psi \tag{9.69}$$

with

$$t \equiv \frac{T - T_c}{T_c} \tag{9.70}$$

and

$$\psi \equiv \frac{\mu - \mu_c}{\mu_c} \tag{9.71}$$

In these equations, $a_{11}, a_{12}, a_{21}, a_{22}$ are constants evaluated at the critical point. The fields g_t and g_ψ are combinations of both t and ψ and the process of defining them is logically referred to as *field mixing*. We also assume that the singular part of the grand free energy Ω_s formally scales with these scaling fields in identical fashion to that used in the Ising case. Hence, the scaling form for Ω_s is given by

$$\Omega_s(g_t, g_\psi) = |g_t|^{2-\alpha}\ \phi_s(g_\psi/g_t^{\Delta}) \tag{9.72}$$

EXERCISE 9.10

What is the value of the exponent Δ in the scaling equation for Ω_s shown in (9.72)?

It is easy to show with these mixed fields that

$$\left(\frac{\partial \Omega_s}{\partial \psi}\right)_t = a_{12}\left(\frac{\partial \Omega_s}{\partial g_t}\right)_{g_\psi} + a_{22}\left(\frac{\partial \Omega_s}{\partial g_\psi}\right)_{g_t} \tag{9.73}$$

$$\left(\frac{\partial^2 \Omega_s}{\partial g_t \partial \psi}\right)_t = \left(\frac{\partial^2 \Omega_s}{\partial g_t^2}\right)\left(\frac{\partial g_t}{\partial \psi}\right)_t + \left(\frac{\partial^2 \Omega_s}{\partial g_\psi \partial g_t}\right)\left(\frac{\partial g_\psi}{\partial \psi}\right)_t \tag{9.74}$$

$$\left(\frac{\partial^2 \Omega_s}{\partial g_\psi \partial \psi}\right)_t = \left(\frac{\partial^2 \Omega_s}{\partial g_\psi \partial g_t}\right)\left(\frac{\partial g_t}{\partial \psi}\right) + \left(\frac{\partial^2 \Omega_s}{\partial g_\psi^2}\right)\left(\frac{\partial g_\psi}{\partial \psi}\right)_t \tag{9.75}$$

EXERCISE 9.11

Prove equations (9.73)–(9.75).

Given equations (9.72)–(9.75), it is also straightforward to show that

$$\begin{aligned}\left(\frac{\partial^2 \Omega_s}{\partial \psi^2}\right)_t = {} & a_{12}\left[a_{12}\left(\frac{\partial^2 \Omega_s}{\partial g_t^2}\right) + \left(\frac{\partial^2 \Omega_s}{\partial g_\psi \partial g_t}\right)a_{22}\right] \\ & + a_{22}\left[\left(\frac{\partial^2 \Omega_s}{\partial g_\psi \partial g_t}\right) + \left(\frac{\partial^2 \Omega_s}{\partial g_\psi^2}\right)a_{22}\right]\end{aligned} \tag{9.76}$$

We now invoke an important property of the function $\Omega_s(g_t, g_\psi)$ based upon the results developed in chapter 3 which imply that $\Omega_s(g_t, g_\psi)$ has *a diagonal Hessian matrix*. From this it follows that

$$\left(\frac{\partial^2 \Omega_s}{\partial \psi^2}\right)_t = a_{12}^2\left(\frac{\partial^2 \Omega_s}{\partial g_t^2}\right)_{g_\psi} + a_{22}^2\left(\frac{\partial^2 \Omega_s}{\partial g_\psi^2}\right)_{g_t} \tag{9.77}$$

Also, from a thermodynamic identity with the grand potential Ω_s, it follows that

$$\left(\frac{\partial^2 \Omega_s}{\partial \psi^2}\right)_t = \mu_c^2\left(\frac{\partial^2 \Omega_s}{\partial \mu^2}\right)_t = -V\mu_c\rho^2 K_T \tag{9.78}$$

We now are in a position to pursue a scaling analysis in fluids. First we define

$$z \equiv \frac{g_\psi}{g_t^\Delta} \tag{9.79}$$

From the scaling equation (9.72) and the use of basic calculus, we find that

$$\left(\frac{\partial \Omega_s}{\partial g_t}\right)_{g_\psi} = (2-\alpha)g_t^{1-\alpha}\phi_s(z) - \Delta g_t^{\beta-1}\left(\frac{\partial \phi_s}{\partial z}\right)g_\psi \tag{9.80}$$

where we have used the exponent relationships $\Delta = \gamma + \beta$ and $\alpha + \beta + \gamma = 2 - \beta$ to get (9.80).

The second derivative yields the result,

$$\left(\frac{\partial^2 \Omega_s}{\partial g_t^2}\right)_{g_\psi} = (2-\alpha)(1-\alpha)g_t^{-\alpha}\phi_s(z) + (2-\alpha)g_t^{1-\alpha}\frac{\partial^2 \phi_s}{\partial z^2}(-\Delta)g_\psi \frac{1}{g_t^{1+\Delta}} + (\beta-1)g_t^{\beta-2}\frac{\partial \phi_s}{\partial z}(-\Delta)g_\psi + \Delta^2 g_\psi^2 g_t^{\beta-1}\frac{1}{g_t^{1+\Delta}} \tag{9.81}$$

It is also easy to show that

$$\left(\frac{\partial^2 \Omega_s}{\partial g_\psi^2}\right)_{g_t} = g_t^{-\gamma}\left(\frac{\partial^2 \phi_s}{\partial z^2}\right) \tag{9.82}$$

Note the divergences of order α and γ in g_t that emerge from (9.81) and (9.82), respectively. Substituting these equations into (9.77) and (9.78) leads to the following scaling equation for K_T along the coexistence line (which we emphasize is now not coincident with the temperature axis):

$$K_T = \Gamma_0 t^{-\alpha} + \Gamma_1 t^{-\gamma} \tag{9.83}$$

where Γ_0 and Γ_1 are constants evaluated at the critical point. Equation (9.83) is an important (and often not recognized) result in scaling theory. It shows that, in the fluid case, because of the inherent asymmetry in the phase boundary, field mixing leads to *both* α-type and γ-type divergences for the compressibility along the coexistence direction. Obviously, for the Ising universality class, since $\gamma > \alpha$, the γ-type divergence predominates as $t \to 0$. This should be compared to the Ising model, where the absence of field mixing leads to only a γ-type of divergence in the susceptibility in this direction.

EXERCISE 9.12

Prove (9.83) and provide the values for the constants Γ_0 and Γ_1. What significance do the functions $\phi_s(z)$, $\partial\phi_s/\partial z$, $\partial^2\phi_s/\partial z^2$, evaluated at the critical point, have in arriving at the result given in (9.83)?

In deriving (9.83), finite-size effects are not considered. Hence, that result is referred to as a bulk-scaling equation. In the next section we investigate how finite-size effects are incorporated into this analysis.

Finite-Size Scaling Analysis in the Neighborhood of the Bulk-Fluid Critical Point

In the situation where the correlation length becomes equal to or larger than the system dimension L, we need to develop a scaling analysis in the fluid that incorporates finite-size effects. From our previous bulk-scaling results, we can define the following susceptibilities:

$$\chi_{g_\psi g_\psi} \equiv \left(\frac{\partial^2 \Omega_s}{\partial g_\psi^2}\right)_{g_t} \tag{9.84}$$

$$\chi_{g_t g_t} \equiv \left(\frac{\partial^2 \Omega_s}{\partial g_t^2}\right)_{g_\psi} \tag{9.85}$$

and we see from (9.77) that

$$K_T = \Gamma_0 \chi_{g_t g_t} + \Gamma_1 \chi_{g_\psi g_\psi} \tag{9.86}$$

The key to a finite-size scaling analysis here is to once again assume that the universal part of the scaling function in (9.72) is taken to be a function of $L/\xi(g_t)$. Since $\xi \approx g_t^{-\nu}$, we have that ϕ_s is a function of $g_t L^{1/\nu}$. In other words, we rewrite our scaling equation for K_T as

$$K_T = \Phi_1 g_t^{-\alpha} f_1(g_t L^{1/\nu}) + \Phi_2 g_t^{-\gamma} f_2(g_t L^{1/\nu}) \tag{9.87}$$

where Φ_1 and Φ_2 are constants evaluated at the critical point. This immediately leads to the following finite-size scaling equation for K_T:

$$K_T = \Phi_1 L^{\alpha/\nu} f_1(g_t L^{1/\nu}) + \Phi_2 L^{\gamma/\nu} f_2(g_t L^{1/\nu}) \tag{9.88}$$

However, very close to the critical point, we expect the γ divergence in (9.87) to dominate the α divergence, in which case,

$$K_T \approx \Phi_2 L^{\gamma/\nu} f_2(g_t L^{1/\nu}) \tag{9.89}$$

If this equation holds true, then data from finite-size simulations showing plots of $\ln K_T$ against $\ln L$ should have a slope equal to γ/ν in the scaling region. Furthermore, at the critical point itself, $\Phi_1 f_1(g_t L^{1/\nu})$ and $\Phi_2 f_2(g_t L^{1/\nu})$ should be constants which we denote by Φ_1^* and Φ_2^*, respectively. We can solve for these constants through the following matrix equation representation of (9.89) at two different values L_i and L_j, respectively:

$$\begin{bmatrix} L_i^{\alpha/\nu} & L_i^{\gamma/\nu} \\ L_j^{\alpha/\nu} & L_j^{\gamma/\nu} \end{bmatrix} \begin{bmatrix} \Phi_1 \\ \Phi_2 \end{bmatrix} = \begin{bmatrix} K_{Ti} \\ K_{Tj} \end{bmatrix} \tag{9.90}$$

The notation K_{Ti} means the compressibility evaluated at L_i. Only at the critical point will the solutions for Φ_1 and Φ_2 from (9.90) yield values for these two variables independent of the system sizes L_i and L_j used; these values for Φ_1 and Φ_2 correspond to Φ_1^* and Φ_2^*, respectively. This result can therefore, in principle, be used to calculate the system's critical temperature through a series of iterative calculations. (see exercise 9.13).

EXERCISE 9.13

Use (9.90) to conceptually construct a computational method for finding the critical point of the system, assuming that GCMC simulations can be done to provide values for the K_{Ti} from fluctuation calculations in the grand canonical ensemble. Sketch a graph showing how your method would provide an estimate of the system's critical temperature.

EXERCISE 9.14

The scaling analysis leading to (9.88) was done for a direction parallel to the coexistence curve. Do the analysis in the perpendicular direction (i.e., $g_t = 0$, g_ψ finite) and comment on which direction of approach would be most useful in finding the system's critical temperature from simulation results.

Thus, as stated earlier, we see that critical scaling in asymmetric systems involves subtle distinctions from the symmetric Ising case. The method followed here could be extended to binary and higher-order fluid mixtures. We have not done so for reasons of economy; the reader choosing to go down this path should be prepared for an algebraic marathon.

9.6 Computer Simulation Methods

The exploitation of finite-size scaling results described in this chapter requires the intensive application of computer simulation methods. These have become indispensable tools for modern research in statistical mechanics, and there are two main approaches used in this area that are now described in many texts on the subject, for example, [7]. The first is *molecular dynamics* (MD) where one solves for the *classical trajectories* of the particles in the system with dynamics governed by *Newton's equations of motion*. MD has the potential for providing both *time-dependent* and equilibrium properties for the system. The second class of techniques is called the *Monte Carlo* (MC) method and is most commonly used to facilitate the calculation of statistical-thermodynamic equilibrium properties. Here we focus upon this latter method illustrating some of the fundamental concepts that underpin this method. In chapter 12, we explicitly present simulation results with this approach in the critical region of a fluid.

A Monte Carlo trajectory is a sequence of configurations for a system where the probability distribution of these configurations is chosen to match that of a known distribution available from fundamental results in statistical mechanics. The purpose of the Monte Carlo method is to effectively sample such states and find averages of properties of interest from these simulation "data". Since the number of states in a real system is extremely large, the efficiency of the Monte Carlo approach depends upon its ability to generate "reasonable" averages, even though it may sample only a small subset of allowable states. In other words, we must often find a way to bias the sampling procedure so that we investigate the most important regions of state space. We now review in a qualitative way some of the main conceptual issues to be confronted in pursuit of these objectives.

Conceptual Outline of the Monte Carlo Sampling Process

1. Decide upon the size of the system and how to characterize its states.
2. Initialize the system configuration.
3. Randomly choose a molecule (subunit) in the system and change its configuration (often referred to as the *move*).
4. Decide whether or not to accept this sample point for use in evaluating the property of interest. It is at this point that we apply statistical tests to the datum to ensure that the state is consistent with the prescribed thermodynamic distribution. If so, evaluate the property of interest in the system at the new configuration. If not, return to step 3.
5. The processes given in steps 3–4 are repeated until some endpoint condition is reached. Based upon these data, averages and error estimates of the quantities of interest are calculated.

Each of the steps described above constitutes an important part of the simulation process, and we refer the reader to many of the texts on the topic for details on each. Here we describe elementary simulation practice together with some practical recommendations that should be of value to newcomers to the area. We attempt to answer some simple but nagging questions, often not explicitly addressed in specialized texts.

Metropolis Monte Carlo

The *Metropolis* algorithm is a celebrated result in the field, first published in 1953 by Metropolis and collaborators who were working at Los Alamos Laboratories [8]. The rationale behind this method can be understood by first looking at the meaning of a probability distribution in the canonical ensemble. In this ensemble, the average value of the system's *configurational energy* $\bar{U}$ can be written as

$$\bar{U} \equiv \frac{\int \cdots \int e^{-\beta U_N} U_N d\boldsymbol{r}_1 \cdots d\boldsymbol{r}_N}{Z_N} \tag{9.91}$$

where Z_N is the *configurational integral*, defined as

$$Z_N \equiv \int \cdots \int e^{-\beta U_N} d\boldsymbol{r}_1 \cdots d\boldsymbol{r}_N \tag{9.92}$$

The molecules are thus free to assume configurations in space with the energy of each configuration weighted by the Boltzmann factor $e^{-\beta U_N}/Z_N$. These weighted configurations are integrated over all positions of the molecules represented by $\boldsymbol{r}_1, \ldots, \boldsymbol{r}_N$.

By analogy with the usual idea of statistical averages, we see that this amounts to finding the average of a property M according to the following formula:

$$\bar{M} = \sum_j P_j M_j \tag{9.93}$$

The index j runs through all configurations sampled, with P_j the probability of configuration j and M_j the value of M in the jth configuration. Evidently, the integral in (9.92) is far too complex for complete enumeration (the integral's coordinate dimensionality is $3N$ with $N \approx 10^{23}$), and the purpose of the Monte Carlo method is to circumvent this problem by sampling only a small fraction of possible configurations using small numbers of molecules. The obvious question arises: how do we know that averages based upon this small sample of state space are meaningful? Furthermore, how does one proceed to sample the configurations that contribute most importantly to the average in equation (9.93). The answers here are intimately connected with the Boltzmann distribution law from statistical mechanics; this result provides that configurations of lower potential energy (negative) are more probable than higher ones. For example, if we denote P_1 and P_2 as the probabilities of the system being in states 1 and 2, respectively, then:

$$\frac{P_2}{P_1} = \exp(-\beta\ \Delta E_{12}) \tag{9.94}$$

where

$$\Delta E_{12} \equiv E_2 - E_1 \tag{9.95}$$

Thus a configuration resulting in a lower energy is always accepted. What should be done, however, if $\Delta E_{12} > 0$? Metropolis et al. [8] proposed the following rules to decide whether or not to accept a move from one configuration to the next in the simulation process. These are

If $\Delta E_{12} < 0$, accept the move
If $\Delta E_{12} \geq 0$, accept move with exponential probability $\exp(-\beta\ \Delta E_{12})$

The latter test ensures consistency with the Boltzmann distribution, and a central question concerns the practical implementation of this rule. How do we sample a random variable so that it obeys a particular statistical distribution—in this case, the exponential distribution? To rigorously answer this question, we first need to look at some basic results in probability theory [9].

Statistical Sampling

For a continuous random variable x, we define a probability density function as the non-negative function f, defined for all real values of $x \in (-\infty, \infty)$, so that the probability P of $a \leq x \leq b$ is given by

$$P\{a \leq x \leq b\} \equiv \int_a^b f(x)dx \tag{9.96}$$

From this definition, we see that

$$\begin{aligned} P\{x \leq a\} &= \int_{-\infty}^a f(x)dx \\ &\equiv F(a) \end{aligned} \tag{9.97}$$

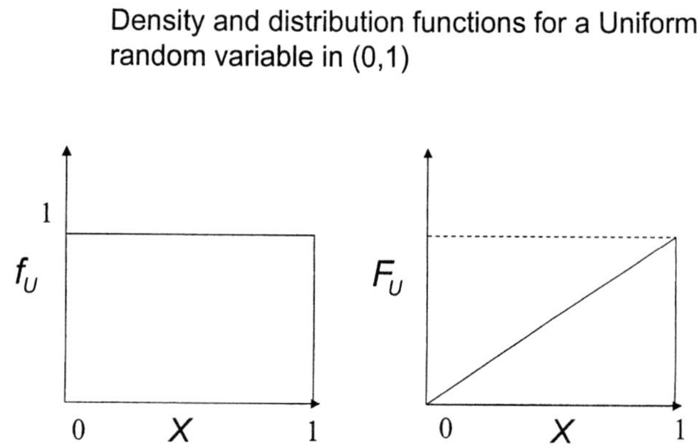

Figure 9.5 Schematic of the probability density and distribution functions for the uniform random variable U

where F is termed the *cumulative distribution function* (*cdf*) with respect to $f(x)$. This distribution function has the following properties:

- *Monotonicity*: if $a \leq b$, then $F(a) \leq F(b)$
- $\lim_{b\to\infty} F(b) = 1$
- $\lim_{b\to-\infty} F(b) = 0$

The Uniform Distribution and its Role in Statistical Sampling of Random Variables

The uniform distribution assumes a special role in the sampling of random variables with prespecified cdf functions. The cdf of a (standard) *uniformly distributed* random variable U is given by

$$F_U(x) = P(U \leq x) = \begin{cases} 0 & \text{if} \quad x < 0, \\ x & \text{if} \quad 0 \leq x < 1, \\ 1 & \text{if} \quad x \geq 1 \end{cases} \tag{9.98}$$

In figure 9.5, we show graphs of both $f_U(x)$ and $F_U(x)$ for the uniform random variable over the interval (0, 1).

To show how the uniform distribution is used to sample other random variables with prespecified cdfs, we now provide simply constructed proofs of some well-known propositions in probability theory.

PROPOSITION 1: Finding a random variable with prespecified cdf from U

Consider that U is a uniform random variable defined on the interval (0,1). If we define another random variable X with a *strictly* monotonic cdf F_X (i.e., $x < y$ implies that $F_X(x) < F_X(y)$) by the equation $X \equiv F_X^{-1}(U)$, then X will be distributed with the cdf F_X.

Proof

Note that the strict monotonicity of F_X ensures the existence of F_X^{-1}. The proof is straightforward and relies upon the monotonicity of F_X.

We have that

$$\begin{aligned} P(X \leq x) &= P(F_X^{-1}(U) \leq x) \\ &= P(U \leq F_X(x)) \\ &= F_X(x) \end{aligned} \tag{9.99}$$

The second step above follows from the monotonicity of F_X and the last statement follows from the property of the uniform distribution. Equation (9.99) shows that the random variable X has the cdf F_X as required.

Exactly what does the function $X \equiv F_X^{-1}(U)$ mean? Mathematically it it is defined by finding the value X such that $F_X(X) = U$. This yields values for X that satisfy the prescribed properties. Say we want to sample a random variable X with a density function given by $p_X(x)$ and cdf $F_X(x)$, then the *realization of this procedure* is to find the value of γ in the following integral equation:

$$F_X(\gamma) = \int_{-\infty}^{\gamma} p_X(x)dx = U \tag{9.100}$$

where U in (9.100) is drawn from the uniform distribution. This is repeated many times and the values of γ found have the probability density $p_X(X)$ and cdf $F_X(X)$. This is a key concept that underlies many ideas in computer simulation of stochastic systems.

Ensuring the Boltzmann Distribution with Metropolis Monte Carlo

How do we ensure that the Boltzmann sampling process is guaranteed by the Metropolis algorithm described above? If the move results in a decrease in ΔE_{12}, then it is evident that it should be accepted. However, what do we do if $\Delta E_{12} > 0$? The recommendation is that we compare the value of $\exp(-\beta \Delta E_{12})$ to the value of a uniform random variable sampled over the interval (0, 1) and *accept the move* if $\exp(-\beta\ \Delta E_{12}) > U$ while rejecting it if the converse is true.

We now show why this criterion ensures that the states accepted this way will follow an exponential distribution.

PROPOSITION 2: Ensuring the Boltzmann distribution in Metropolis Monte Carlo sampling

Consider a random variable $Z \in [0, Z_{max}]$ chosen with a uniform probability density function. Assume a test is applied to the Z and a new set of variables $\tilde{Z}$ constructed consisting of the set of Z that satisfy the test condition $e^{-z} \geq U$ (where U is uniform on [0, 1]). We will show that the set $\tilde{Z}$ is exponentially distributed.

Proof

The proof is similar to that for the rejection statistical simulation method described in [9]. We are interested in finding the form of the probability density function $f_{\tilde{Z}}$ in the following cdf:

$$P\{\tilde{Z} \leq a\} = \int_0^a f_{\tilde{Z}}\, d\tilde{Z} \tag{9.101}$$

This cdf can be expressed using the following conditional probability expression:

$$P\{\tilde{Z} \leq a\} = P\{Z \leq a \text{ and } U \leq e^{-z}\}/P\{U \leq e^{-z}\} \tag{9.102}$$

This conditional probability can, furthermore, be expressed as follows:

$$P\{\tilde{Z} \leq a\} = \left(\int_0^a \int_0^{e^{-z}} f_{ZU}(Z,U)dU\, dZ\right) \Big/ P\{U \leq e^{-z}\} \tag{9.103}$$

where f_{ZU} is the joint probability density function of the random variables Z and U, respectively. It follows that, since Z and U are independent (why?), we can write

$$f_{ZU}(Z,U) = f_Z(Z).f_U(U) = 1/Z_{\max} \tag{9.104}$$

with

$$f_z = \frac{1}{Z_{\max}} \tag{9.105}$$

by assumption. It follows, that

$$P\{Z \leq a \text{ and } U \leq e^{-z}\}$$
$$= \frac{1}{Z_{\max}} \int_0^a \int_0^{e^{-z}} dU\, dZ \tag{9.106}$$

Equations (9.102)–(9.106) imply that

$$P\{\tilde{Z} \leq a\} = \frac{1}{Z_{\max}\Omega} \int_0^a \int_0^{e^{-z}} dU\, dZ \tag{9.107}$$

where we have defined

$$\Omega \equiv P\{U \leq e^{-z}\} \tag{9.108}$$

Doing the first integral in equation (9.107), we get that

$$P\{\tilde{Z} \leq a\} = \frac{1}{Z_{\max}\Omega} \int_0^a e^{-Z}\, dZ \tag{9.109}$$

or

$$P\{\tilde{Z} \leq a\} = \frac{1}{Z_{\max}\Omega}(1 - e^{-a}) \tag{9.110}$$

Furthermore, from a basic statistical identity,

$$\left(\frac{\delta P\{\tilde{Z} \le a\}}{\delta a}\right) = f_{\tilde{Z}} = \frac{e^{-a}}{Z_{\max}\Omega} \tag{9.111}$$

Equation (9.111) implies that $\tilde{Z}$ is exponentially distributed, which completes the proof. We can normalize its probability density by defining a new density function $g_{\tilde{Z}}$:

$$g_{\tilde{Z}} \equiv Z_{\max}\Omega \cdot f_{\tilde{Z}} \tag{9.112}$$

EXERCISE 9.15

Prove that the random variable Y with probability density function $P_Y(y) = \alpha \cdot e^{-\beta y}$ has mean $\langle Y \rangle = \alpha/\beta$. What is its variance? Prove that Y satisfies the following conditional probability:

$$P\{Y > a + b | Y \ge b\} = P\{Y > a\}$$

What does this latter result tell you about the energy states sampled in Boltzmann sampling?

Determining a Sufficiently Large Sample Size for the Simulations

One of the most basic questions to be answered with simulations is how to determine whether the number of data points used to find property averages is sufficient to get statistically meaningful results. These averages are usually obtained from data points that themselves are averages calculated within a given block of simulation results. If the block sample size is too small, then the overall average will be poor since each block average will not be *statistically independent*; conversely, if the block size is too large, then many of the calculations will be redundant. We now consider this important issue in more detail with the use of the following results from probability theory.

PROBABILITY PROPOSITION: the variance of sums of random variables

If $X_1, \ldots, X_n$ are all pairwise independent random variables that belong to the same probability density function, then

$$\operatorname{var}\sum_{i=1}^{n} X_i = \sum_{i=1}^{n} \operatorname{var} X_i \tag{9.113}$$

and

$$\operatorname{var}(aX_i + b) = a^2 \operatorname{var} X_i \tag{9.114}$$

where a and b are constants.

Consider that, within a simulation block, there are T data points that are averaged to yield a single "average" point for that block, which we denote as x_i. If N such blocks are used, and all of the x_i obey the same probability distribution and are statistically independent, then the overall average of the property of interest y is found simply from the equation:

$$y = \frac{\sum_{i=1}^{N} x_i}{N} \tag{9.115}$$

with the *standard deviation* of y, defined as σ_y, related to that for each of the x_i by the following equation (which comes form the use of equations (9.113) and (9.114)):

$$\sigma_y = \sigma_x/\sqrt{N} \text{ (block size } T) \tag{9.116}$$

If we double the original block size so that it incorporates twice the number of data points (i.e., $2T$), and if we define a new block average z_i by the equation

$$z_i \equiv (x_i + x_{i+1})/2 \tag{9.117}$$

since

$$2z_i = x_i + x_{i+1} \tag{9.118}$$

we have

$$\sigma_{z_i} = \sigma_x/\sqrt{2} \tag{9.119}$$

and

$$y = \frac{\sum_{i=1}^{N/2} z_i}{N/2} \tag{9.120}$$

From these equations (statistical independence assumed) it is easy to show, using equations (9.113) and (9.114), that

$$\sigma_y = \sigma_x/\sqrt{N} \text{ (block size } 2T) \tag{9.121}$$

Equation (9.121) demonstrates that, if the original ensemble size of T measurements is large enough to generate independent sets of averages (i.e., the original x_i), then doubling the sample size does not enhance the accuracy of error estimates for the desired quantity y. This is shown by the equivalence between equations (9.116) and (9.121) for the two different block sizes. In this case, T is a large enough block sample size—an important issue in ensuring the integrity of the simulation results.

Confidence Limits: Their Use for Bracketing Averages Found from Simulation Data

A different sort of question from the one answered above asks how to determine the interval in which the probability of finding the true sample mean is located—given a set of data from a stochastic process. Say we have a sample of a random variable, say

$\{x_1, \ldots, x_n\}$ with mean value $\bar{x}$ and known standard deviation (of each measurement) σ_x. We ask the question: if the *true sample mean* is μ, what is the interval around $\bar{x}$ in which we can be, for example, 95% sure that we will find the true sample mean? This interval is designated the 95% *confidence limit.*

We analyze this question using normal random variables, although the arguments carry over to other types of random variable. Define the standard normalized variable z as [9]

$$z \equiv \frac{\bar{x} - \mu}{\sigma_x / \sqrt{n}} \tag{9.122}$$

For this random variable, the equation

$$P\{z \geq z_\alpha\} = \alpha \tag{9.123}$$

can be used to find the value for z_α, so that the probability of finding a larger value for z is equal to some prespecified value α. We can find this particular value for z_α from the cdf of the standard normal variable, which is tabulated in most elementary probability texts. From these previous two equations, it follows that

$$P\left\{\frac{\sqrt{n}}{\sigma_x}|\bar{x} - \mu| \leq z_{\alpha/2}\right\} = 1 - \alpha \tag{9.124}$$

EXAMPLE 9.16

Prove equation (9.124).

with the equivalent result being

$$P\left\{\bar{x} - z_{\alpha/2} \cdot \frac{\sigma_x}{\sqrt{n}} \leq \mu \leq \bar{x} + z_{\alpha/2} \cdot \frac{\sigma_x}{\sqrt{n}}\right\} = 1 - \alpha \tag{9.125}$$

In other words, in equation (9.125), we have a description of the interval around $\bar{x}$ in which we are $(1 - \alpha) \cdot 100\%$ sure the true mean lies. This is the so-called "$(1 - \alpha) \cdot 100\%$ confidence limit." If the variance of each measurement is unknown, then the analysis is similar but the confidence limit is now given by the interval

$$\bar{x} \pm t_{n-1,\alpha/2} \cdot \frac{S}{\sqrt{n}} \tag{9.126}$$

where S is the sample standard deviation and $t_{n,\alpha}$ the t random variable tabulated, once again, in most statistics textbooks. These tests are useful in determining how large to make the sample size so that the specified confidence interval falls below a certain size.

Simulation Test for Thermal Equilibration in Monte Carlo Sampling

We finish our discussion on statistical issues in simulation by focusing upon how we might ensure that the states sampled are indeed representative of equilibrium states. While this is generally difficult to do we present one method that has been proposed for doing this. It is termed the *replica method* and described here in the context of the Ising model [4]. It is generally applicable, however, to any physical system.

At given conditions, we take two replicas of the system being studied, at identical thermodynamic conditions, and start the Monte Carlo simulations from two different initial random spin configurations. For each system, we follow the progress of the Metropolis calculations through a large number of Monte Carlo steps, keeping track of all the lattice spin configurations. We define S_{1i} as the value of the spin in the ith position of replica 1, S_{2i} the similar property in replica 2, N the number of spins, t the beginning point from which the equilibration test is used, and τ the estimated differential in the number of steps sufficient to ensure equilibration. We then calculate the following correlation quantities:

$$a_1 = \frac{1}{N}\left\langle \sum_i S_{1i}(t)S_{1i}(t+\tau) \right\rangle \tag{9.127}$$

$$a_2 = \frac{1}{N}\left\langle \sum_i S_{2i}(t)S_{2i}(t+\tau) \right\rangle \tag{9.128}$$

$$a_c = \frac{1}{N}\left\langle \sum_i S_{1i}(t)S_{2i}(t) \right\rangle \tag{9.129}$$

The question we want to answer is whether τ is large enough to ensure equilibrium sampling. If the quantities a_1, a_2, and a_c are equivalent, within error-bar estimates, then we assume that the value of τ used is large enough to ensure thermal equilibration. This conclusion follows from the fact that, in this situation, both replicas—although started from entirely independent configurations—must be tracking through the same regions of phase space: a result we take to mean that equilibrium sampling is occurring. If, for example, $a_1 \neq a_2$ for a given τ, it would imply that both replicas are mapping out different areas of phase space, separated by free-energy barriers that have yet to be surmounted.

Thus, we have described a number of basic issues in using computer techniques for Monte Carlo simulations: (1) sampling according to a required statistical distribution, (2) deciding when a sufficient number of data points have been taken to assure statistical accuracy in the results, and (3) ascertaining whether or not the system is thermally equilibrated. These issues cover some of the basic requirements in any set of equilibrium simulations, although getting them to work satisfactorily takes persistence and attention to detail.

9.7 Chapter Review

In this chapter, we discussed scaling ideas and their integral relationship to critical phenomena in the Ising universality class which includes simple fluids with short-range intermolecular potentials. The fundamental idea behind scaling is that certain divergent thermodynamic properties approach the critical point with a singularity described by specific power laws and characteristic critical exponents. The origin of this property divergence arises from the singular part of the system's free energy at the critical point, a manifestation of the fact that the free energy is not analytic there.

There is extensive experimental evidence showing that these scaling exponents appear to be the same for many superficially different systems (like simple fluids and Ising magnets). Members of the Ising universality class have scalar order parameters and short-range interactions between subunits that constitute the system. This similarity in exponents gives rise to the notion of universality in the critical behavior of systems belonging to a given universality class. This universality can be expressed mathematically by formulating the free energy as a general homogeneous function of the thermodynamic field variables pertinent to the system at hand. This approach has led to well-known phenomenological scaling theories like those proposed by Kadanoff and Widom in which the critical exponents naturally appear. These theories also introduce the idea of the correlation length—a key concept in modern theories of critical phenomena.

Since the correlation length diverges as the critical point is approached, it is difficult, if not impossible, to calculate divergent critical properties by computer simulation using finite-size systems. This leads to the idea of finite-size scaling theories which describe the scaling behavior of divergent properties in terms of the finite length scale of the system considered. These approaches often lead to elegant computer simulation methods for determining important critical properties, like the critical temperature, by exploiting the size dependence of divergent properties in the critical region. These ideas were discussed in the context of both Ising and fluid systems. In fluids, field mixing was shown to lead to critical scaling behavior that is considerably more complex than in the Ising case, because of the inherent asymmetry of the coexistence boundary in fluids. We developed detailed results showing how field mixing can be accounted for using a pure-fluid system as a paradigm.

Finally, we discussed basic ideas related to computer simulation in these systems describing the fundamental rationale behind the well-known Monte Carlo technique. This method plays a central role in computer simulation research in critical systems.

9.8 Additional Exercises

Finite-Size Scaling Analysis

1. Consider $G(x, y) = x^2 + y^2$, which we express as a homogeneous function as follows:

$$G(x, y) = \phi G(\phi^p x, \phi^q y)$$

If this relationship is true for all values of ϕ, then show that this must imply that $p = q = -1/2$.

2. Suppose that the free energy is taken to scale as

$$G(t, h) = \phi G(\phi^p t, \phi^q h)$$

where ϕ, p, q are real numbers. Show that

$$\gamma = \frac{2q+1}{p}, \qquad \beta = -\left(\frac{q+1}{p}\right), \qquad \alpha = \frac{2p+1}{p}, \qquad \delta = -\left(\frac{q}{q+1}\right)$$

[Hint: Look at scaling for the properties $\chi(t, 0)$, $M(0, h)$, $M(t, 0)$, and $C_H(t, 0)$.]

3. If the specific heat in a 2-d Ising system scales logarithmically with temperature, i.e., $C_H \propto \ln t$, show that this implies the finite-size scaling relationship $C_H \propto 1/v \ln L$.
4. Show that, if the scaling of the free energy is taken to be the same as in question 2, then the lattice order parameter can be expressed as

$$M(t, h) = |t|^{\beta} M\left(+1, \frac{\mathrm{h}}{|t|^{\beta+\gamma}}\right)$$

5. Prove that the susceptibility χ_{M} in the Ising system can be computed from the fluctuation formula

$$\chi_{\mathrm{M}} \equiv \frac{L^d}{T}(\langle M^2\rangle - \langle M\rangle^2)$$

6. For any finite-size system L, the Ising order parameter vanishes, namely $\langle M\rangle = 0$. If the singular part of the free energy scales as

$$G_{\mathrm{s}}(t, h, L) = L^{-d} G_{\mathrm{s}}(tL^{y}, hL^{x}, 1)$$

show that this results in a scaling law for the order parameter squared, at zero field, as follows:

$$M^2 = L^{2(x-d)} T G_{hh}(tL^{y}, 0)$$

Expand the scaling function $G_{hh}(tL^{y})$around T_{c}(for small $tL^{1/v}$) and show that

$$M^2(T, L) = L^{-2\beta/v}\left[\phi_0 + \phi_1 L^{1/v}(T - T_{\mathrm{c}}) + \cdots\right]$$

to first order in $L^{1/v}$. What are the definitions of ϕ_0 and ϕ_1? Describe how you could use this approach to estimate the reciprocal $1/v$ of the correlation length exponent in this system.

7. Using the Barber–Selke finite-size scaling method discussed in this chapter, show how you would develop a strategy to calculate the critical exponent ratio β/v in a 2-d Lennard–Jones fluid.
8. How does one calculate C_{M} and C_H from fluctuation formula in the lattice model using computer simulation?
9. In a lattice of finite size L, where $T_{\mathrm{c}}(\infty)$ represents the critical temperature of the infinite system, prove that

$$T_{\mathrm{c}}(L) = T_{\mathrm{c}}(\infty) + bL^{-1/v}$$

where b is some constant.

10. Derive the equations showing how you would expect the susceptibility χ_{M} to scale with L in any finite-size scaling calculation for the Ising system at the critical point?

11. How would you use finite-size scaling results to estimate the critical exponent β for the Ising system? Do you have to know the critical temperature for these calculations?
12. Show that the heat-bath simulation algorithm which gives the acceptance probability of a Monte Carlo move as

$$P_{1\to 2} = \frac{1}{1 + e^{\beta(E_2 - E_1)}}$$

obeys the principle of detailed balance, that is,

$$\frac{P_{1\to 2}}{P_{2\to 1}} = e^{\beta(E_1 - E_2)}$$

Show, using perturbation ideas like those discussed in this chapter, that the difference in free energy of an Ising system at two temperatures β and β', respectively, is given by the expression

$$A(N, \beta) - A(N, \beta') = -\frac{1}{\beta} \ln\langle \exp(-E_\upsilon(\beta - \beta'))\rangle'$$

Write a formal expression for the quantity $\langle \exp(-E_\upsilon(\beta - \beta'))\rangle'$ in terms of the partition function $Q'(N, \beta)$. Consider an Ising model at temperature $\beta = 1/k_B T$. Show how the difference in the free energy between a temperature β and β' can be computed by averaging the quantity

$$\exp[-E_\upsilon(\beta - \beta')]$$

over a Monte Carlo trajectory at either β or β'.
13. Umbrella sampling is a computer simulation method that artificially imposes an external potential on a system's Hamiltonian to facilitate either efficient barrier crossing or the sampling of rare events. If the "real" system energy is E_ν and the imposed potential is V_ν (a constant), show that the property G is given by

$$\frac{\langle G\rangle^\circ}{\langle G\rangle} = e^{\beta V_\nu}$$

where $E_\nu^\circ \equiv E_\nu - V_\nu$. What represents the umbrella sampling step in the above expression?
14. It is possible, at least in principle, to keep track of the frequency of visited states of a specific magnetization, for example, during Monte Carlo simulations of an Ising system at a given temperature. The free energy of a state at magnetization M, $A(M)$, is then formally given by the expression

$$e^{-\beta A(M)} = \sum_{\nu \text{ states at } M} e^{-\beta E_\nu}$$

From a histogram of the frequency of these states, show how ΔA_{12} can be found where $\Delta A_{12} = A(M_2) - A(M_1)$. Below the critical temperature, sketch the function $A(M)$.

15. Adsorption data often conform to the general form shown in figure 9.6 for measurements of X (density) versus Y (pressure) at given values of Z (temperature). Note that the data for X plateaus at large values of Y for each Z. Can you conceive of trying to use a scaling function for representing these data that would (possibly) reduce all of these curves to a single universal curve? [Hint: Consider the logarithm of

$$\left(\frac{X}{Z^{\alpha}}\right) \text{ versus logarithm of } \left(\frac{Y}{Z^{\beta}}\right)$$

where α and β are constants]. How would you assess the success of this approach in a given experimental situation?

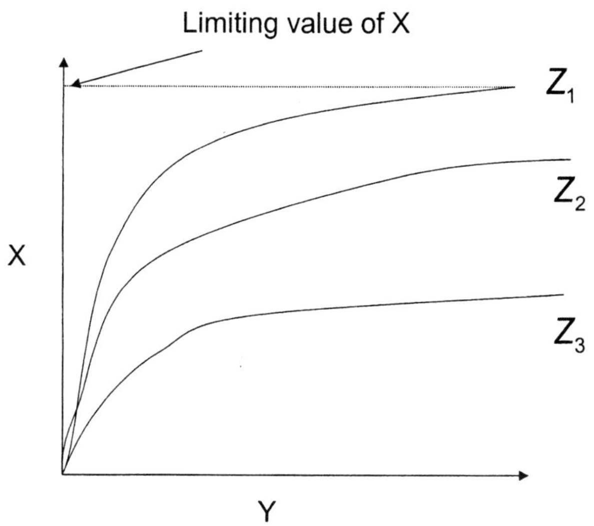

Figure 9.6 Data for doing a scaling analysis

Analysis of Random Variables

16. The cumulative distribution function (cdf) for a normal random variable v with mean μ and standard deviation σ is given by the integral

$$F(x) = \frac{1}{\sigma\sqrt{2\pi}} \int_{-\infty}^{x} e^{-\frac{1}{2}\left(\frac{v-\mu}{\sigma}\right)^2} dv$$

Using the transformation $y = (v - \mu)/\sigma$, show that

$$F(x) = \Phi(z)$$

where $z \equiv (x - \mu)/\sigma$ and

$$\Phi(z) \equiv \frac{1}{\sqrt{2\pi}} \int_{-\infty}^{z} e^{-y^2/2}\, dy$$

which is the cumulative distribution function of a normal variable with zero mean and unit standard deviation—sometimes referred to as a *standard* (or unit) normal distribution. What fraction of the observed values of a normal random variable will lie in the range $\mu \pm \sigma$?

17. Using the definitions in the previous problem prove that,

$$P\{z_1 \leq y \leq z_2\} = P\{\sigma z_1 + \mu \leq v \leq \sigma z_2 + \mu\}$$

This shows how to use a normalized random variable to sample a random variable of the same type with any specified values for the mean and variance.

18. If x^* is a normalized random variable with any type of probability density (mean 0, variance 1) and we define $x = \sigma_x x^* + \langle x \rangle$, then show that the density functions of x and x^* are related as follows:

$$p_x(x) = \frac{1}{\sigma_x} p_{x^*}\left(\frac{x - \langle x \rangle}{\sigma_x}\right)$$

19. Prove that, if

$$\Phi(z) \equiv \frac{1}{\sqrt{2\pi}} \int_{-\infty}^{z} e^{-y^2/2} dy$$

then $\Phi(-z) = 1 - \Phi(z)$. This equation states that $P\{y \leq -z\} = P\{y > z\}$.

20. Show that, if the random variable x is normally distributed with parameters μ and σ^2 (mean and variance), then the variable $y \equiv \alpha x + \beta$ is normally distributed with parameters $\alpha\mu + \beta$ and $\alpha^2\sigma^2$. [Assume $\alpha, \beta > 0$.]
21. Show that if the random variable x is normally distributed with parameters μ and σ^2 (mean and variance), then $P\{x > 0\} = P\{z < \mu/\sigma\}$, where $z \equiv (x - \mu)/\sigma$.
22. Show that, if the random variable x is normally distributed with parameters μ and σ^2 then $P\{x > \mu + \sigma\} = 1 - \Phi(1)$.
23. If x is a unit normal random variable, show that the random variable x^n, where n is any integer $\neq 1$, is *not* a unit normal random variable. What does the density function of x^2 look like?
24. The *variance* and *covariance* of two random variables x and y are defined as

$$\begin{aligned} &\text{var } x \equiv \langle (x - \langle x \rangle)^2 \rangle = \sigma_x^2 \\ &\text{cov}(x, y) \equiv \sigma_{xy} \equiv (x - \langle x \rangle)(y - \langle y \rangle) \\ &= \langle xy \rangle - \langle x \rangle \langle y \rangle \end{aligned}$$

If the covariance of x and y is zero, then these variables are said to be *uncorrelated*. Prove the following results:

(a) $\text{var}(ax + b) = a^2 \text{ var } x$.

(b) $\text{var}(x + y) = \text{var } x + \text{var } y + 2 \text{ cov } (x, y)$.

25. If the mean value of a series random numbers is defined by

$$\bar{x} = \frac{1}{n} \sum_{i=1}^{n} x_i$$

show that

$$\text{var } \bar{x} = \frac{1}{n^2} \sum_{i=1}^{n} \sigma_{xi}^2 + \frac{2}{n^2} \sum_{j<k} \sigma_{x_j x_k}$$

If the various x_i are uncorrelated, each with variance σ_x^2, then

$$\text{var } \bar{x} = \frac{\sigma_x^2}{n}$$

What does this tell us about the role of n in establishing $\bar{x}$? How can this result be gainfully used in computer simulations?

26. If the uniform distribution is defined as

$$f(x) = \begin{cases} 1/(b-a) \text{ for } a \leq x \leq b, \\ 0 \text{ otherwise} \end{cases}$$

then show that

$$\langle x \rangle = \frac{b+a}{2}, \qquad \text{var } x = \frac{(b-a)^2}{12}$$

27. Prove that, if

(a) U is a uniformly distributed random variable over the interval [0, 1], then 1–U is also uniformly distributed over [0, 1].

(b) U is a uniformly distributed random variable over the interval [0, 1], then –ln U is an exponentially distributed random variable with a mean value equal to 1.

28. Consider a "composite" canonical ensemble consisting of two subsystems, designated α and β with volumes and numbers of molecules V_α, V_β, N_α, N_β, respectively. If simulations are done in this ensemble, allowing particles to move both within and between the subsystems, what would the appropriate criterion be for deciding whether or not to accept a particular move? This approach leads to a direct technique for determining the coexistence boundary of a system with α and β, called the *Gibbs ensemble simulation method* [11].

29. Prove that if $x_i \quad (i = 1, \ldots, n)$ are independent random variables that are normally distributed with parameters μ_i and σ_i^2, then their sum $\sum_{i=1}^n x_i$ is also normally distributed with mean $\sum_{i=1}^n \mu_i$ and variance $\sum_{i=1}^n \sigma_i^2$.

Functional Equations [13]

30. If

$$M(tx, ty) = t^m M(x, y)$$

and

$$N(tx, ty) = t^m N(x, y)$$

for $x, y, t > 0$ and $m = 2$, is it necessary that $M(x, y) = N(x, y)$? Can you think of any functions for which these conditions hold and $M(x, y) \neq N(x, y)$?

31. Prove that if

$$M(tx, ty) = t^m M(x, y)$$

for all $x, y \neq 0$, then it follows that any solution to this equation takes on the form

$$M(x, y) = (\sqrt{xy})^n \phi\left(\frac{y}{x}\right)$$

where ϕ is some general function. Express ϕ in terms of x, y and $M(x, y)$.

32. If

$$M(tx, ty) = t^m M(x, y)$$

and

$$N(tx, ty) = t^m N(x, y)$$

with $M \neq N$, find the conditions so that the solution to the differential equation

$$M(x, y)dy + N(x, y)dx = 0$$

is a hyperbola with $xy =$ constant for all values of x, y.

33. Is $f(x, y) = xy$ a solution to the following functional equation?

$$f(x_1 + x_2, y_1 + y_2) = f(x_1, y_1) + f(x_2, y_2)$$

How about $f(x, y) = x + y$? If $f(x, y)$ is homogeneous, is the same true for its Legendre transforms?

34. If $f(x)$ satisfies the following functional (scaling) equation:

$$f(zx) = z^n f(x), \quad z, n > 0$$

show that the solution to the following differential for $x(t)$:

$$\frac{df}{dt} + bx^a = 0, \quad x(0) = 0$$

admits a solution of the form

$$x(t) \sim t^{1/(1-c)}$$

where b and c are both finite constants with $c \neq 1$.

Bibliography

Useful references dealing with finite-size scaling theory, especially in the context of computer simulation, can be found in the publications in [1, 4, 5, 6]. Reference [12] contains simulation results for critical properties in both pure and binary Lennard–Jones systems. Reference [14] provides an excellent overview of field mixing and corrections-to-scaling ideas in the context of fluid systems.

[1] M. Plischke and B. Bergersen, *Equilibrium Statistical Physics*. Singapore: World Scientific, 1994.

[2] H. J. Maris and L. P. Kadanoff, "Teaching the renormalization group," *Am. J. Phys.*, vol. 6, p. 652, 1978.

[3] D. ben-Avraham and S. Havlin, *Diffusion and Reactions in Fractals and Disordered Systems*. Cambridge: Cambridge University Press, 2000.

[4] S. De, "Near-Critical Behavior of Fluids Embedded in Porous Media," Ph.D. thesis in Physics. Rochester, NY: University of Rochester, 2002.

[5] M. N. Barber and W. Selke, "Phenomenological renormalization of Monte-Carlo data," *J. Phys. A*, vol. 15, p. L617, 1982.

[6] R. Lee and S. Teitel, "Phase transitions in classical two-dimensional lattice Coulomb gases," *Phys. Rev.* B, vol. 46, p. 3247, 1992.

[7] D. Frenkel and B. Smit, *Understanding Molecular Simulation: From Algorithms to Applications*, vol. 1, second ed. Academic Press, 2002.

[8] N. Metropolis, A. Rosenbluth, M. Rosenbluth, A. Teller, and E. Teller, "Introduction of the Metropolis algorithm for molecular dynamics simulations," *J. Chem. Phys.*, vol. 21, p. 1087, 1953.

[9] S. Ross, *A First Course in Probability*. New York: Macmillan, 1988.

[10] K. Binder, "Some Recent Progress in the Phenomenological Theory of Finite Size Scaling and Application to Monte Carlo Studies of Critical Phenomena," in *Finite Size Scaling and Numerical Simulation of Statistical Systems*, V. Privman (ed.). Singapore: World Scientific, 1990, p. 173–221.

[11] A. Z. Panagiotopoulos, "Direct determination of phase co-existence properties of fluids by Monte Carlo simulation in a new ensemble," *Mol. Phys.* vol. 61, p. 813, 1987.

[12] J. J. Potoff and A. Z. Panagiotopoulos, "Critical point and phase behavior of the pure fluid and a Lennard–Jones mixture," *J. Chem. Phys.*, vol. 109, p. 10914, 1998.

[13] E. Castillo and M. R. Ruiz-Cobo, *Functional Equations and Modeling in Science and Engineering*, Monographs and Textbooks in Pure and Applied Mathematics, vol. 161, E. J. Taft and Z. Nashed (eds.). New York: Marcel Dekker, 1992.

[14] N. B. Wilding, "Simulation studies of fluid critical behavior," *J. Phys.: Condens. Matter*, vol. 9, p. 585, 1997.

10 The Renormalization-Group Method

The renormalization-group (RG) method discussed in this chapter has assumed a pivotal role in the modern theory of critical phenomena. It attempts to relate the partition function of a given system to that of a "similar system" with decreased degrees of freedom through a process referred to as *renormalization*. Exactly how these degrees of freedom are removed from the system, what we mean by a "similar" system, and how successive systems are coupled to one another are essentially the questions we take up in the introductory treatment given in this chapter. The RG method is a topic with large scope and found widely disseminated in an extensive physics literature on the topic; however, it is seldom found in engineering journals. Our purpose here is to try and make sense of some basic ideas with the RG approach so that it is more accessible to this wider community. For this we often rely upon some prior exposes of the subject in more specialized settings [1, 2, 3, 5].

In its complete sense, the RG method has only been made to work, at least analytically, for a few simple statistical-mechanical models. But aside from these numerical results, many important and quite general insights about critical phenomena can be developed from studying this approach to the problem, especially the central role played by length scale as a factor in describing the phenomenology. These ideas have significantly enhanced our understanding of ideas like scale invariance, universality classes, *relevant scaling fields* (as opposed to *irrelevant* ones), Hamiltonian renormalization, and so on; these and related concepts lie at the center of modern discourse on the subject.

The essential concepts of the approach can be well illustrated using the Ising system since, with this model, lattice spins are fixed in space, which makes the analytical work quite transparent. This approach, called *real space renormalization*, is the RG method studied in this chapter.

10.1 The One-Dimensional Ising Model and RG Approach ($h = 0$)

The partition function for this form of the Ising model with nearest neighbor interactions is of the form [1]

$$Q(\beta, N) = \sum_{S_1, S_2, \ldots, S_N = \pm 1} \cdots \sum \exp[\beta J(S_1 S_2 + S_2 S_3 + \cdots + S_N S_1)] \tag{10.1}$$

Invoking the block-spin idea of Kadanoff illustrated in an earlier chapter, one might ask how this could be used in simplifying the finding of $Q(\beta, N)$. Is it possible to group spins into effective blocks in the Ising chain and reformulate the summation expression given in (10.1)? As a start, we might partition (10.1) as

$$\begin{aligned} Q(K, N) = \sum_{S_1, \ldots, S_N} & \exp(K(S_1 S_2 + S_2 S_3)) \exp(K(S_3 S_4 + S_4 S_5)) \cdots \\ & \cdots \exp(K(S_5 S_6 + S_6 S_7)) \cdots \exp(K(S_{N-1} S_N + S_N S_1)) \end{aligned} \tag{10.2}$$

where we have defined the spin *coupling parameter* $K \equiv \beta J$. Summing this expression over only the even spins, we find that the summation takes on the form

$$Q(K, N) = \sum_{S_1, S_3, \ldots} [\exp(K(S_1 + S_3)) + \exp(-K(S_1 + S_3))] \cdots \tag{10.3}$$

where the S_1–S_3 term in (10.3) is obtained from considering the values ± 1 for S_2. We now see that the sum in (10.3) involves half the spins, or degrees of freedom, present in the original summation for $Q(\beta, N)$ given by (10.1). This *thinning-out* process is akin to collapsing pairs of spins into single groups; essentially this can also be seen as doubling the length scale over which the remaining spins interact.

The next important idea in the RG approach is to cast the thinned-out sum for $Q(K, N)$ into an expression that formally looks identical to the original summation expression. For example, in (10.3) we take the term containing adjacent spins i and j, with i and j odd numbers given by

$$\exp(K(S_i + S_j)) + \exp(-K(S_i + S_j)) \tag{10.4}$$

and ask how this may be replaced by a term of the form

$$q(K)\exp(K' S_i S_j) \tag{10.5}$$

In this expression, we presume that the coupling constant of the "new" expression, represented by K', is changed to accommodate the new i–j interaction in the blocked system. Since the terms in (10.3) are all formally identical, the summation for $Q(K, N)$ in terms of the "pruned" system thus becomes

$$Q(K, N) = q(K)^{N/2} Q(K', N/2) \tag{10.6}$$

where $Q(K', N/2)$ is the partition function of the reduced lattice containing $N/2$ spins.

The interesting thing about this result is that it relates the partition function for a system of size $N/2$ to that of one of size N. As stated earlier, this process amounts to changing the length scale governing spin interactions and leads, if continued, to successively larger blocked spin system configurations. This is sometimes referred to as a *rescaling process*, and the transformation given by (10.6) is called a *Kadanoff relation* [2].

The next question is how to find K' in the rescaled system. This is easily accomplished by solving the equations

$$\exp(K(S_i + S_j)) + \exp(-K(S_i + S_j)) = q(K)\exp(K' S_i S_j) \tag{10.7}$$

for $S_i, S_j = \pm 1$ and $S_i = -S_j = \pm 1$. This solution leads to the equations

$$K' = \tfrac{1}{2}\ln\cosh 2K \tag{10.8}$$

$$q(K) = 2\sqrt{\cosh 2K} \tag{10.9}$$

These results show why we needed at least two terms in equation (10.5), since equation (10.7) provides two independent equations that have to be satisfied.

EXERCISE 10.1

Prove equations (10.8) and (10.9).

These equations are the *RG recursion expressions* for the one-dimensional Ising chain. In general these recursion equations may possess a *fixed point* K^*, which we define as the condition where $K^* = f(K^*)$. The RG equations thus allow successive rescaling of the system's partition function (and hence free energy) through the use of (10.6) in conjunction with (10.8) and (10.9). We see from (10.8) that

$$K' = \tfrac{1}{2}\ln\tfrac{1}{2}(e^{2K} + e^{-2K}) \tag{10.10}$$

with $K > 0$. Since we have that

$$\frac{dK'}{dK} = \frac{1}{e^{2K} + e^{-2K}}(e^{2K} - e^{-2K}) \tag{10.11}$$

and $0 < dK'/dK < 1$; this implies that, in this situation, $K' < K$ for all K (why?). This result means that, as we block the system, the coupling parameter gets smaller, which amounts to weakened spin interactions between the residual lattice spins.

We now proceed to study equations (10.6)–(10.9) by starting out with $K=\infty$, which corresponds to the $T \rightarrow 0$ limit. The partition function for the thinned-out system is found using equations (10.8)–(10.9). We notice that successive calculations of K allow one to progress from the high end of the interval $K \in [0, \infty]$ to the other in an uninterrupted fashion. This sequential pattern of K values is often referred to as *flow* in coupling-parameter space. At either extreme, these recursion relations have so-called *fixed points*, implying that the flow stops at this value of K. We could also begin the flow process from the other extreme of the interval where $K \approx 0$. In this case, the RG equations are given by

$$K = \tfrac{1}{2}\cosh^{-1} e^{2K'} \tag{10.12}$$

$$f(K) = 2\exp K' \tag{10.13}$$

10.2 Further Examples of RG Calculations

The One-Dimensional Ising model at Finite Field Strength ($h \neq 0$)

In this section, we consider an extension to our introductory example by considering the 1-d Ising model at finite field strength. The partition function is given by

$$Q(K, N) = \sum_{\{S_i\}=\pm 1} \exp \sum_{i-1}^{N} (KS_i S_{i+1} + \beta h S_i) \tag{10.14}$$

We can express the term

$$\sum_{i=1}^{N} \beta h S_i$$

as

$$\beta h \sum_{i=1}^{N} S_i = \frac{\beta h}{2} \sum_{i=1}^{N} (S_i + S_{i+1}) \tag{10.15}$$

EXERCISE 10.2

Prove the relationship given in (10.15).

In this case, (10.14) becomes

$$Q(K, N) = \sum_{\{Si\}=\pm 1} \exp\left[\sum_{i-1}^{N}(KS_i S_{i+1}) + \frac{\beta h}{2}(S_i + S_{i+1})\right] \tag{10.16}$$

As in the zero-field case, we can factorize the terms of this sum and sum them over odd spins. The partition function in (10.16) is more completely given as

$$\begin{aligned} Q(K, N) = \sum_{S_i,\ldots,S_N} &\exp\left[K(S_1 S_2 + S_2 S_3) + \frac{\beta h}{2}(S_1 + S_2) + \frac{\beta h}{2}(S_2 + S_3)\right] \\ &\times \exp\left[K(S_3 S_4 + S_4 S_5) + \frac{\beta h}{2}(S_3 + S_4) + \frac{\beta h}{2}(S_4 + S_5)\right]\cdots \end{aligned} \tag{10.17}$$

The term containing S_3 in this sum is

$$\exp[K(S_2 S_3 + S_3 S_4) + \beta h S_3] \tag{10.18}$$

Every odd-numbered spin in (10.17) has a corresponding form which yields, in the case of $S_3 = \pm 1$, the following terms in the summation:

$$\exp[K(S_2 + S_4) + \beta h] + \exp[(-K(S_2 + S_4) - \beta h] \tag{10.19}$$

By now, the analogy with the zero-field case studied earlier should be clear, and we may ask how terms like those in (10.19) can be replaced by a Kadanoff term of the form

$$q(K)\exp\left[K' S_2 S_4 + \frac{\beta h'}{2}(S_2 + S_4)\right] \tag{10.20}$$

where K' and h' are the parameters in the renormalized system. The Kadanoff transformation in this case is given by the equation (written here for spin 3)

$$\begin{aligned} &\exp[K(S_2 + S_4) + \beta h] + \exp[-K(S_2 + S_4) - \beta h] \\ &= q(K)\exp\left[K' S_2 S_4 + \frac{\beta h'}{2}(S_2 + S_4)\right] \end{aligned} \tag{10.21}$$

which, for $S_2, S_4 = \pm 1$, leads to the RG results for this model:

$$K' = \tfrac{1}{4}\ln\frac{\cosh(2K + h)\cosh(2K - h)}{\cosh^2 h} \tag{10.22}$$

$$h' = h + \tfrac{1}{2}\ln\frac{\cosh(2K + h)}{\cosh(2K - h)} \tag{10.23}$$

$$q = \exp 2g \tag{10.24}$$

where

$$g = \tfrac{1}{8}\ln[16\cosh(2K+h)\cosh(2K-h)\cosh^2 h] \tag{10.25}$$

EXERCISE 10.3

Prove the RG equations for K' and h' in the 1-d Ising model ($h \neq 0$) given by equations (10.22)–(10.23). Sketch the flows of K and h for the Ising model given by these equations, and comment on the nature of the fixed-point conditions.

We point out that each odd spin term in the partial function sum is identical to the S_3 term described above, which means that the partition function can be expressed by the equation

$$\begin{aligned} Q(K,N) &= \sum_{\{S_i\}} \exp\left[K\sum_{i=1}^{N}(S_iS_{i+1}) + \frac{\beta h}{2}(S_i+S_{i+1})\right] \\ &= q^{N/2}\sum_{\{Sj\}}\exp\left(K'\sum_j S_jS_{j+2} + \beta h'\sum_j S_j\right) \end{aligned} \tag{10.26}$$

where j is the summation index over the remaining even-numbered spins in the chain (i.e., $j = 2, 4, \ldots, N$). What is of note in this expression for the sum over j is that it appears identical in form to the original partition-function summation in equation (10.14), in keeping with the RG approach. Thus again we see that the recursive relationship for the partition function between the successive renormalized systems is similar to (10.6):

$$Q(K,N,h) = q(K,h)^{N/2}Q(K', N/2, h') \tag{10.27}$$

This latter expression displays the so-called *group property* whereby successive partition functions are expressed in identical functional forms with the parameters K and h changing to reflect the change in scale from one iteration (thinning-out procedure) to the next.

EXERCISE 10.4

The RG approach as formulated here is called the space-RG method because it involves scale changes of the lattice model whose successive structures are similar except for the scale change of the lattice dimension. Can you geometrically show this type of transformation for a triangular lattice and motivate Kadanoff expressions, at zero field, where only nearest-neighbor interactions are considered?

The RG Approach in the Two-Dimensional Ising Model ($h = 0$)

In this section, we discuss the two-dimensional Ising model with only nearest neighbor interactions [1, 2]. This model involves a topography similar to the structure in figure 10.1. This problem is mainly of pedagogical interest, because it shows how simply

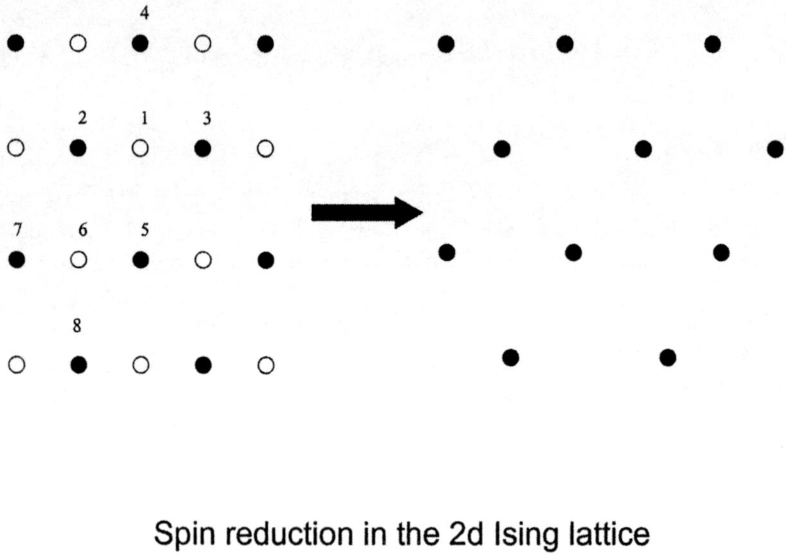

Figure 10.1 Schematic of RG spin reduction in the 2-d Ising system

increasing the system dimension leads to difficulties in formulating an exact RG analysis. Nearest-neighbor interactions in the lattice are of now of the form

$$-\frac{J}{2}\sum{}' S_i S_j \tag{10.28}$$

except that now there are four such contributions to each spin, and this particular type of summation is denoted by the prime on the summation sign. We now proceed with the thinning-out process, as done in the prior case, and attempt to evaluate the partition function

$$Q = \sum_{\{S_i S_j\}} \exp\left(K \sum{}' S_i S_j\right) \tag{10.29}$$

By eliminating (summing) alternate spins in the two-dimensional lattice, we end up with a structure of the form shown in figure 10.1. In order to remove spins from the lattice, we need to find their contributions to Q. Using figure 10.1, we see that, in summing over spins 1 and 6, the following terms enter into the summation for Q:

$$\begin{aligned} Q &= [\exp(K(S_2 + S_3 + S_4 + S_5)) + \exp(-K(S_2 + S_3 + S_4 + S_5))] \\ &\quad \cdot [\exp(K(S_2 + S_5 + S_7 + S_8)) + \exp(-K(S_2 + S_5 + S_7 + S_8))] \\ &\quad \cdots \text{ rest of terms in } Q \end{aligned} \tag{10.30}$$

If we proceed before, as in the one-dimensional case, and seek Kadanoff transformations for these terms, we end up (for the first term by way of example) seeking to establish $f(K)$ and K' that satisfies the following equation:

$$\begin{aligned} &\exp[K(S_2 + S_3 + S_4 + S_5)] + \exp[-K(S_2 + S_3 + S_4 + S_5)] \\ &\quad = f(K)\exp[K'(S_2S_3 + S_3S_4 + S_4S_5 + S_2S_5)] \end{aligned} \tag{10.31}$$

where all the possible values of S_2, S_3, S_4, S_5 are considered. Note that the topographic connectivity in the thinned-out lattice connects spins that are not nearest neighbors, and this is an early indication that something is amiss with this particular RG scheme, as will now be looked at in more detail (the reader is recommended to first do exercise 10.5).

EXERCISE 10.5

The number of combinations of k-upward and $(n-k)$ downward pointing spins in n-spin system can be found from the combinatorial factor

$$\binom{n}{k} = \frac{n!}{k!(n-k)!}$$

For the RG scheme shown in equation (10.31), there are 16 combinations of the 4 spin variables S_2, S_3, S_4, S_5, where each spin can take on the values ± 1. However, *only* four independent equations result from considering these combinations for use in equation (10.31). Develop expressions for each of these four equations.

From the results of exercise 10.5, we see that we have an overdetermined system of equations, since we are faced with having to determine only two independent parameters $f(K)$ and K' in equation (10.31)! This should be contrasted with the result for the one-dimensional Ising model, where a fully determined set of Kadanoff transformations emerged from the analysis. One way to get around this problem is to increase the terms (and number of parameters) in the exponential argument in (10.31). A possibility is to look at the following more complex Kadanoff transformation [1], now containing at least four *coupling parameters* to make the problem fully determined:

$$\begin{aligned} &e^{K(S_2+S_3+S_4+S_5)} + e^{-K(S_2+S_3+S_4+S_5)} \\ &= f(K)\exp\left[\frac{K_1}{2}(S_2S_4 + S_3S_4 + S_3S_5 + S_2S_5)\right. \\ &\qquad \left. + K_2(S_2S_3 + S_4S_5) + K_3 S_2S_3S_4S_5\right] \end{aligned} \tag{10.32}$$

Note that the argument in (10.32) structurally represents various terms relating spins on the four corners of the square lattice in the thinned-out system. These were all nearest neighbors to spin 1 in the original lattice which was removed in the thinning process. The first (K_1) term represents nearest-neighbor sums, the K_2 term *next*-nearest-neighbor sums, and the K_3 term consists of all four spins. If the four independent spin states for $S_2S_3S_4$ and S_5 are substituted into (10.32), it is now possible to solve for K_1, K_2, K_3, and $f(K)$. These results, presented by Chandler, are given by

$$K_1 = \tfrac{1}{4}\ln\cosh 4K \tag{10.33}$$

$$K_2 = \tfrac{1}{8}\ln\cosh 4K \tag{10.34}$$

$$K_3 = \tfrac{1}{8}\ln\cosh 4K - \tfrac{1}{2}\ln\cosh 2K \tag{10.35}$$

with

$$f(K) = 2(\cosh 2K)^{1/2}(\cosh 4K)^{1/8} \tag{10.36}$$

The complete partition function $Q(K, N)$ can now be found in the thinned-out lattice by a complex summation which includes nearest-neighbor, next-to-nearest-neighbor, and quaternary (four-point) sums using the respective coupling parameters K_1, K_2, K_3.

EXERCISE 10.6

Prove equations (10.33)–(10.36). Show how $Q(K, N)$ can be formally expanded in terms of the $\frac{1}{2}N$ (thinned-out lattice) spins in the 2-d Ising model to yield a recursive RG result with the structure

$$Q(K, N) = f(K)^{N/2} \sum \exp\left[K_1 \sum_{i,j}{}' S_i S_j + K_2 \sum_{m,n}{}'' S_{m,n} + K_3 \sum_{a,b,c,d} S_{a,b,c,d}\right] \tag{10.37}$$

In this expression, the i–j terms, are nearest-neighbor ones, the m–n terms are next-to-nearest neighbors and the a–b–c–d terms are the 4-spin ones. In an $N \times N$ lattice how many next-to-nearest-neighbor terms are present in the thinned-out lattice's Hamiltonian (size $\frac{1}{2}N \times \frac{1}{2}N$)?

The problem with the increased complexity in the connectivity in the structure of the thinned-out system's Hamiltonian is that it makes an exact RG calculation impossible to do as it stands. One should recall that, in a RG calculation, lattice Hamiltonians should all have the same structural form in order for an equation like (10.26) to pertain. We have already pointed out a fundamental difference in topography between connected spins in the original and thinned-out lattice. This illustrates how what appears to be a simple increase in dimensionality makes for an impossible situation to solve exactly. How can we attempt to circumvent these problems? Once again, the approximate approach given in [1] is instructive. By ignoring the four-point sum in the Hamiltonian, the idea is to convert the nearest-neighbor and next-nearest-neighbor terms into an effectively nearest-neighbor expression and thereby restore topographic consistency to the analysis. This can be accomplished by approximating these terms using the expression

$$K_1 \sum_{i,j}{}' S_i S_j + K_2 \sum_{m,n}{}'' S_{m,n} = K'(K_1, K_2) \sum_{i,j}{}' S_i S_j \tag{10.38}$$

in which case, we now can recover the familiar recursive form for the RG formulation of the problem:

$$Q(K, N) = f(K)^{N/2} Q(K', N/2) \tag{10.39}$$

This, however, creates another problem, namely how to come up with a way of relating K' to K_1 and K_2. If we approximate K' by the equation

$$K' = K_1 + K_2 \tag{10.40}$$

then we get the following RG equation linking K' to K:

$$K' = \tfrac{3}{8} \ln \cosh 4K \tag{10.41}$$

which has a *fixed point* at

$$K_c = 0.50698$$

which should be compared to the exact value for the 2-d Ising model given by $K_c = 0.4407$.

EXERCISE 10.7

Rationalize the approximation given in (10.40) which says that

$$K' \approx K_1 + K_2$$

In a given lattice, at what temperatures would you expect this approximation to be most accurate (why?) Also, plot the flow in coupling-parameter space given by equation (10.41).

We ought to be clear about how the RG process leading to equation (10.38) works. The original Hamiltonian involves only nearest-neighbor terms. However, after a single step of the RG scheme (spin "decimation"), a more complex Hamiltonian structure is required to represent the renormalized lattice. This Hamiltonian requires at least 3 spin coupling parameters that provide values for the renormalized lattice's parameters K_1, K_2, K_3 in terms of the original coupling parameter K. However, this lattice's Hamiltonian is different to its antecedent's structure, and to restore this consistency, an effective nearest-neighbor coupling parameter K' is defined by (10.40). This *uncontrolled approximation* is only one of many that could easily be conceived for this purpose. What this highlights is the difficulty faced when trying to do an exact analytic RG scheme for this model. For this reason, hybrid approaches involving both analytic methods and computer simulation are now finding favor.

All of these processes are called *renormalization*. Naturally they invite several interesting questions which we now discuss.

EXERCISE 10.8

Starting out with a weakly interacting system (i.e., $K \approx 0$) compare the exact free energy of the 1-d Ising model with that given by the RG equations in (10.6)–(10.9). What do you conclude from these results?

EXERCISE 10.9

Can you qualitatively identify either of the extreme values of K (i.e., $K = 0$ and $K = \infty$) with order and/or disordered states in the system? If so, which is the ordered point and why?

EXERCISE 10.10

The Hamiltonian $H(K,N)$ for the Ising system is related to its partition function by

$$Q(K, N) = \sum_{\{S_i, S_j\}} e^{H(K,N)} \tag{10.42}$$

In the 1-d Ising model, at zero field, the simplest form of this Hamiltonian function is given by the expression $-K \sum_{\{S_i, S_j\}} S_i S_j$. However, if the lattice energy function were to contain more than simple nearest-neighbor interaction terms (e.g., next-to-nearest-neighbor ones), the Hamiltonian would contain a *vector* $\boldsymbol{k}$ of coupling constants. If the RG recursion equations have a fixed point $\boldsymbol{k}^*$ (i.e., they converge to a solution $\boldsymbol{k}^*$), show that a Taylor expansion of the RG equations around $\boldsymbol{k}^*$ yields a matrix equation linking $\delta\boldsymbol{k}'$ to $\delta\boldsymbol{k}$ of the form

$$\delta\boldsymbol{k}' = \boldsymbol{M}\delta\boldsymbol{k} \tag{10.43}$$

where δ denotes a small linear departure from the fixed-point condition. Note that $\delta\boldsymbol{k} \equiv \boldsymbol{k} - \boldsymbol{k}^*$, $\delta\boldsymbol{k}' \equiv \boldsymbol{k}' - \boldsymbol{k}^*$, $\boldsymbol{k}' = \boldsymbol{r}(\boldsymbol{k})$ with $\boldsymbol{M}$ the Jacobian matrix of the function $\boldsymbol{r}(\boldsymbol{k})$.

Some Generic Questions about the RG Method

The RG method, as used above for the Ising model calculations, invites several questions, some of which we now state.

- What, if any, relationship exists between the RG method and the scaling theories given in the previous chapter?
- Can one always reconstitute the thinned-out system Hamiltonian in terms of a functional form identical to its immediate predecessor Hamiltonian?
- How does one "prune" (block) the system Hamiltonian from one RG step to another? The procedure used in the introductory example is called *spin decimation* (for obvious reasons). However, it is clear that blocking can occur in a variety of ways. In the 1-d example, just given, spins were blocked in groups of two. If larger-size blocking were done, would this change the results reached in a fundamental way? In other words, how does the manner in which blocking is done affect the outcome of the RG calculations?
- Do the intermediate lattice structures have any physical significance? What is the meaning of the lattice's thermodynamic properties such as the susceptibility from one iteration of the RG analysis to the next? In other words, how do these properties change with the length scale of the rescaled lattices?
- What, if any, physical significance can be attached to fixed-point conditions in the RG process?
- What constraints exist on the values taken on by the spin variables in the thinned-out lattice?

These and related questions now take up our attention, beginning with an analysis of the analogies between RG and scaling theories. However, we first provide a very brief discourse on fractal objects, since these have come to play an important role in RG ideas.

DISCOURSE

Reducing the lattice so that the mother structure and its offspring have similar topography with different length scales is intimately linked to the concept of a *fractal structure*. Such objects are called *self-similar* if the rescaling is isotropic (i.e., the same in all directions). These find wide use in many branches of physics and engineering where scaling and complex structure formation are central to the phenomena of interest [4]. One idea central to these objects is that, if we have a property g of the lattice that depends upon x, then it is natural to relate g in both structures by the equation

$$g(x) = b^{-a} g(bx)$$

where b is a rescaling factor and a is the dimension of the system. Doesn't this remind you of the Kadanoff blocking equation for the free energy in the previous chapter? If $a = 1$, what is special about the situation? [Hint: Consider extensive thermodynamic variables.]

10.3 The Relationship of RG to Scaling of the Free Energy

We now generalize the previous analysis and show how the RG equations lead to scaling forms for the free energy identical to those set out earlier using the scaling hypothesis. In this sense, the RG theory provides a firm theoretical underpinning for the intuitive scaling ideas that historically were used to describe critical behavior.

Renormalized Hamiltonians with the Ising Model ($h = 0$)

The Hamiltonian of the N lattice is related to its partition function and free energy through the equation

$$Q(K, N) = \sum_{\{S_i\}} e^{H(K)} = \mathrm{Tr}\{e^{H(K)}\} = e^{-\beta A(K,N)} \tag{10.44}$$

where the trace operator Tr means a summation over all possible states. From the previous example in the 1-d Ising model, we see that the essential goal of the RG approach is to relate the partition function of successive lattices, exemplified by the result given in equation (10.6). That result suggests that we relate the partition function of the starting lattice to a renormalized Hamiltonian of its successor lattice by a generic equation of the form

$$Q(K, N) = \sum_{\{S_i\}} e^{H(K)} = \sum_{\{S_j\}} e^{H(K')} \tag{10.45}$$

The renormalized Hamiltonian $H(K')$ usually contains a spin term as well as a prefactor term, as was illustrated with the 1-d Ising chain in the introductory example.

To proceed further, we look at an intensive property of the lattice, a convenient one being the free energy per site defined by the following function f:

$$f(K) \equiv -\frac{\beta A(K, N)}{N} \tag{10.46}$$

Using (10.44)–(10.46) applied to the Ising lattice, we have that

$$e^{Nf(K)} = e^{Nf(K')/2 + Ng(K)} \tag{10.47}$$

where the free energy of the thinned-out lattice is given by

$$\frac{N}{2} f(K') = -\beta A\left(K', \frac{N}{2}\right) \tag{10.48}$$

The $N/2$ factor in the term $Nf(K')/2$ accounts for the reduction in lattice sites that occurs in the $(K', N/2)$ Ising chain. In general though, this factor is equal to b^d, where b is the lattice rescaling parameter chosen in the thinning-out process, and d the system dimensionality; the parameter b is simply the factor that is required to shrink the rescaled lattice to its antecedent's scale and form. In the 1-d Ising model used in the introduction, for example, $b = 2$, but in general b can be any value.

For example, in the case of the 1-d Ising model, if we define $H(K')$ as

$$H(K') \equiv K' \sum_j S_j S_{j+2} + N g(K) \tag{10.49}$$

and the reduced $N/2$ lattice partition function by

$$Q(K', N/2) \equiv \sum_{\{S_j\}} e^{K' \sum_j S_j S_{j+2}} \tag{10.50}$$

then, to be consistent with equation (10.6), the prefactor function $g(K)$ is related to $q(K)$ by the equation

$$g = \tfrac{1}{2} \ln q \tag{10.51}$$

General Hamiltonians

The analysis leading to (10.47) for the Ising lattice is easily generalized to other Hamiltonians [3]. We see that the more general analogue to (10.47) is given by

$$f(K) = \frac{f(K')}{b^d} + g^*(K) \tag{10.52}$$

If we focus on a Hamiltonian with two coupling parameters, this equation can be rewritten in the form

$$f(K_1, K_2) = b^{-d} f(K_1', K_2') + g^*(K_1, K_2) \tag{10.53}$$

At this point, we make use of the RG expansion about the fixed point en route to a transformation of the variables in equation (10.53).

The RG equations can be recast into the following form (see exercise 10.10):

$$\boldsymbol{k}' = \boldsymbol{M}\boldsymbol{k} \tag{10.54}$$

where $\boldsymbol{M}$ is the Jacobian matrix of the RG equations evaluated *at the fixed-point condition*. If we define the matrix of eigenvectors of $\boldsymbol{M}$ as the matrix $\boldsymbol{P}$, then a fundamental theorem in linear algebra (first introduced in chapter 1), yields

$$\boldsymbol{\Lambda} = \boldsymbol{P}^{-1}\boldsymbol{M}\boldsymbol{P} \tag{10.55}$$

where $\boldsymbol{\Lambda}$ is a diagonal matrix of eigenvalues of $\boldsymbol{M}$. If we now define the following (2×1) vector quantities $\boldsymbol{u}$ and $\boldsymbol{u}'$ by

$$\boldsymbol{u} = \boldsymbol{P}^{-1}\boldsymbol{k} \tag{10.56}$$

$$\boldsymbol{u}' = \boldsymbol{P}^{-1}\boldsymbol{k}' \tag{10.57}$$

then we see from (10.54)–(10.57) that

$$\boldsymbol{u}' = \boldsymbol{\Lambda}\boldsymbol{u} \tag{10.58}$$

These transformations imply that we may rewrite the free-energy equation (equation (10.53)) in terms of a new function of elements of the vectors $\boldsymbol{u}$ and $\boldsymbol{u}'$ as follows:

$$f(U_1, U_2) = b^{-d} f(U_1', U_2') + g(U_1, U_2) \tag{10.59}$$

where

$$\boldsymbol{u} \equiv \begin{bmatrix} U_1 \\ U_2 \end{bmatrix} \text{ and } \boldsymbol{u}' \equiv \begin{bmatrix} U_1' \\ U_2' \end{bmatrix} \tag{10.60}$$

From equation (10.58), we can also write that

$$U_1' = \lambda_1 U_1 \equiv b^{y_1} U_1 \tag{10.61}$$

$$U_2' = \lambda_2 U_2 \equiv b^{y_2} U_2 \tag{10.62}$$

In (10.61) and (10.62) λ_1 and λ_2 are the eigenvalues of $\boldsymbol{M}$ evaluated at the fixed-point condition, and y_1 and y_2 are referred to as the RG exponents.

From these definitions, it follows that

$$f(U_1, U_2) = b^{-d} f(b^{y_1} U_1, b^{y_2} U_2) + g^*(U_1, U_2) \tag{10.63}$$

Equation (10.63) is remarkable. It emerges from a rich combination of concepts: lattice rescaling, fixed-point conditions, and eigenvalue transformations. It leads directly to a scaling form for the free energy already familiar to us from phenomenological scaling theory discussed in the previous chapter. To see this, we define $\lambda \equiv b^{-d}$, $t \equiv U_1$, and $h \equiv U_2$, and substitute these into (10.63), which leads to the equation

$$f(t, h) = \lambda f(\lambda^{-y_1/d} t, \lambda^{-y_2/d} h) + g^*(t, h) \tag{10.64}$$

Now, since λ is arbitrary, we may choose $\lambda = |t|^{d/y_1}$, in which case we find that

$$f(t, h) \propto |t|^{d/y_1} f\left(\frac{t}{|t|}, |t|^{-y_2/y_1} h\right) \tag{10.65}$$

Equation (10.65) is a wonderful result since it independently provides a scaling form for the free energy identical to that postulated by phenomenological scaling theory in a previous chapter.

Note that, from (10.65), we can relate the RG exponents y_1 and y_2 to the more familiar scaling exponents α, β, δ, and so on. In (10.63), scaling fields identified with positive RG exponents y_i are called *relevant* fields since they assume increasing significance as the RG process unfolds toward a fixed point. Conversely, fields with a negative exponent are termed *irrelevant* scaling fields.

EXERCISE 10.11

Prove equation (10.65) and use it to show that the RG approach leads to the following equation relating one of its exponents y_1 to both d and α:

$$\frac{d}{y_1} = 2 - \alpha \tag{10.66}$$

Show that this equation is equivalent to Josephson's law, also known as the hyperscaling equation. What are the signs of y_1 and y_2 in (10.65)?

Expansion of the RG Parameters in Coupling-Parameter Space

If we denote the Hamiltonian flow by a vector $\boldsymbol{k}$ representing the trajectory of the coupling parameters in the Hamiltonian as the RG analysis proceeds, we can expand $\boldsymbol{k}$ around the fixed-point condition, represented by $\boldsymbol{k}^*$. We do this in terms of the eigenvectors, a basis for the transformation $\boldsymbol{M}$ described earlier, which relates $\boldsymbol{k}^{j+1}$ to $\boldsymbol{k}^j$, successive vectors of the coupling parameters. We have that [3]

$$\boldsymbol{k} = \boldsymbol{k}^* + \alpha_1 \boldsymbol{p}_1 + \alpha_2 \boldsymbol{p}_2 \tag{10.67}$$

In (10.67), $\boldsymbol{p}_1$ and $\boldsymbol{p}_2$ are the eigenvectors of $\boldsymbol{M}$ and α_1 and α_2 are nonzero constants. In the region where (10.54) is applicable, n iterations of the RG equations applied to (10.67) lead to the result

$$\boldsymbol{M}^n(\boldsymbol{k} - \boldsymbol{k}^*) = \boldsymbol{M}^n \alpha_1 \boldsymbol{p}_1 + \boldsymbol{M}^n \alpha_2 \boldsymbol{p}_2 \tag{10.68}$$

Since, at the fixed-point condition,

$$\boldsymbol{M}^n \boldsymbol{k}^* = \boldsymbol{k}^* \tag{10.69}$$

and

$$\boldsymbol{M}^n \boldsymbol{p}_1 = \lambda_1^n \boldsymbol{p}_1 \tag{10.70}$$

$$\boldsymbol{M}^n \boldsymbol{p}_2 = \lambda_2^n \boldsymbol{p}_2 \tag{10.71}$$

we see that

$$\boldsymbol{M}^n \boldsymbol{k} = \boldsymbol{k}^n = \boldsymbol{k}^* + \alpha_1 \lambda_1^n \boldsymbol{p}_1 + \alpha_2 \lambda_2^n \boldsymbol{p}_2 \tag{10.72}$$

Since, from equation (10.61) and (10.62), $y_1 > 0$, $y_2 < 0$, $|\lambda_1| \geq 1$, and $|\lambda_2| \leq 1$. In the limit as $n \to \infty$, we would expect $\boldsymbol{k}^n \to \boldsymbol{k}^*$ (the fixed-point condition). In this case, from (10.72), we see that α_1 must *approach zero* as the critical point is approached along the eigendirection given by $\boldsymbol{p}_1$ which defines the temperature or thermal scaling field. Note that the importance of the eigenvector directions was first pointed out in chapter 3! These conditions also show that the RG function relating flow in coupling-parameter space has a saddle point at this type of fixed point. This can be used to develop the scaling law for the correlation length as the critical temperature is approached (see the additional exercises).

Summary of General Insights from this Analysis

The remarkable outcome of all of this analysis is that the RG approach leads to a scaling form of the free energy consistent with that provided earlier using intuitive phenomenological scaling ideas. We summarize some of these main findings:

- The eigenvalues of the linearized RG equations are seen to be related to the various scaling exponents. The signs of the RG exponents determine whether or not the corresponding scaling fields are *relevant* ones as the critical point is approached. Relevant fields have positive exponents, while *irrelevant fields* have negative exponents (one should be careful to distinguish between the usual critical scaling exponents and the RG exponents).
- Another important consequence of these results is their intimation of a *universality property* for the given Hamiltonian structure, leading to so-called *universality classes* describing scaling in various systems. In deriving (10.63), the only assumption used is that the system Hamiltonian can be reexpressed (renormalized) in a form similar to its antecedent expression as the thinning-out process moves forward. At each point, what changes in each respective Hamiltonian are coupling parameters such as K. As a result, the scaling form for the free energy that results from this process (e.g., equation (10.64)) is anticipated to be universal for the given Hamiltonian structure. We, therefore, would expect identical scaling exponents with Hamiltonian structures that fall into this class. In other words, systems with scalar order parameters, the same dimensionality, and short-order intermolecular interactions will show the same scaling behavior.

Having developed these general results we now look more closely at the question of lattice reduction strategies different to the spin decimation method used in the introductory example used for this chapter and how these may affect the RG approach.

10.4 Another "Generic" Lattice Reduction Strategy with the RG Approach

Lattice reduction refers to the process of removing spins from the lattice as the RG analysis proceeds. Thus far we have used the spin decimation approach for doing this, however, there are many ways to thin the lattice. Here we look at one such alternative strategy referred to as the *linear renormalization scheme* (LRS) and follow closely the analysis given by Binney et al. [5]. This general analysis leads to very specific results about how certain properties such as the susceptibility and magnetization change from one step of the RG analysis to the next.

We begin by focusing upon the correlation length ξ and how it changes from step n to step $n+1$ in any lattice reduction process. We must have that

$$\xi^{(n+1)} = \frac{\xi^{(n)}}{b} \tag{10.73}$$

where b is the rescaling parameter. However, how do other properties like the susceptibility change from one iteration of the RG approach to the next using LRS?

In the LRS approach, the spins in a given block K in the lattice at step n are represented by the set $\{S_i^{(n+1)}\}, i \in \mathcal{S}_K$. These spins are subsequently combined, in some fashion, to give a resultant spin $S_K^{(n+1)}$ in the successor $(n+1)$ lattice using the formula:

$$S_K^{(n+1)} \equiv A^{(n)} \sum_{i \in \mathcal{S}_K} S_i^{(n)} \tag{10.74}$$

where $A^{(n)}$ is some *renormalization constant*. In equation (10.74), the linear character of the RG scheme is obvious, and the subscript integer refers to the spin number, while the superscript refers to the step in the lattice reduction process. Equation (10.74) provides that the new lattice variable S_K at step $n+1$ is simply the sum of the old spin variables in the Kth block at step n. The purpose of the renormalization constant $A^{(n)}$ is to maintain feasible values for the $S_K^{(n+1)}$. For example, in the Ising model, the variables $S_K^{(n+1)}$ must lie between -1 and 1; without the flexibility afforded by the use of $A^{(n)}$, it is easy to see that this might not occur (why?).

One logical convention for $A^{(n)}$, discussed by Binney et al., is to define it in terms of the rescaling parameter b as follows:

$$A^{(n)} \equiv b^{-d\omega} \tag{10.75}$$

where ω is, in principle, a function of temperature. It can be chosen to maintain certain properties for the $S_i^{(n)}$ values from one iteration to the next, as previously alluded to, and alters the more naturally conceived b^{-d} term we have repeatedly seen in prior chapters. For instance, the mean square value of the spin could be kept constant by adjusting ω at each iteration. Given this LRS scheme, several results immediately follow. For instance, we can now can study the magnetization m of the lattice at successive steps of the RG analysis.

Magnetization with Linear Renormalization

The lattice magnetization $m^{(n)}$ per spin at the nth stage of the RG analysis is given by the definition

$$m^{(n)} \equiv \langle S_i^{(n)} \rangle \tag{10.76}$$

where the average is taken over all spin values in the lattice at the nth step given by (10.74) and (10.75). Using the linear renormalization scheme, it follows that,

$$m^{(n+1)} = \langle S_K^{(n+1)} \rangle = \left\langle b^{-d\omega} \sum_{i \in \mathcal{S}_K} S_i^{(n)} \right\rangle \tag{10.77}$$

or

$$m^{(n+1)} = b^{-dw} b^d \langle S_i^{(n)} \rangle \tag{10.78}$$

since there are b^d spins in each block. Thus we have the following equation linking $m^{(n)}$ to $m^{(n+1)}$:

$$m^{(n+1)} = b^{d(1-\omega)} m^{(n)} \tag{10.79}$$

There are some important qualitative issues worth pointing out with this result. In the LRS scheme, the number of spins removed is simply controlled by adjusting b, while in (10.79) the transformation of the lattice property m is controlled by the choice of ω. How ω is chosen, though, is not obvious at this point. This should be contrasted with the results shown earlier, where lattice transformations were accomplished through Kadanoff relationships linking the coupling parameters of successive lattices. Here we obviously renormalize the lattice in quite a different way.

EXERCISE 10.12

Think about the magnetization values $m^{(n+1)}$ and $m^{(n)}$ in terms of the 2-d Ising model with the LRS and $b = 2$. How would you find $m^{(n)}$ and $m^{(n+1)}$ using computer simulation and Boltzmann sampling of the larger lattice? Would this approach yield insight into the value of ω in equation (10.79)?

The Pair Correlation Function Γ_{ij}

The pair correlation function Γ_{ij} defined in the previous chapter can also be analyzed using the results introduced in this section. We ask the question: how is Γ_{ij}^{n+1} related to Γ_{ij}^{n} in the LRS renormalization scheme? To answer this question one first needs to carefully define the spins involved in the pair correlation function at each stage of the RG analysis. We define the spin sets $\mathcal{S}_K$ and $\mathcal{S}_l$ which refer to the spins at the nth stage that are *blocked into spins* S_i and S_j through LRS at the $(n+1)$th stage. The pair correlation function of the *blocked* spins i and j at the $(n+1)$th RG iteration is defined by the equation

$$\Gamma_{ij}^{(n+1)} \equiv \langle S_i^{(n+1)} S_j^{(n+1)} \rangle \tag{10.80}$$

From the LRS formulation, we have that

$$\Gamma_{ij}^{(n+1)} = b^{-2d\omega} \sum_{\substack{k \in \mathcal{S}_k \\ l \in \mathcal{S}_l}} \langle S_k^{(n)} S_l^{(n)} \rangle \tag{10.81}$$

which can be approximated by the equation

$$\Gamma_{ij}^{(n+1)} \approx b^{2d(1-\omega)} \Gamma_{kl}^{n} \tag{10.82}$$

Once again, the ubiquitous ω appears, now in equation (10.82).

EXERCISE 10.13

What is the basis for the approximation given above in equation (10.82)?

The Lattice Susceptibility χ_M

Another important property to analyze is the lattice susceptibility. The definition of the lattice susceptibility at the $(n+1)$th stage of the RG analysis is given by the equation

$$\chi_M^{(n+1)} = \frac{1}{N^{(n+1)}} \sum_{i,j} \langle S_i^{(n+1)} S_j^{(n+1)} \rangle \tag{10.83}$$

From the LRS scheme, we have that

$$\sum_{i,j} \langle S_i^{(n+1)} S_j^{(n+1)} \rangle = b^{-2d\omega} \sum_{k,l} \langle S_k^{(n)} S_l^{(n)} \rangle \tag{10.84}$$

where we use the abbreviated notation $\sum_{k,l}$ to denote a summation over all spins in respective blocks K and l.

Since

$$N^{(n+1)} = N^{(n)}/b^d \tag{10.85}$$

it immediately follows that

$$\chi_M^{(n+1)} = b^{d(1-2\omega)} \chi_M^{(n)} \tag{10.86}$$

EXERCISE 10.14

Prove equation (10.86).

The consistent result we see from (10.78), (10.82), and (10.86) is that the choice of ω affects the relative magnitudes of m, Γ_{ij}, and χ_M from one iteration of the RG to the next. For $\omega = 1$, the values of m and Γ_{ij} are preserved while χ_M decreases. However, this is *not* the case for the critical exponents which are *invariant to the length scale* in LRS, as now discussed.

10.5 Critical Exponents from the RG Analysis

In the study of critical behavior, the critical exponents assume a special significance because of their universal character, within broad classes of systems. In this section, we show how the exponents arise from the LRS and how, in contrast to the results of the previous section, they are universal with respect to the system rescaling.

The Exponent δ

A significant reason why the LRS scheme is of theoretical interest is a consequence of the external magnetic field in the Ising Hamiltonian coupling linearly to the spins through a term of the form

$$h^{(n)} = \sum_i S_i^{(n)} \tag{10.87}$$

Using the LRS scheme, it is easy to show that the structure of this term in the Hamiltonian is preserved if

$$h^{(n+1)} = b^{d\omega} h^{(n)} \tag{10.88}$$

which, coupled with the equation for $m^{(n+1)}$, namely

$$m^{(n+1)} = b^{d(1-\omega)} m^{(n)} \tag{10.89}$$

allows us to show that the scaling exponent δ, which describes the scaling of m with h, follows as

$$m \approx h^{1/\delta} \quad \text{at } T = T_c \tag{10.90}$$

and is independent of n in the RG scheme. We emphasize the importance to the RG scheme of preserving the structure of the Hamiltonian from one iteration to the next, and hence the significance of (10.87) and (10.88).

EXERCISE 10.15

Prove that δ is independent of n in equation (10.90).

This type of result is generally true with regard to all the other critical exponents. For example, this can be seen with the pair correlation exponent η, introduced in the previous chapter, which characterizes the scaling of $\Gamma(r)$ at the *critical temperature*:

$$\Gamma(r) \approx r^{-(d-2+\eta)} \tag{10.91}$$

In (10.91), r refers to the distance separating two spins. That this exponent can be found from any stage of the RG analysis can be deduced from the following relationships:

$$\begin{aligned}\Gamma^{(n)}(r) &\approx \Gamma^{(n-1)}(br) \approx \cdots \approx \Gamma^{(0)}(b^n r)\\ &\approx (b^n r)^{-(d-2+\eta)}\\ &\approx b^{-n(d-2+\eta)} r^{-(d-2+\eta)}\end{aligned} \tag{10.92}$$

in conjunction with (10.82).

Equation (10.92) shows that, at each lattice size, the scaling of Γ has the same exponent. Proving results for other exponents is left as an exercise for the reader (see the additional exercises at the end of this chapter).

The Relationship between η and ω_c

We have seen that establishing ω controls the relative magnitude of the of various properties in the LRS between RG iterations. Is there a way to determine the value of ω at the critical point? Here we look at this question and show that indeed the value of ω at the critical point is related to specific critical exponents.

If the distance between two points on the lattice is r, then at $T = T_c$ the pair correlation function scaling relationship in the limit $r \to \infty$ is given by the scaling law

$$\Gamma(r) \approx r^{-(d-2+\eta)} \tag{10.93}$$

Using equation (10.82), which relates Γ in the unrenormalized and renormalized lattices, it is straightforward to show that

$$\eta = d + 2 - 2d\omega_c \tag{10.94}$$

where ω_c is the value of ω at the critical point.

EXERCISE 10.16

Prove equation (10.94).

Further practice in deriving the relationship between ω and other critical exponents can be found from doing exercise 10.17.

EXERCISE 10.17

Using the results developed earlier,

$$m^{(n+1)} = b^{d(1-\omega)} m^{(n)} \tag{10.95}$$

and the scaling relationship for m, namely

$$m \approx h^{1/\delta} \quad \text{at } T = T_c \tag{10.96}$$

show that

$$\delta = \frac{\omega_c}{1 - \omega_c} = \frac{d + 2 - \eta}{d - 2 + \eta}$$

10.6 The Scaling Laws and the RG Approach

We now provide another view on how scaling laws can be derived from the RG perspective we have just been following. The approach is to formulate properties in terms of the RG scheme and then compare the results obtained this way with the regular scaling equations presented in chapter 9.

For the coexistence curve, we have the order parameter scaling law

$$m \approx (T - T_c)^{\beta} \tag{10.97}$$

and, since $\xi \approx |t|^{-\nu}$, we have that

$$m \approx \xi^{-\beta/\nu} \tag{10.98}$$

which implies that

$$\frac{m^{(n+1)}}{m^{(n)}} \approx \left(\frac{\xi^{(n+1)}}{\xi^{(n)}}\right)^{-\beta/\nu} = (1/b)^{-\beta/\nu} \tag{10.99}$$

Also, from equation (10.79), we have that

$$\frac{m^{(n+1)}}{m^{(n)}} = b^{d(1-\omega)} \tag{10.100}$$

Comparing (10.99)–(10.100) immediately leads to the result

$$\beta/\nu = d(1 - \omega) \tag{10.101}$$

At the critical point $\omega = \omega_c$, and using (10.94) in (10.101), we get the following *scaling law*:

$$\beta = \frac{\nu}{2}(d + \eta - 2) \tag{10.102}$$

If we look at the scaling of χ_M with temperature defined by the scaling relationship

$$\chi_M \approx |T - T_c|^{-\gamma} \tag{10.103}$$

then it is easy to derive the result

$$\gamma = \nu(2 - \eta) \tag{10.104}$$

which was previously referred to as Fisher's law.

EXERCISE 10.18

Prove equation (10.104) and derive the hyperscaling equation from the RG perspective.

These results serve the useful objective of demonstrating how a RG analysis can be used to provide a basis for the scaling laws. From a historical standpoint, this was an important advance over the scaling hypothesis which, as the name suggests, provided the scaling laws as hypothesized relationships. However, from a practical point of view, the RG approach can become quite intractable, as we previously demonstrated in the 2-d Ising model. To further exploit the RG approach, we need to consider its use with computer simulation methods.

10.7 Computational Techniques in Conjunction with RG Results

Even in the 2-d nearest-neighbor Ising model, the connectivities of the spins in the renormalized Hamiltonian are much more complex than in the original system, leading to insurmountable barriers for an exact analytical approach. Given these drawbacks, we now consider how computer simulation might be used, in some situations, to numerically solve for RG parameters once a renormalization scheme has been suitably identified.

We take up this issue using the 1-d Ising model as a basis for illustrating the ideas, before generalizing to more complicated Hamiltonians.

The 1-d Ising Model and Spin Tracing (Summing)

In figure 10.2 we show a schematic of the 1-d Ising model with alternate spins designated by symbols • and ∘, respectively. If we renormalize the lattice by removing spins designated by ∘ from the lattice, we must somehow account for all their possible states,

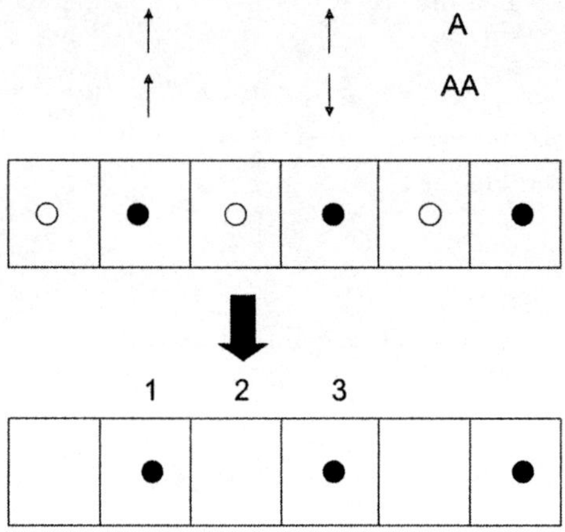

Figure 10.2 Spin tracing schematic in the Ising system

consistent with the remaining spins, as we attempt to develop the new lattice coupling parameters in the renormalized lattice. This spin tracing process requires that we trace (sum) over the states of the spins to be removed with the remaining spins pinned to specific configurations. This leads to a numerical way of estimating the new coupling parameters in the renormalized lattice.

We illustrate the concepts using the 1-d Ising model. As given previously, a partial summation of the partition function in this model can be viewed as follows:

$$\begin{aligned} Q(K,N) &= \sum_{S_1,S_3,\ldots} [\exp(K(S_1+S_3)) + \exp(-K(S_1+S_3))] \\ &\quad \times [\exp(K(S_3+S_5)) + \exp(-K(S_3+S_5))]\cdots \\ &= q(K)^{N/2} \sum_{S_1,S_3\ldots} \exp(K'S_1S_3)\exp(K'S_3S_5)\cdots \end{aligned} \tag{10.105}$$

where we have shown the terms arising from the trace over spins 2 and 4, respectively. Now imagine that we specify two particular configurations for the • spins 1 and 3 with the only difference being the relative alignment of spin 3 to spin 1 in either an aligned (A) or antialigned position (AA)—see figure 10.2. If, instead of analytically tracing over spin 2 as represented in equation (10.105), we did this trace numerically over all ∘ spins, we end up with two numerically evaluated quantities $Q_{\mathrm{AA}}(K,N)$ and $Q_{\mathrm{A}}(K,N)$, respectively. The quantity Q_{AA} is the sum with 1 and 3 • spins in the AA configuration designated by the symbol $\sum_{\circ,\bullet=\mathrm{AA}}$ while Q_{A} is the sum with these • spins in the A configuration. From (10.105), we have the following relationships: For $S_3=-1$ (the antialigned position)

$$Q_{\mathrm{AA}}(K,N) = q(K)^{N/2}\exp(-K')\sum_{S_5,\ldots}\exp(-K'S_5)\exp(K'S_5S_7)\cdots \tag{10.106}$$

while for $S_3=+1$ (the aligned position)

$$Q_{\mathrm{A}}(K,N) = q(K)^{N/2}\exp(K')\sum_{S_5,\ldots}\exp(K'S_5)\exp(K'S_5S_7)\cdots \tag{10.107}$$

from which it follows that

$$\frac{Q_{\rm AA}}{Q_{\rm A}} = \exp(-2K') \tag{10.108}$$

or

$$K' = \tfrac{1}{2}\ln\frac{Q_{\rm A}(K,N)}{Q_{\rm AA}(K,N)} \tag{10.109}$$

EXERCISE 10.19

Show that equation (10.109) follows exactly from (10.106) and (10.107).

Finding the RG coupling parameter in this manner using Monte Carlo simulation to find $Q_{\rm AA}(K,N)$ and $Q_{\rm A}(K,N)$ is, in principle, a powerful way of attacking the RG problem. We now generalize some of these ideas to lattices of higher dimension with more complicated Hamiltonians.

Use of Monte Carlo Simulation in the Renormalization of General Ising-Type Hamiltonians

If we want to extend these ideas to systems with more general Hamiltonians, it helps to develop a formalism appropriate to this problem. In this case, we consider Hamiltonians with more than nearest-neighbor interaction terms, and choose to define a general Hamiltonian

$$H = \sum_{i=1}^{m} K_i f\{S_i\} \tag{10.110}$$

where the function $f\{S_i\}$ within the summation describes the expressions for all the spins interacting in the ith-size spin cluster. The coupling parameter for the spin cluster of size i is given by the term K_i in the Hamiltonian. For example, in the *standard* 1-d Ising model, the nearest-neighbor interaction terms correspond to spin clusters with size $i=2$, and the function $f\{S_i\}$ takes on the form

$$f\{S_{i=2}\} = S_1S_2 + S_2S_3 + \cdots + S_NS_1 \tag{10.111}$$

We now prove some basic mathematical properties of these spin-cluster terms as they appear in standard Ising-like models. To do this, we first define a particular spin-cluster product of size n by the quantity t_n:

$$t_n \equiv \prod_{i\in\mathcal{S}_n} S_i \tag{10.112}$$

where $\mathcal{S}_n$ refers to the set of spins contained in the particular cluster product. For example, $t_1 = S_i$ for any single spin i, $t_2 = S_iS_j$ for any two spins i and j, $t_3 = S_iS_jS_k$ etc. The *trace* of t_n is defined as the following summation:

$$\mathrm{Tr}\{t_n\} \equiv \sum_{S_1,\ldots,S_n=\pm1}\prod_{i=1}^{n} S_i \tag{10.113}$$

from which it follows that

$$\mathrm{Tr}\{t_1\} \equiv \sum_{S_i=\pm 1} S_i = 0 \tag{10.114}$$

$$\mathrm{Tr}\{t_2\} \equiv \sum_{S_i=\pm 1} S_i \sum_{S_j=\pm 1} S_j = 0 \quad (i \neq j) \tag{10.115}$$

More generally it can be shown that

$$\mathrm{Tr}\{t_l t_k\} = 0 \quad (l \neq k) \tag{10.116}$$

$$\mathrm{Tr}\{t_l t_k\} = 2^N \quad (l = k) \tag{10.117}$$

in a system containing N spins.

EXERCISE 10.20

Prove equations (10.116) and (10.117).

The results given in equations (10.114)–(10.117) hint at the possible use of an *orthogonality* property for evaluating the system Hamiltonian if it can be expressed in terms of the various t_i. For example, consider the Hamiltonian of the original lattice given by

$$H = \sum_{i=1}^{m}{}' K_i t_i \tag{10.118}$$

where the prime on the summation in (10.118) implies that all nearest-neighbor spin clusters of size i are to be included in the sum; that is,

$$\sum_{i=1}^{m}{}' t_i = \sum_{i} S_i + \sum_{i,j} S_i S_j + \sum_{i,j,k} S_i S_j S_k + \cdots \tag{10.119}$$

It follows that

$$\mathrm{Tr}\left\{\sum_{n=1}^{m}{}' K_i t_i t_n\right\} = \mathrm{Tr}\left\{\sum_{n=1}^{m}{}' H t_n\right\} \tag{10.120}$$

and so

$$K_n = 2^{-N}\mathrm{Tr}\left\{\sum_{n=1}^{m}{}' H t_n\right\} \tag{10.121}$$

in a lattice of size N. Equation (10.121) is key and provides, in some sense, the reverse of the usual information we have been accustomed to developing thus far in this chapter.

In this equation, the coupling parameter K_n, associated with a spin-cluster product of size n, is given as a function of the Hamiltonian H instead of vice versa.

What does the complicated-looking term $\mathrm{Tr}\{\sum_{n=1}^{m}{}' Ht_n\}$ mean? If we look, for example, at the Standard Ising case when $n = 2$, then we have that

$$H\{S_i\} = \sum_{i=1}^{2}{}' K_i t_i \tag{10.122}$$

$$= \sum{}' K_1 t_1 + \sum{}' K_2 t_2$$

$$= K_1 \sum_{i=1}^{n} S_i + K_2 \sum_{i,j}{}' S_i S_j \tag{10.123}$$

which can be expanded into the more familiar expression:

$$H\{S_i\} = K_1(S_1 + S_2 + \cdots + S_N) + K_2(S_1 S_2 + \cdots + S_N S_1) \tag{10.124}$$

Thus we find that

$$\mathrm{Tr}\left\{\sum_{n=1}^{2}{}' Ht_n\right\} = \sum_{\{S_i=\pm 1\}} \left(\sum{}' t_1\right) \cdot H\{S_i\} + \sum_{\{S_i=\pm 1\}} \left(\sum{}' t_2\right) \cdot H\{S_i\} \tag{10.125}$$

and, from (10.121),

$$K_2 = 2^{-N}\left[\sum_{\{S_i=\pm 1\}} \left(\sum{}' t_1\right) \cdot H\{S_i\} + \sum_{\{S_i=\pm 1\}} \left(\sum{}' t_2\right) \cdot H\{S_i\}\right] \tag{10.126}$$

Equation (10.126) is implicit in the parameter K_2, though, since it enters into both sides of the equation.

To formulate a general RG process using this approach, we need to be able to relate the parameters from a Hamiltonian H with variables $\{S_i\}$ as we proceed to its renormalized counterpart H' with variables $\{S_i'\}$. To do this, we employ a previous result that relates H' to H, namely

$$\mathrm{Tr}\{H'\{S_i'\}\} = -g + \sum_{\bullet,\circ} e^{-H\{S_i\}} \tag{10.127}$$

where $\sum_{\bullet,\circ}$ represents a summation of states in the $\{S_i\}$ system consistent with a given state in the $\{S_i'\}$ system; that is, the spins $\{S_i'\}$ are pinned at a given state while the remaining ones in the $\{S_i\}$ system are allowed to fluctuate, during which period the sampling occurs and the required properties calculated via Monte Carlo simulation, for example. Given this, it is easy to show [5] that combining (10.127) with (10.121) leads to the result

$$K_n' = 2^{-N'} \sum_{\{S_i'\}} t_n' \sum_{\bullet,\circ} e^{-H(S_i)} \tag{10.128}$$

where the term $\sum_{\{S_i'\}} t_n'$ is calculated from the pinned states' configuration and N' is the number of spins in the renormalized lattice. Equation (10.128) shows how one can relate the coupling constants of the renormalized Hamiltonian to that of its predecessor using numerical procedures. To be consistent with the idea of an invariant Hamiltonian structure from iteration to iteration during the RG process, one should carry through a given number, say n, of the K_n parameters at each step. Of course, this does not necessarily imply that the values of any parameters K'_j, $(j > n)$, calculated using (10.128) would turn out to be zero. In fact, this is unlikely to be the case and represents a compromise that arises from the inherently approximate nature of the RG process. We are forced to accept this situation notwithstanding the conceptually attractive features of the RG process, a point discussed in detail by Binney et al. [5].

EXERCISE 10.21

Prove equation (10.128).

EXERCISE 10.22

In a 2-d Ising model (zero field, nearest-neighbor interactions only), conceptually structure a computer algorithm using Monte Carlo simulation to find K' and h' in an RG treatment using majority rule as the lattice-reduction RG technique. Describe carefully how you would find $\sum_{\bullet,\circ} e^{-H(S_i)}$ in this calculation.

10.8 Chapter Review

The RG approach provides an indispensable contribution toward a better understanding of the theory of critical phenomena. In particular, it provides a firmer theoretical basis for the concept of universality than does phenomenological scaling theory, which we described in chapter 9. Critical universality hypothesizes that the critical exponents for apparently many superficially different systems may be identical and, if so, these systems are said to belong to the same critical universality class (U class). There are three fundamental requirements for systems to belong to a given universality class. It is widely held that they must have the same spatial dimensionality, similar order-parameter character (e.g., a scalar property in the case of simple fluids and Ising magnets), and possess similar structures for their Hamiltonians. For example, in the case of both simple fluids and Ising magnets, the Hamiltonians contain only terms that represent short-range interactions. All of these issues are clearly represented in the RG description of critical behavior in these systems.

The essential idea in the RG approach is to renormalize the Hamiltonian for a given system. This is achieved by isotropically rescaling the system's linear length scale from one RG iteration to the next, all the while maintaining the same structure for the Hamiltonian. During this process, recursive relationships are derived that relate coupling parameter values in the renormalized Hamiltonian to those in its antecedent Hamiltonian. The *fixed point* of these recursive relationships defines a critical point and leads to RG predictions for the critical exponents in the system, as well as other critical properties of interest.

We illustrated these general concepts and showed how all of the well-known scaling laws for the free energy can be derived from the RG viewpoint in the Ising U class. We also looked at specific RG schemes for some Ising systems and showed how RG calculations could be carried out using the given lattice reduction schemes. This analysis quickly pointed out the essentially approximate nature of the RG approach, even in a system as familiar as the 2-d Ising model. Because of the increased lattice connectivity in the 2-d system compared to its 1-d counterpart, lattice reduction in the 2-d case leads to an indeterminate set of RG relationships for the Hamiltonian parameters. As a result, a purely analytic RG approach can only be worked out in an approximate way for this model. This drawback pointed out the need for hybrid analytic–numerical RG approaches which lead to a more exacting use of the RG method in systems with more complex topologies. We discussed the basis of one such approach which requires the use of Monte Carlo computer simulation techniques to carry out the renormalization process.

10.9 Additional Exercises

1. Using the susceptibility χ_M, show how the critical exponent ratio γ/ν can be related to the lattice normalization constant ω.
2. From the results of the previous problem, prove Fisher's law:

$$\gamma/\nu = \eta - 2$$

3. Prove the following mathematical identity:

$$x^n = [y^n]^{\ln x/\ln y}$$

4. The eigenexpansion for the coupling parameter is given by

$$\boldsymbol{M}^n\boldsymbol{k} = \boldsymbol{k}^n = \boldsymbol{k}^* + \alpha_1\lambda_1^n\boldsymbol{p}_1 + \alpha_2\lambda_2^n\boldsymbol{p}_2$$

Since $|\lambda_1| \geq 1$, we must have that $\alpha_1 = 0$ at the critical point. We can express this condition mathematically [5]:

$$\alpha_1 = at$$

with $\alpha_1\lambda_1^n = p \ll 1$. The correlation length can be expressed as

$$\xi(t) = qb^n$$

with $q < 1$. Use the identity proved in the previous problem to show that

$$\xi \propto t^{-\ln b/\ln \lambda_1}$$

Earlier in this chapter we defined $\lambda_1 \equiv b^{y_1}$, which implies that

$$\xi \propto t^{-\nu}$$

if we define $\nu \equiv 1/y_1$ as was done earlier in this chapter.
5. By rescaling the lattice dimension by a factor of two, show that the free energy of the Ising model can be expressed as

$$-\beta A(K, N, h) = \tfrac{1}{2}N\ln q(N, K, h) + \ln Q(K', \tfrac{1}{2}N, h')$$

If we define $q \equiv e^{2g(N,K,h)}$, show that the free energy can be expressed by the series expansion:

$$\frac{-\beta A(K,N,h)}{N} = \sum_{i=1}^{\infty} g_i\left(\frac{N}{2^{i-1}}, K_i, h_i\right)$$

where the K_i, h_i refer to the lattice coupling parameters at each successive reduction in scale.

6. In their paper, Maris and Kadanoff [1] show that the specific-heat scaling exponent for the 2-d Ising model defined by the renormalization equations set out in this chapter is given by the result:

$$\alpha = 2 - \ln 2/\ln\left(\frac{dK'}{dK}\right)_{K=K_c}$$

Prove this result and show that it yields the value $\alpha = 0.13$.

7. Show, for the 1-d RG results for the Ising model given by equation (10.6), that the free energies per spin ψ in the RG lattices are related by

$$\psi(K) = \tfrac{1}{2}\ln 2 + \tfrac{1}{2}K' + \tfrac{1}{2}\psi(K')$$

8. If one chose to simply ignore K_2 and K_3 in the 2-d RG equations given in this chapter, what would the coupling-parameter flow equations look like? Is there a phase transition in this system?

RG, Fractals, and Scaling

RG, fractals, and scaling analyses often draw upon a core set of similar ideas. Here are a few problems in this direction.

9. (a) The fractal dimension, d_f, of an object is defined as [4]

$$d_f = \lim_{r\to 0}\frac{\ln(n(r))}{\ln(1/r)}$$

where r is the measuring scale length and $n(r)$ the "number density" of elements covering the object (as a result, the product $n(r)\cdot r^{d_f}$ is sometimes colloquially referred to as the object's mass). If $d_f \neq d$, where d is the imbedding (Euclidean space) dimension, the object is said to follow fractal geometry. Does the 2-d lattice reduction scheme illustrated by figure 10.1 lead to a fractal structure?

(b) What would you conclude if one started out with a triangular lattice and adopted the same lattice reduction strategy referred to in part (a)?

(c) Does the 2-d lattice reduction scheme illustrated by figure 10.1 lead to a fractal structure? If so, what is d_f?

(d) What happens in 3-d structures?

10. Imagine that through simulation (or other means) you are able to calculate a property G in a (size L) lattice structure that is thought to follow the scaling relationship

$$G(Lx) = L^d G(x)$$

with x being some appropriate scaling field. If L and x obey a fractal relationship, how would you find the fractal dimension that characterizes this relationship between these two variables?

11. Random walks on a lattice structure can lead to walk trajectories that are fractal objects. Consider the square lattice shown in figure 10.1. If a random walker begins at some point in the lattice and, at each node, has an equal probability of moving one lattice spacing in any available direction, show that

$$\langle r^2(n)\rangle \approx n$$

where r is walker's displacement and n the number of steps taken. However, if the walk took place along a fractal lattice structure, show that in this case

$$\langle r^2(n)\rangle \approx n^{2/d_f}$$

where d_f is the fractal dimension of the lattice. This situation is often referred to as *anomalous diffusion* [5]—why the use of this term?

12. If we have a property that scales as

$$G(yx) = y^z G(x)$$

prove the following two identities:

$$x\left(\frac{\partial G}{\partial x}\right) = y\left(\frac{\partial G}{\partial y}\right) \tag{a}$$

and

$$xG'(x) - zG(x) = 0 \tag{b}$$

where

$$G'(x) \equiv \left(\frac{\partial G}{\partial x}\right)$$

[Note: $G(xy)$ should not be construed as being the same as $G(x, y)$!]

13. Throughout this chapter we have often observed the importance of terms like $e^z \pm e^{-z}$ in RG analyses of the Ising system. Show that these functions also enter into the solution of second-order homogeneous differential equations of the form

$$f''(x) - f(x) = 0$$

with boundary conditions $f(0) = 0,\ f(a) = a$.

14. We can get ordinary differential equations, like that given in exercise 13, from functional equations [7]. For example, show that D'Alembert's equation

$$f(x+y) + f(x-y) = 2f(x)f(y)$$

gives rise to the following ordinary differential equation:

$$f''(x) - kf(x) = 0$$

where k is constant. What is the expression for k in terms of f? Find general solutions for this differential equation for finite values of k (both positive and negative) as well as for the situation where $k = 0$. Is there a relationship that could be developed further between these results and the RG analysis given in equations (10.1)–(10.9)?

15. Are all the solutions to D'Alembert's equation odd, even, or neither type of function? We define an even function here as one where

$$f(x+y) = f(-x-y)$$

16. In the RG analyses given in this chapter (e.g., as shown in equations (10.1)–(10.9)), one often is attempting to force functions into an abstract procrustean bed (of sorts). This is epitomized by trying to satisfy functional equations (cf. equations (10.4) and (10.5)) of the form

$$f(xy) = f(x+y) + f(-(x+y))$$

Can you find a general solution to this equation and, if not, how does the RG approach attempt to resolve this dilemma in an approximate way?

17. If a lattice property G scales as follows:

$$G(Lx) = L^z G(x)$$

show that if the Fourier transform of $G(x)$ is denoted by $F_x(k)$, then

$$F_x\left(\frac{k}{L}\right) = L^{(z+1)} F_x(k)$$

What happens if $z < -1$?

18. If a property follows the familiar scaling form given by

$$G(Lx) = L^z G(x)$$

show that the following relationship holds:

$$\frac{G_L(b)}{G_L(a)} = \left(\frac{b}{a}\right)^z$$

where $G_L(a)$ denotes the property's value in a system of size L when $x = a$, etc.

19. Consider an RG analysis that leads to the following "flow" equation for the reduced temperature T in a system, away from some initial temperature T_1:

$$\frac{K(T)}{K(T_1)} = e^{\phi(T,T_1)}$$

with

$$\phi(T, T_1) = a \int_{T_1}^{T} \frac{(b + cx^q)}{x(1 + x^q)} dx$$

where constants $a \neq 0, b, c > 0$, $q = a(b - c)$. Prove that this particular transformation *always has a unique fixed point*, which we denote by T_c. Find T_c for $a = 1, q = 2$. Also show that, in general, given such a flow equation from an RG analysis, we must have

$$\left(\frac{dK(T)}{dT}\right)_{T=T_c} = l^{1/\nu}$$

where l is the rescaling factor used and ν the correlation length exponent.

20. If a property $G(x, y)$ scales with lattice size L as follows:

$$G(L(x + y)) + G(-L(y - x)) = L^z G(x)H(y)$$

(where $H(y)$ is a function of y only), then find the fixed point of the equation

$$x = G(x)$$

with $G(0) = -1$, $G'(0) = 2$.

Bibliography

Elementary [1, 2] and more advanced [3, 5] treatments of the RG approach were used as a basis for many of the ideas in this chapter.

[1] H. J. Maris and L. P. Kadanoff, "Teaching the renormalization group," *Am. J. Phys.*, vol. 6, p. 652, 1978.

[2] D. Chandler, *Introduction to Modern Statistical Mechanics*. New York: Oxford University Press, 1987.

[3] M. Plischke and B. Bergersen, *Equilibrium Statistical Physics*. Singapore: World Scientific, 1994.

[4] A.-L. Barabasi and H. E. Stanley, *Fractal Concepts in Surface Growth*. Cambridge: Cambridge University Press, 1995.

[5] J. J. Binney, N. J. Dowrick, A. J. Fisher, and M. E. J. Newman, *The Theory of Critical Phenomena*. Oxford: Clarendon Press, 1992.

[6] D. ben-Avraham and S. Havlin, *Diffusion and Reactions in Fractals and Disordered Systems*. Cambridge: Cambridge University Press, 2000.

[7] E. Castillo and M. R. Ruiz-Cobo, *Functional Equations and Modeling in Science and Engineering*, Monographs and Textbooks in Pure and Applied Mathematics, vol. 161, E. J. Taft and Z. Nashed (eds.). New York: Marcel Dekker, 1992.

11 Critical Behavior in Confined Systems

The prediction of properties in complex materials is a problem of importance in many applications in chemical and materials engineering; by the term "complex material" we mean a heterogeneous substance, like a porous material containing a confined fluid. Such materials appear in many technological applications, including: (1) processes using supercritical fluids to dry porous aeorogels and thin films [1], (2) physical adsorption of trace components from gaseous effluents, (3) gas storage using microporous materials [2], and (4) chemical separation using inorganic membranes [3]. Inorganic membranes are often highly porous and randomly structured materials with large surface areas available for adsorption, a property that makes them useful in chemical separation and as catalyst supports.

In addition to their heterogeneity, complex materials have another distinguishing characteristic that relates to the structure of the heterogeneity itself. Is it periodic, or is it dispersed throughout in some random fashion? These two situations are quite distinct and may, in each instance, show critical behavior for a confined fluid belonging to entirely different universality classes, an issue that to the present time is still unsettled in the literature.

In this chapter, we investigate the critical properties of fluids confined in randomly structured host materials like that found in porous silicon. The main question we address is: how does confinement in a porous structure affect the critical point or phase behavior of a fluid mixture? Before investigating some of the more advanced ideas in this area, we look at the basic thermodynamics of interfaces, and the phenomenon of capillarity in a single idealized pore structure. This simple example provides the impetus for a more detailed study of confinement effects.

11.1 Interfaces and Capillarity

The Laplace Equation

Consider two phases in equilibrium separated by an interface. The total energy of the composite system is the sum of the energy of each phase plus the energy associated with the interface. In formulating the fundamental thermodynamic equation for energy in this system, we presume that the formation of an interface requires energy; therefore, the energy equation must reflect this fact. Thus, we use the following well-known form for E in the system:

$$dE = T\,dS - p\,dV + \mu\,dN + \gamma\,d\sigma \tag{11.1}$$

where γ is the intensive variable associated with the formation of the interface's surface area denoted by the variable σ. The property γ is usually referred to as the *surface tension* and it must be positive (why?).

Now for simplicity imagine a spherical interface in which a spherical vapor bubble or liquid droplet is surrounded by liquid or vapor, respectively. The film surrounding this spherical shape is the interface. At equilibrium, since $dE = 0$, we have that the incremental reversible work associated with the formation of this shape is given by

$$p\,dV \equiv \gamma\,d\sigma \tag{11.2}$$

This equation shows that the driving mechanical force for establishing the interface is pressure, an intuitively appealing result. Hence we consider a pressure differential over the interface, with p_{i} the interior pressure in the sphere and p_{o} the external pressure. In an expanding interface, $p_{\mathrm{i}} > p_{\mathrm{o}}$ and so

$$(p_{\mathrm{i}} - p_{\mathrm{o}})dV = \gamma\,d\sigma \tag{11.3}$$

For a spherical shape, $dV = 4\pi r^2 dr$, $d\sigma = 8\pi r\, dr$, and hence

$$4\pi(p_i - p_o)r^2\, dr = 8\pi\gamma r\, dr$$

or

$$p_i = p_o + \frac{2\gamma}{r} \tag{11.4}$$

This is the *Laplace equation* for mechanical equilibrium in the system of the vapor sphere in bulk fluid. When $r \to \infty$, we have that $p_i \to p_o$ (i.e., pressures are equal on either side of a plane interface). With a curved interface, a pressure differential exists with the pressure larger on the convex side of the interface. For a given r, this pressure differential increases linearly with the surface tension. Note that the system is not homogeneous, because pressure varies from one side of the interface to the other.

The Kelvin Equation

In the situation where we have a droplet dispersed in its own vapor, the droplet's chemical potential at saturation, corrected for the increase in pressure due to the interface, is found from the thermodynamic equation

$$\int_{\mu_o}^{\mu_i} d\mu = \int_{p_o}^{p_i} v\, dp \tag{11.5}$$

Thus

$$\mu(p_i) = \mu(p_o) + v(p_i - p_o) \tag{11.6}$$

From the Laplace equation, we have that

$$p_i - p_o = \frac{2\gamma}{r}$$

Therefore

$$\mu(p_i) = \mu(p_o) + \frac{2\gamma}{\rho r} \tag{11.7}$$

which, at low pressures (where $d\mu = RT d \ln p$), leads to the following variant of the *Kelvin equation*:

$$p_i = p^{\text{sat}} \exp \frac{2\gamma}{\rho RTr} \tag{11.8}$$

where we have identified p_o in equation (11.7) with p^{sat} in the two-phase system. One consequence of this result is that the pressure in the droplet is larger than the external saturated pressure.

Capillarity and Young's Equation

The phenomenon of capillarity can be understood using force balances like those just used for deriving the Laplace equation; however, the thermodynamic approach we use here is also very instructive.

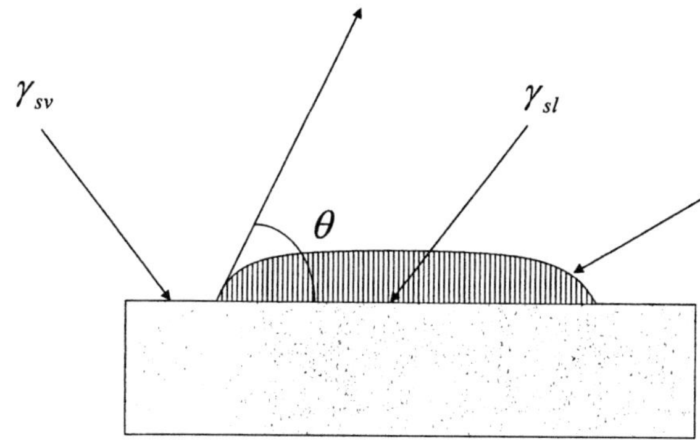

Figure 11.1 Schematic for droplet formation on a solid surface

Consider a planar solid surface in contact with a reservoir of gas. If the effect of the solid is to induce the formation of a liquid droplet on the surface (see figure 11.1), then a simple force balance leads to *Young's equation* relating the three relevant surface tensions in the system as follows:

$$\gamma_{\mathrm{lv}} \cos\theta = \gamma_{\mathrm{sv}} - \gamma_{\mathrm{sl}} \tag{11.9}$$

where θ is the angle of contact of the tangent to liquid droplet surface going through the 3-phase point of contact. Now consider a control volume as shown in figure 11.2. In this control volume, the grand potential energy at constant T and μ is given by

$$\Omega_{\mathrm{g}} = -pV + \gamma_{\mathrm{sv}} A_{\mathrm{s}} \tag{11.10}$$

where V is the volume of the system, p the pressure, and A_{s} the solid area exposed to the gas. We assume that the system is undersaturated, so that $p < p_{\mathrm{sat}}$ and $\mu < \mu_{\mathrm{sat}}$ at the given conditions. Now consider that the two semi-infinite planes, placed perpendicular to the horizontal plane, are replaced by solid surfaces at a separation H. As H decreases, one can imagine that the gas might condense (within the domain between the two walls) if the walls are sufficiently attractive. This situation is tantamount to *condensation* in

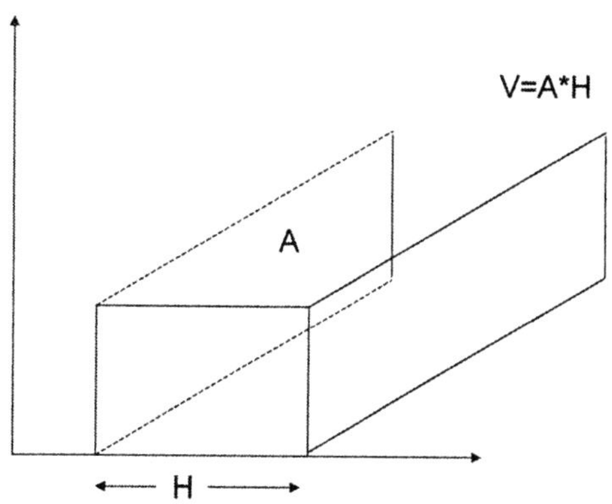

Figure 11.2 Schematic for a slit pore

a slit-pore geometry. The corresponding grand potential for the condensed system is given by the equation [4]

$$\Omega_{\mathrm{l}} = -p^{*}V + \gamma_{\mathrm{sl}}A_{\mathrm{s}} \tag{11.11}$$

where p^* is the pressure of the metastable liquid at the same T and μ. (Why is p^* not equal to p?) A first-order phase transition between the gas and liquid phases will occur if $\Omega_{\mathrm{g}} = \Omega_{\mathrm{l}}$. These results, when combined with Young's equation, provide that

$$p - p^{*} = \frac{2\gamma_{\mathrm{lv}}\cos\theta}{H} \tag{11.12}$$

This is an interesting equation since it implies that the difference between the pressures in the liquid phase and in the reservoir (bulk) can be substantial. It depends on the forces between the surface and liquid molecules as well as on the slit width H. Note that the conditions of saturation in this system, which apply when $H \to \infty$, appear nowhere in (11.12). We designate the (bulk) saturation chemical potential and pressure at this temperature by μ^{sat} and p^{sat}, respectively. If the entire system is not too far from saturation, then

$$p^{\mathrm{sat}} - p = \rho_{\mathrm{v}}(\mu^{\mathrm{sat}} - \mu) \tag{11.13}$$

$$p^{\mathrm{sat}} - p^{*} = \rho_{\mathrm{l}}(\mu^{\mathrm{sat}} - \mu) \tag{11.14}$$

from which it follows that

$$p^{*} - p = RT(\rho_{\mathrm{v}} - \rho_{\mathrm{l}})\ln\frac{p^{\mathrm{sat}}}{p} \tag{11.15}$$

Substituting (11.15) in (11.12) leads to the well-known *Kelvin equation*:

$$p = p^{\mathrm{sat}}\exp\left[-\frac{2\gamma_{\mathrm{lv}}\cos\theta}{RTH(\rho_{\mathrm{l}} - \rho_{\mathrm{v}})}\right] \tag{11.16}$$

for predicting the *capillary vapor–liquid coexistence line* in a single-slit capillary system. If a cylindrical pore were to be analyzed, the final equation would be similar to that shown in (11.16) with the radius of the pore replacing H.

At a given temperature, the Kelvin equation predicts a first-order transition in the pore at a bulk pressure p lower than p^{sat}. However, it erroneously predicts that the capillary coexistence curve ends at a critical point identical to that of the bulk fluid (why?), in disagreement with experimental observation—see figure 11.3, for example, which was adapted from data for SF_6 in a controlled-pore-size glass [5].

We now discuss the implications of some of the above results in the supercritical processing of porous materials.

11.2 Supercritical Drying of Nanoscale Porous Materials

An example of a problem where the critical behavior of a confined fluid is important involves the removal of the electrolytic solution from electrochemically etched *porous*

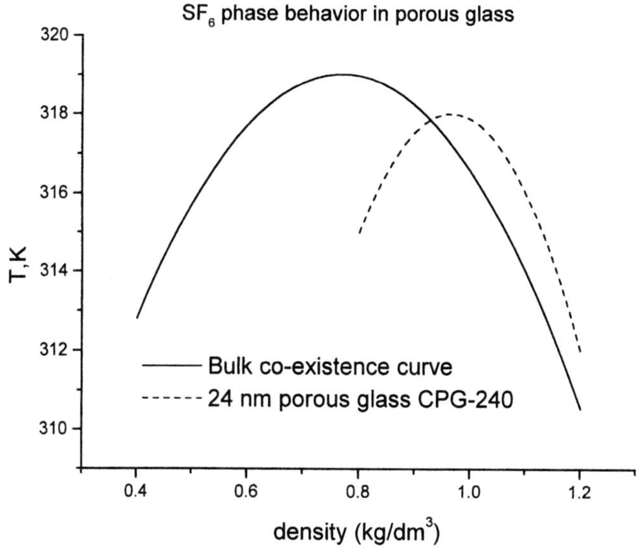

Figure 11.3 Bulk and pore-confined coexistence curves for SF_6 in porous glass

nanostructures. This is often an important step in the manufacture of porous materials like porous silicon [1] where straightforward air-drying leads to the formation of vapor–liquid interfaces inside the material. These interfaces, as one might expect, produce capillary-induced tensile stresses which often destroy the fragile pore structure, especially in highly porous materials with porosities exceeding 80%. Highly porous silicon structures are the most desirable. For example, in optical applications, they provide high quantum yields when photoluminescence (PL) is a desired property of the material. To avoid capillary-induced stresses, the use of supercritical fluids as drying agents has been proposed with the purpose of avoiding the intrapore fluid's vapor–liquid phase boundary during drying [1].

This can be illustrated with the schematic shown in figure 11.4, where the coexistence phase boundaries of two fluids are shown ending at the bulk-fluid critical points A and C, respectively. The dotted lines show how the critical point of the confined fluid changes at progressively smaller pore sizes—a feature we shall say something about in subsequent sections of this chapter. Drying of the porous material is accomplished by first expelling etchant or any other liquid with a liquid, like carbon dioxide, at low temperature and high pressure. This is followed by increasing the temperature and decompression in the system. Two such thermodynamic trajectories designated E and F are shown in figure 11.4. The trajectory E is always *subcritical* for fluid C regardless of the shift in T_c caused by confinement effects. For fluid A, this is true for only a specific range of pore sizes. Trajectory F on the other hand is always *supercritical* for A and subcritical for C. As such, A is likely to be successful as a drying fluid for this system while the opposite is true for C.

This process is just one illustration of the importance of understanding the critical behavior of fluids in pore-confined systems. Another one of increasing interest is the proposed use of solid materials for hydrogen storage for fuel-cell-powered devices

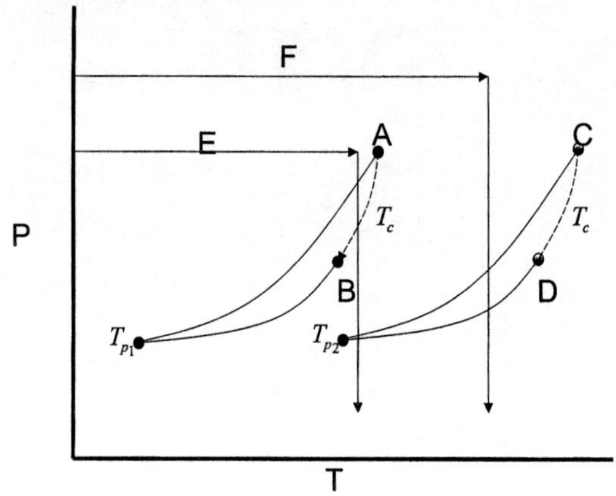

Figure 11.4 Schematic of a supercritical-fluid drying process

[6], [13]. This continues to be an active area of research, since the phase behavior of the confined fluid can be quite different to that of its bulk-fluid counterpart, especially in small (nanoscale) pores. In this respect, the Kelvin equation developed above (for an idealized pore geometry) is not accurate, and more sophisticated approaches are required.

11.3 Critical Scaling for Systems Confined between Parallel Walls

Consider a slit pore consisting of noninteracting parallel plates separated by a distance D in between which we find the confined fluid. What is the nature of the confined fluid's critical point as D decreases? At wide separations, it seems reasonable to assume that bulk 3-d scaling behavior should pertain; at lower separations, however, when the slit width approaches the order of a few molecular spacings, one might expect the critical scaling behavior to resemble that of a 2-d system. Given the knowledge gleaned from studying chapters 8 and 9, the temperature at which this *crossover* might be expected to occur would be when the correlation length ξ becomes approximately equal to D. This would imply a confined-fluid critical temperature dependent upon D among other things.

Fisher and Nakanishi [7] considered this problem in an Ising system. They proposed a reasonably familiar form for the singular part of the free energy, given by

$$f_s(T, D, H, H_1) \approx |t|^{2-\alpha} W(D|t|^{\nu},\ H/|t^{\Delta}|,\ H_1/|t|^{\Delta_1}) \qquad (11.17)$$

where W is an unknown, but universal, scaling function. The variable H_1 is new in this text and represents a field due to the walls with an associated surface-field scaling exponent Δ_1. It accounts for the influence of the slit wall on the adjacent row of spins. Note the close analogy between (11.17) and the scaling function for bulk fluids given in chapter 9. In (11.17), the (dimensionless) temperature t is defined as $t \equiv [T_c(D) - T_c(\infty)]/T_c(\infty)$, which we point out is not zero at the critical point of the slit-confined fluid. Also, in the limit as t, H, $H_1 \to 0$ and $D \to \infty$, we correctly get the usual bulk-fluid scaling form.

The Hamiltonian that accompanies equation (11.17) is given by

$$H = -J \sum_{i,j}{}' S_i S_j - H \sum_i S_i - H_1 \sum_i{}' S_i \tag{11.18}$$

where the superscript 1 on the last term in (11.18) signifies that the surface field is assumed to act only on the first layer of spins adjacent to the walls.

EXERCISE 11.1

Show that a mean-field theory satisfies the scaling function in (11.17) with exponents given by $\alpha = 0$, $\nu = \frac{1}{2}$, and $\Delta_1 = \frac{1}{2}$.

The central question now is to find expressions for the *critical-point shifts* caused by confinement in the slit pore. These occur in both temperature and field variables and are defined as follows:

$$\begin{aligned} \Delta T_c &\equiv T_c(D, H_1) - T_c^{\infty} \\ \Delta H_c &\equiv H_c(D, H_1) \end{aligned} \tag{11.19}$$

where T_c^{∞} is the bulk ($D \to \infty$) critical temperature. At the critical point, $f_s = 0$; and since $t \neq 0$, we must have $W = 0$ there. If we rewrite H/t^{Δ} as

$$\frac{H}{t^{\Delta}} = \frac{HD^{\Delta/\nu}}{(Dt^{\nu})^{\Delta/\nu}} \tag{11.20}$$

we can express W in terms of the variables $W(D|t|^{\nu}, HD^{\Delta/\nu}, H_1 D^{\Delta_1/\nu})$ which, together with the result that $W = 0$ at the slit's critical point, implies that [7]

$$D^{1/\nu}|t| = -\beta_T X_c(H_1 D^{\Delta_1/\nu}) \tag{11.21}$$

and

$$HD^{\Delta/\nu} = -\beta_H X_c(H_1 D^{\Delta_1/\nu}) \tag{11.22}$$

where β_T, β_H, and X_c are functions of the shown arguments. From equation (11.21):

$$|t| \equiv -\beta_T D^{-1/\nu} X_c(H_1 D^{\Delta_1/\nu}) \tag{11.23}$$

For the *weak-wall* limit $H_1 \to 0$,

$$|t| = -\beta_T D^{-1/\nu} X_c(0) \tag{11.24}$$

which implies that

$$\Delta T_c = T_c(D) - T_c(\infty) = -\beta_T T_c(\infty) D^{-1/\nu} X_c(0) \tag{11.25}$$

Therefore, the critical-point shift with D predicted by the scaling form given in equation (11.17) is

$$\Delta T_c \propto D^{-1/\nu} \tag{11.26}$$

This is the first equation presented in this text that suggests a specific relationship between the shift in T_c and geometric confinement. It says that the temperature shift decreases according to a power law, but does not provide any indication as to when crossover from 3-d to 2-d behavior occurs. This analysis is for a single idealized pore geometry; however, in practice, the more interesting cases will concern critical behavior in more realistic substrates.

EXERCISE 11.2

How would you study crossover in a slit pore using equation (11.26)?

11.4 Critical Phenomena in Disordered Systems

Spinodal Decomposition

A common way of generating more realistic substrate structures relies upon the use of computer simulation via a process referred to as *spinodal decomposition*. The structure of porous *Vycor glass*, for example, can be represented using this technique. It mimics the mixing of two types of molten glass that are subsequently allowed to phase-separate for a short time, followed by cooling and chemical etching of one of the two component glasses, leaving behind a porous substrate.

The Ising model can be used to simulate this process with the now familiar Hamiltonian given by

$$\beta H_{sp} = -J_{sp} \sum{}' S_i S_j \tag{11.27}$$

where J_{sp} is the spin–spin interaction parameter. Spins are located at positions r_i on a simple cubic lattice, and each can assume either of the values $S_i = \pm 1$ representing the two types of glass. At the beginning of the simulation, a spin is located at each site in the lattice, each randomly allocated to one of these two spins states. The system is then simulated using spin-exchange Monte Carlo kinetics as described in chapter 9. After a given number of steps, the process is halted and upward-pointing spins are removed, which models quenching of the structure, while the remaining downward-pointing spins constitute the model of the remaining Vycor glass substrate. The structure factor $S(k)$of the porous-medium model created this way is defined in the usual manner as

$$S(k) = \left| \sum_j e^{ikr_j} s_j \right|^2 \tag{11.28}$$

From the structure factor, a mean pore diameter is defined as [8]

$$D = \frac{\pi}{\langle k \rangle} \tag{11.29}$$

where

$$\langle k \rangle = \frac{\int_0^\infty k S(k) dk}{\int_0^\infty S(k) dk} \tag{11.30}$$

Finite-Size Scaling in Spinodally Decomposed Porous Media: Ising Simulations

MacFarland et al. [8] studied the critical behavior of a spin system confined in such a spinodally decomposed medium using finite-size-scaling computations. Their model consisted of spins in the "holes" left after spinodally decomposing the material as just described. This type of model is often referred to as a *randomly site-diluted* Ising model for obvious reasons.

The form of the scaling function for the remaining spins in these random porous structures is not precisely known, but a logical starting point is the scaling hypothesis for finite-size homogeneous systems described in chapter 9, where we postulated that in such a system the susceptibility χ of the system scales as [9]

$$\chi \propto L^{\gamma/\nu} \bar{\chi}(tL^{1/\nu}) \tag{11.31}$$

where $\bar{\chi}$ is a universal scaling function. As in the homogeneous case, finite-size simulation can be used to generate data for the quantity $\chi L^{-\gamma/\nu}$ as a function of $tL^{1/\nu}$, varying the parameters γ/ν, t, and $1/\nu$ to see if all the results are observed to collapse onto a single curve. If so, this would lend credence to the scaling function used. The results of MacFarland et al. [8], however, are inconclusive in this regard, showing considerable scatter about the "single curve" graph. This casts some doubt on the appropriateness of equation (11.31) for this system. To the present time, this remains an unsolved problem in critical phenomena.

Nevertheless, these authors proceeded to find the order-parameter exponent β. It can be found by appealing to the finite-size-scaling relationship for $M(t)$ given in the homogeneous case by

$$M_L(T_c) \propto L^{-\beta/\nu} \tag{11.32}$$

from which it follows that, for two distinct system sizes L_1 and L_2, we may define a function $\phi_{L_1,L_2}(T)$ as

$$\phi_{L_1,L_2}(T) = \log_{L_1/L_2} \frac{M_{L_1}(T)}{M_{L_2}(T)} \tag{11.33}$$

with the property that, at T_c,

$$\phi_{L_1,L_2}(T_c) = -\beta/\nu \tag{11.34}$$

EXERCISE 11.3

Prove equation (11.34).

Therefore we see that, when sets of curves $\phi_{L_1,L_2}(T)$, found from simulation data, are graphed for different choices of L_1 and L_2 at various temperatures, they should all intersect at the point $T = T_c$ where $\phi_{L_1,L_2}(T_c) = -\beta/\nu$. This approach, once again, in principle provides a method for locating both the critical temperature and the exponents β and ν in a porous system with average pore size D. The results of

MacFarland et al. [8] yielded a critical-temperature dependence on average pore size of the form

$$T_c(D) = 4.512 - 3.6\,D^{-1} \tag{11.35}$$

Since $T_c(\infty) \approx 4.512$ is the currently accepted value of the critical temperature for the pure 3-d Ising model, it is a boundary condition in (11.35) for the limiting case $D \to \infty$. We note that this function is similar (but not identical) to equation (11.26) given earlier in respect of scaling phenomena for an idealized single slit pore. That result suggests a function of the form

$$T_c(D) = 4.512 - a D^{-1/\nu} \tag{11.36}$$

with a some constant. Equation (11.36) also provided a reasonable fit to simulation data. This ambiguity underscores the current uncertainty in our knowledge of critical behavior in such systems. We point out that, in the work of MacFarland et al., which represents one of the few critical phenomena simulation studies with the equivalent of pore-confined systems, no spin–surface interactions were accounted for—which is obviously a very significant omission! This means that only the effects of physical confinement could possibly be present in their results.

EXERCISE 11.4

If the exponent $1/\nu$ were to be calculated using (11.36) for Ising spins confined in a spinodally decomposed material, which critical universality class would you expect it to fall within?

The critical exponents provided by MacFarland et al. in comparison with various other well-known Ising models are shown in table 11.1.

The *Gaussian RFIM* is an Ising model with spins located at each site, each one of which is subject to a local field, whose values are drawn randomly from a Gaussian distribution. The *site-diluted random-field Ising model* (*SDRFIM*) is an Ising system for which a number of spins have been randomly removed from the structure, thereby diluting the density of spins in the lattice as described in the previous section. Both of these models have been proposed as models of the critical universality class for confined fluids. The random fields in the RFIM are considered to be representative of

Table 11.1 Comparison of critical exponents in various Ising models

	Gaussian RFIM [10] $h/T = 0.35$, 3-*d*	*Site-Dilute* [15] *Ising Model* 3-*d*	*Bulk Ising Model* 3-*d*	*System of MacFarland et al.* [8]
α	−0.500	−0.0517	0.125	
β	0.250	0.3546	0.326	0.60 ± 0.13
γ	1.700	1.3420	1.239	1.40 ± 0.1
ν	1.100	0.6837	0.63	0.81 ± 0.1
η	0.505	0.0374	0.040	

the interactions a fluid particle would have with a porous medium, but in its standard form implies that every particle interacts with a pore surface; this is clearly not the case in a realistic substrate. The SDRFIM wards off certain regions of space (through the dilution process) to particles but does not account for particle–surface interactions, so the correspondence to the fluid-in-porous-medium problem for both of these models is not even qualitatively correct, an issue we discuss in the next section. What is evident from the results shown in table 11.1, though, is that none of these situations appear to be consistent with the simulation data of MacFarland et al. The correspondence between these standard Ising models and the fluid-in-porous-medium problem remains an open question in the extant literature to the present.

While computer simulation results as just described are useful for this complex fluid structure, analytical theories are valuable in their own right. In confined-fluid systems, there has been little published on this topic for realistic substrates. However, in the next section we describe a recent approach to this problem in quenched random porous structures.

11.5 Fluid Phase Transitions in Quenched Random Porous Structures

The Hamiltonian for a Confined Lattice Gas

In porous systems, the host material consists of void space with complementary regions taken up by solid matrix (adsorbate) material. In the void space, one will generally find isolated molecules, those with nearest neighbors, molecules adjacent to pore surfaces, and simply empty regions. Furthermore, pore surface heterogeneities will generally be present, and all of these effects should be captured in any realistic analytical theory.

De et al. [11] considered a lattice gas excluded from a fixed matrix of solid sites randomly assigned throughout the structure (see figure 11.5 for a 2-d schematic), with the remaining lattice sites being void spaces that may, or may not, be occupied by fluid particles. Nearest-neighbor fluid particles interacted through a constant interaction energy while solid-matrix–fluid energetic interactions were represented by a probability density function that represents the effect of binding-site energy heterogeneities on the

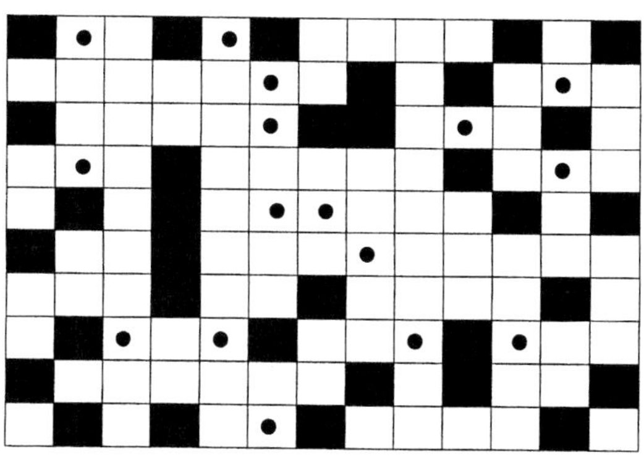

Figure 11.5 Schematic of a random quenched pore structure

solid surface. They considered a simple cubic lattice in d dimensions and, for each site, defined a "quenched" random variable ε_i, where ε_i can take on either the values 0 or 1. The value of 0, found with probability p, implies the existence of a solid matrix particle at that position in the lattice, while the value 1 designates a void space. In addition, at each void site they assigned an "annealed" variable n_i which can assume either the value 0, denoting the absence of a fluid particle, or 1 which represents the presence of a fluid particle. The variable n_i is the usual density variable associated with the lattice-gas partition function. For a completely random structure (like highly porous aerogels), the variables ε_i are assumed to be uncorrelated. The Hamiltonian for such a system is given by the equation

$$-H_{\mathrm{LG}} = 4\Im \sum_{\langle i,j\rangle} \varepsilon_i \varepsilon_j n_i n_j + \Gamma \sum_{\langle i,j\rangle} [\varepsilon_i n_i(1-\varepsilon_j) + \varepsilon_j n_j(1-\varepsilon_i)] + \mu \sum_i \varepsilon_i n_i \tag{11.37}$$

where $4\Im$ is the coupling constant between two adjacent fluid particles, Γ the coupling constant between a fluid and solid particle, and μ the chemical potential of the confined fluid. The summation in (11.37) (represented by the notation $\langle i, j\rangle$) is over all (i, j) pairs, but the effect of the ε_i and ε_j variables is to ensure that the first term in the Hamiltonian captures fluid–fluid interactions and the second term represents fluid–solid ones while the third term is the standard field term. By using the relationship connecting lattice-gas variables n_i and Ising spin variable s_i, namely

$$s_i \equiv 2n_i - 1 \tag{11.38}$$

we can easily transform (11.37) into its equivalent Ising Hamiltonian (up to a constant term) given by

$$-H_{\mathrm{I}} = J \sum_{\langle i,j\rangle} \varepsilon_i \varepsilon_j s_i s_j + K \sum_{\langle i,j\rangle} [\varepsilon_i s_i(1-\varepsilon_j) + \varepsilon_j s_j(1-\varepsilon_i)] + B \sum_i \varepsilon_i s_i + F(\varepsilon_i, \ldots, \varepsilon_j) \tag{11.39}$$

where

$$F(\varepsilon_i, \ldots, \varepsilon_j) = \left[\frac{\mu}{2} + \frac{\Gamma z}{2}\right] \sum_i \varepsilon_i - \Gamma \sum_{\langle i,j\rangle} \varepsilon_i \varepsilon_j \tag{11.40}$$

The variable z in (11.40) represents the lattice coordination number, while the function $F(\varepsilon_i, \ldots, \varepsilon_j)$ is dependent only upon the quenched variables ε_i and the parameters shown. At a specific value of the chemical potential it is constant for a given realization of the disorder in the system, and hence can be omitted from subsequent consideration.

The relationships between the various parameters in the two Hamiltonians given in equations (11.37) and (11.39) are given by

$$J = \Im \tag{11.41}$$

$$K = \tfrac{1}{2}\Gamma - \Im \tag{11.42}$$

$$B = \tfrac{1}{2}\mu + \Im z \tag{11.43}$$

Mean-Field Treatment: The Confined Lattice Gas

We now develop a mean-field approximation for the exact Hamiltonian given in (11.37). The "exact" thermal average of molecular density at site i, which we define as $\langle n_i \rangle \equiv \rho$, is given rigorously by the statistical-mechanical equation

$$\langle n_i \rangle \equiv \rho = \frac{\sum_{n_i=0,1} \cdots \sum_{n_k=0,1} n_i \exp[\beta H_{\text{LG}}]}{\sum_{n_i=0,1} \cdots \sum_{n_k=0,1} \exp[\beta H_{\text{LG}}]} \tag{11.44}$$

where H_{LG} is given by equation (11.37) and $\beta \equiv 1/kT$.

In the mean-field regime, however, we can simplify equation (11.44) considerably. Consider a site i surrounded by z sites each of which may be blocked by the solid, empty, or contain another molecule. The fundamental mean-field assumption is that molecules in any of these surrounding sites must be at their respective mean densities. If, among these z neighboring sites, N of them are blocked, then the probability of observing this configuration is given by the quantity Ω_N defined as

$$\Omega_N = \frac{z!}{(z-N)!N!}(1-p)^{z-N}p^N \tag{11.45}$$

with the total number of all possible configurations designated as $\Omega_{\text{total}} = \sum_{N=0}^{z} \Omega_N = 1$. Using this approach, equation (11.44) can be approximated in the mean-field situation by summing over the values $n_i = 0, 1$ to get

$$\langle n_i \rangle = \frac{1}{\Omega_{\text{total}}} \sum_{N=0}^{z} \frac{\Omega_N \exp[\beta(y_N + \mu)]}{\exp[\beta(y_N + \mu)] + 1} \tag{11.46}$$

or

$$\langle n_i \rangle = \frac{1}{\Omega_{\text{total}}} \sum_{N=0}^{z} \frac{\Omega_N \exp[\frac{1}{2}\beta(y_N + \mu)]}{\exp[\frac{1}{2}\beta(y_N + \mu)] + \exp[-\frac{1}{2}\beta(y_N + \mu)]} \tag{11.47}$$

which can be simplified to yield

$$\langle n_1 \rangle = \frac{1}{2\Omega_{\text{total}}} \sum_{N=0}^{z} \Omega_N \left[1 + \tanh \tfrac{1}{2}\beta(y_N + \mu)\right] \tag{11.48}$$

where, in equations (11.46)–(11.48), we have made use of a quantity y_N defined as

$$y_N \equiv 4\Im\langle n_i \rangle(z - N) + N\Gamma \tag{11.49}$$

Substituting equation (11.45) in equation (11.48) yields the following result for the density of the confined fluid:

$$\rho = \tfrac{1}{2}\left\{1 + \sum_{N=0}^{z} \frac{z!}{(z-N)!N!}(1-p)^{z-N}p^N \cdot \tanh[\tfrac{1}{2}\beta(4\Im\rho(z-N) + N\Gamma + \mu]\right\} \tag{11.50}$$

which to first order in p yields the equation:

$$\rho = \tfrac{1}{2}\{1 + (1 - zp)\tanh[\beta'(2z\rho + \tfrac{1}{2}\mu')] + zp\tanh[\beta'(2\rho(z-1) + \tfrac{1}{2}\Gamma' + \tfrac{1}{2}\mu')]\} \tag{11.51}$$

where we have used the following dimensionless quantities:

$$\beta' \equiv \beta\Im \tag{11.52}$$

$$\mu' \equiv \mu/\Im \tag{11.53}$$

$$\Gamma' \equiv \Gamma/\Im \tag{11.54}$$

$$T' \equiv 1/\beta' = kT/\Im \tag{11.55}$$

Equation (11.51) is the *mean-field equation of state for a confined fluid in a highly porous quenched random structure*. Developing the corresponding equations for the Ising system is given as an exercise in the additional exercises. In this case, the $K = 0$ solution is equivalent to the well-known *site-diluted Ising model* in the mean-field regime.

The Critical Point for the Model

The conditions for the critical point in this system are defined by the following two thermodynamic equations:

$$\left(\frac{\partial \mu'}{\partial \rho}\right)_{\beta'_c} = 0 \tag{11.56}$$

$$\left(\frac{\partial^2 \mu'}{\partial \rho^2}\right)_{\beta'_c} = 0 \tag{11.57}$$

To establish these derivatives, we rewrite equation (11.51) more generally as

$$\rho = f(\rho, \beta', \mu') \tag{11.58}$$

which together with (11.56) and (11.57) give rise to the following equations at the critical point:

$$\rho_c = f(\rho_c, \beta'_c, \mu'_c) \tag{11.59}$$

$$\left(\frac{\partial f}{\partial \rho}\right)_{\beta'_c, \mu'_c} = 1 \tag{11.60}$$

$$\left(\frac{\partial^2 f}{\partial \rho^2}\right)_{\beta'_c, \mu'_c} = 0 \tag{11.61}$$

The Low-p Limit: for use in Highly Porous Aerogels

The conditions given by (11.59)–(11.61) provide three equations that can be solved for the critical values of T'_c, μ'_c, and ρ_c; however, these are implicit equations, and for the low-p limit we use an analytic approximation for (11.51). This is done by defining *perturbation functions* χ and ϕ as follows:

$$\mu' - \mu'_{p=0} \equiv \chi(\beta', \Gamma')p \tag{11.62}$$

$$\rho - \rho_{p=0} \equiv \phi(\beta', \Gamma')p \tag{11.63}$$

The functions χ and ϕ are, as yet, unknown. However, solutions of (11.51) for $p = 0$ yield the values $\mu'_{p=0} = -2z$ and $\rho_{p=0} = \frac{1}{2}$. Using these results in equations (11.60) and (11.61), we find that

$$2\phi(1 - \beta' z) = 2\beta'\chi + z\tanh[\beta'(\tfrac{1}{2}\Gamma' - 1)] \tag{11.64}$$

and

$$2z^2\beta'(z\phi + \chi) + (z-1)^2 z\tanh[\beta'(\tfrac{1}{2}\Gamma' - 1)]\,\mathrm{sech}^2[\beta'(\tfrac{1}{2}\Gamma' - 1)] = 0 \tag{11.65}$$

Solving both of these equations simultaneously for χ and ϕ leads to the results

$$\phi = \frac{1}{2z}\tanh[\beta'(\tfrac{1}{2}\Gamma' - 1)] \cdot \{z^2 - (z-1)^2\,\mathrm{sech}^2[\beta'(\tfrac{1}{2}\Gamma' - 1)]\} \tag{11.66}$$

and

$$\chi = \frac{1}{2z\beta'}\tanh[\beta'(\tfrac{1}{2}\Gamma' - 1)] \cdot [z\beta'\{(z-1)^2\,\mathrm{sech}^2[\beta'(\tfrac{1}{2}\Gamma' - 1)] - z^2\} - (z-1)^2\,\mathrm{sech}^2[\beta'(\tfrac{1}{2}\Gamma' - 1)]] \tag{11.67}$$

When (11.66) and (11.67) are combined with equations (11.51) and (11.60)–(11.63), they yield the following mean-field results for the critical properties ρ_c, T'_c and μ'_c in the confined fluid, to first order in p:

$$\rho_c = \frac{1}{2} + \frac{1}{2}\left[\tanh\frac{\Gamma' - 2}{2z}\right]\left[z^2 - (z-1)^2\,\mathrm{sech}^2\frac{\Gamma' - 2}{2z}\right]p \tag{11.68}$$

$$T'_c = z\left[1 - p\left(z - (z-1)\,\mathrm{sech}^2\frac{\Gamma' - 2}{2z}\right)\right] \tag{11.69}$$

$$\mu'_c = -2z^2\left(\tanh\frac{\Gamma' - 2}{2z}\right)p - 2z \tag{11.70}$$

Note that equation (11.69) predicts that the critical temperature of the confined fluid will always be less than that of the bulk fluid, a result consistent with available data.

At the $p = 0$ limit, we may use equations (11.68)–(11.70) to get the critical properties of the pure 3-d lattice-gas model within a mean-field approximation. These results are given by

$$\rho_c(p=0) = \tfrac{1}{2} \tag{11.71}$$

$$T'_c(p=0) = z \tag{11.72}$$

$$\mu'_c(p=0) = -2z \tag{11.73}$$

and conform with the established results for this system given in chapter 8.

EXERCISE 11.5

Using equations (11.68)–(11.70), establish the critical properties of the site-diluted Ising model in the mean-field regime. [Hint: Use equations (11.37)–(11.43).]

Energy Heterogeneity in the Fluid–Solid Interaction

These results can be extended to the situation where the fluid–solid interaction term is given by a statistical distribution—that is, where the surface is composed of sites with energetically heterogeneous binding energies. In this case, we assume that Γ' is given by a probability distribution $\eta(\Gamma')$.

The mean-field equation of state can be modified so that the average of the right-hand side in (11.69) is now taken, leading to a critical temperature of the confined fluid given by

$$T'_c = z[1 - p\{z - (z-1)\gamma\}] \tag{11.74}$$

where γ is the "average" of $\operatorname{sech}^2(\Gamma'/z)$, defined by the integral

$$\gamma = \int \eta(\Gamma')\operatorname{sech}^2\frac{\Gamma'-2}{2z}d\Gamma' \tag{11.75}$$

For a narrow distribution around some average energy Γ'_0, we derive a simple expression for γ by Taylor expansion of $\operatorname{sech}^2[(\Gamma'-2)/2z]$. This yields the result

$$\gamma = \operatorname{sech}^2\frac{\Gamma'_0-2}{2z} + \frac{1}{2}\left[\frac{d^2\operatorname{sech}^2\frac{\Gamma'-2}{2z}}{d\Gamma'^2}\right]_{\Gamma'=\Gamma'_0}\int(\Gamma'-\Gamma'_0)^2\eta(\Gamma')d\Gamma' \tag{11.76}$$

For a narrow Gaussian distribution with variance $\sigma^2 \ll \Gamma'_0$, we find that

$$\gamma = \operatorname{sech}^2\frac{\Gamma'_0-2}{2z} + \frac{1}{2}\left[\frac{d^2\operatorname{sech}^2\frac{\Gamma'-2}{2z}}{d\Gamma'^2}\right]_{\Gamma'=\Gamma'_0}\sigma^2 \tag{11.77}$$

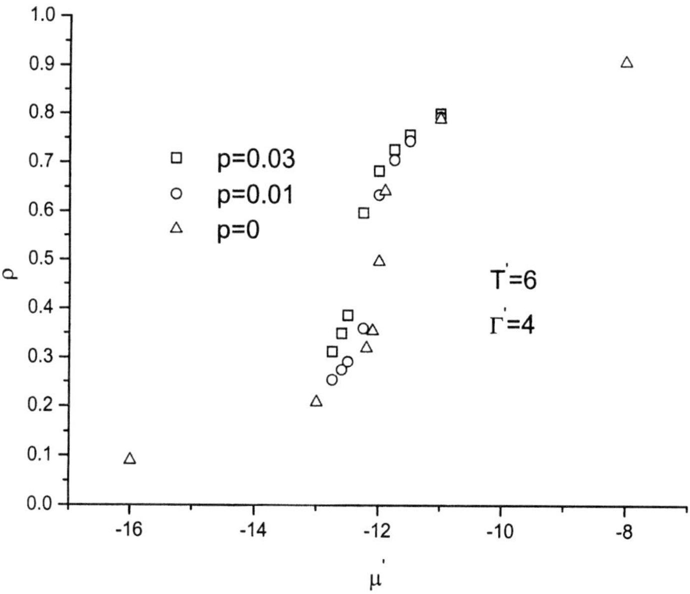

Figure 11.6 Mean-field equation of state prediction of adsorption isotherms for a fluid confined in a random quenched pore structure

Model Predictions

We now illustrate some of the predictions available from this mean-field theory. In figure 11.6, we show two adsorption isotherms in a 3-d system at various values of p calculated at a dimensionless temperature $T' = 6$ with the value of the fluid–pore-surface interaction parameter $\Gamma' = 4$. This temperature is the critical one for the zero-p case, as derived in (11.72), and the infinite slope of the $p = 0$ adsorption isotherm is evident at the critical chemical potential ($\mu'_c(p = 0) = -2z = -12$). As p increases, the system moves further away from criticality (into a supercritical temperature regime), and the slopes of the various isotherms become finite throughout the region shown.

The variation of the system's critical temperature with the fluid–pore-surface interaction parameter Γ', at various values of p, is presented in figure 11.7, where we observe that the critical temperature decreases with increasing values of p, a result that follows mathematically from equation (11.69).

The effect of porosity on the *density enhancement* of the adsorbed fluid, relative to the bulk fluid in thermodynamic equilibrium with it, is also seen in the results given in figure 11.6. In the technologically interesting case (attractive fluid–pore-surface interactions, i.e., $\Gamma' > 0$), the adsorbed fluid density is enhanced with increasing p at a given value of both Γ' and temperature. This enhancement is particularly pronounced in the region of lower pressure (i.e., lower chemical potential)—a result that forms the basis for practical applications involving light-gas storage in porous media. One would also expect that increasing the fluid–pore-surface interaction parameter Γ', at a given temperature and value of p, should increase the magnitude of the adsorbed fluid's density. This is

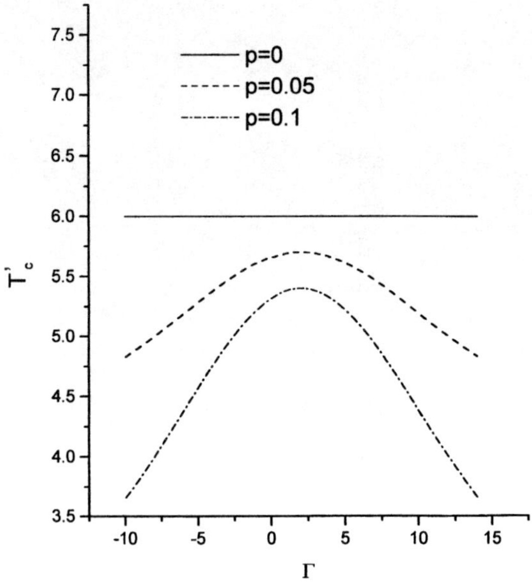

Figure 11.7 Mean-field equation-of-state prediction for the critical properties of a fluid confined in a random quenched pore structure

the case, as shown in the results presented in figure 11.8 for the situation $p = 0.03$ and a dimensionless temperature $T' = 6$.

In any real material, the fluid–pore energy interaction parameter Γ' is likely to be heterogeneous. For the results shown in figure 11.9 for $p = 0.1$, the distribution of the energy heterogeneity was assumed to follow a Gaussian form with various values for the *mean-to-variance* ratio. The effect of this heterogeneity on the critical temperature of the system is to effectively widen the distribution of the $T_c'(\Gamma')$ function about Γ'_{max}.

Comparison with Computer Simulation Results

Some comparisons between this mean-field model and GCMC computer simulation results in a confined system have been made [11]. These are given in figures 11.10 and 11.11. The *reducing parameters* for all the thermodynamic properties in these comparisons are the corresponding critical values in the respective pure systems (i.e., $p = 0$). These are known exactly for the lattice gas and used here for the simulation results, while those for the mean-field model were found from equations (11.71)–(11.73).

Figure 11.10 shows a comparison of adsorption isotherms of both the mean-field model and simulation results in a highly porous system ($p = 0.01$) at a slightly supercritical temperature corresponding to a reduced temperature of 1.01. In figure 11.11, comparisons are shown at a lower porosity and two different temperatures. The average proportional absolute deviation between the mean-field model and simulation results at the lower (closer to critical) temperature is about 10.3%, while at the higher temperature it is 7.4%.

The model can also be used in conjunction with experimental data. Adsorption data for methane in a silica aerogel at 308 K are shown with the model representations

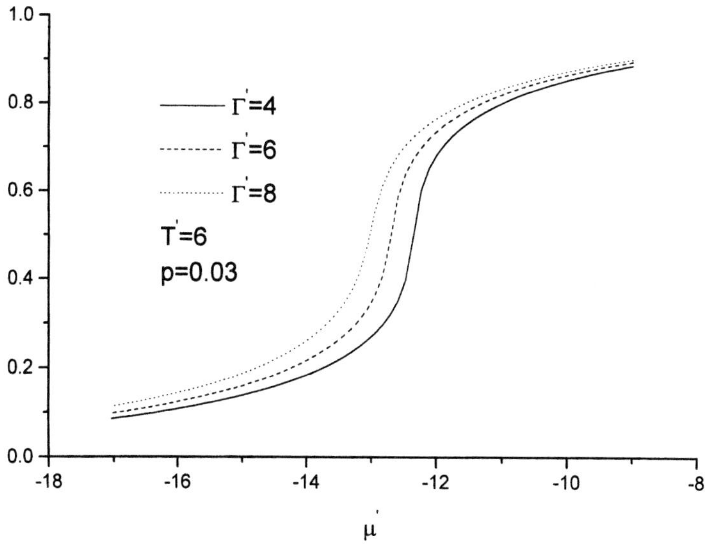

Figure 11.8 Mean-field equation-of-state prediction-of-density-adsorption isotherms for a fluid confined in a random quenched pore structure

in figure 11.12. Such data are usually represented as pressure–density (or chemical-potential–density) adsorption isotherms. The mean-field equation of state given earlier in equation (11.51) provides a way of calculating the variation of density with chemical potential along the experimental isotherm—given assumed values for the model parameters Γ' and $\Im$, at a given value of porosity, which is often known from independent measurements. Integration of the equation of state provides the pressure change in the system between any two chosen states 1 and 2, using the following thermodynamic identity:

$$\Delta P_{1-2} = \int_{\mu_1'}^{\mu_2'} \rho(\Gamma', \Im, \beta')d\mu' \tag{11.78}$$

This series of calculations is repeated in an attempt to optimize the values of Γ' and $\Im$, thereby establishing the values that provide the best fit with the data. These results are shown in figure 11.12 for $p = 0.05$ and illustrate very good agreement between the model and these data—especially at the higher pressures, where this agreement is often accurate to within 1% of the experimental data.

We now summarize the broad characteristics of the effects of confinement on the critical behavior of a confined fluid as revealed by the analysis just given:

- Critical temperatures are lowered by confinement, while critical densities increase.
- The confined-fluid density is substantially enhanced, relative to that of the bulk fluid in equilibrium with it at given thermodynamic conditions. This underpins the use of porous materials as gas storage media, in particular, for hydrogen in fuel-cell applications.

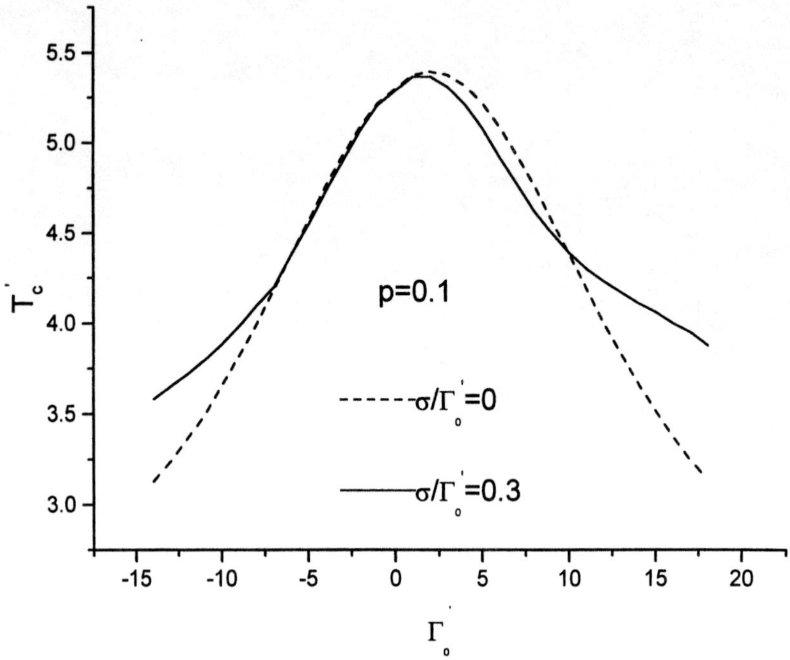

Figure 11.9 Predictions of the effects of surface-energy heterogeneity on the critical temperature of a fluid confined in a random quenched pore structure

- The critical universality class of these systems remains an open question to the present time. The critical exponents for a pore-confined fluid in random heterogeneous structures have not been accurately calculated to the present time.

11.6 Chapter Review

In many applications of supercritical fluids, particularly in materials processing, it becomes important to understand the effects of confinement on the critical behavior of confined fluids. A well-known manifestation of the effects of confinement on phase behavior is the phenomenon of capillary condensation, whereby confinement induces a phase transition in the capillary fluid, at thermodynamic conditions that would be considered undersaturated in the bulk state. This phenomenon is predicted by a famous result in this area, the Kelvin equation, which predicts the pressure drop over curved interfaces as well as the capillary fluid's condensation line. However, the Kelvin equation predicts that the confined fluid's coexistence boundary ends at a capillary critical point identical to that of the pure bulk fluid. This is known to be incorrect from experimental observation. Thus, critical behavior in confined structures requires a more sophisticated analysis than the classical analysis leading to the Kelvin equation.

We explored ideas in this regard, first using scaling ideas in idealized geometries like the slit pore. This abstract scaling analysis showed the potential importance of confinement effects in the scaling region. It suggests that, the smaller the region of

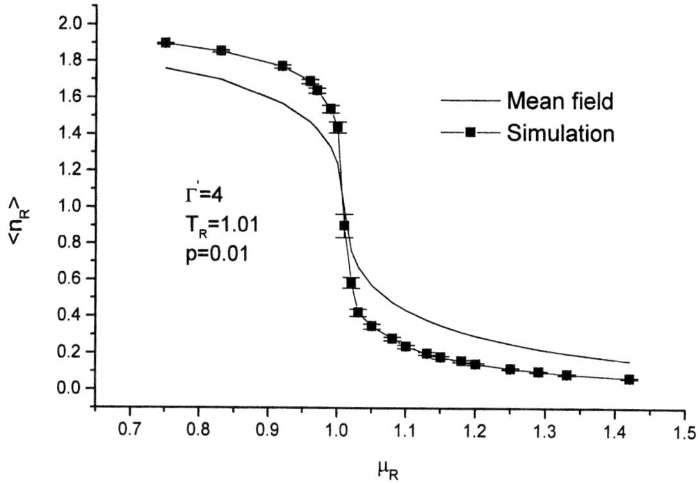

Figure 11.10 Comparison between mean-field predictions and GCMC simulation data for the density isotherm of a fluid confined in a random quenched pore structure

confinement, all other things being equal, the larger should be the shift in critical properties of the confined fluid, relative to bulk-fluid critical values.

Approaches for describing the phase behavior of fluids confined in complex three-dimensional porous structures were discussed, using both mean-field ideas and computer simulation. Both methods provide routes to thermodynamic properties of interest and reflect the influence of the porous matrix on the confined fluid's properties. To the present time, however, the exact nature of the critical universality class of pore-confined fluid mixtures remains an open question in the research literature.

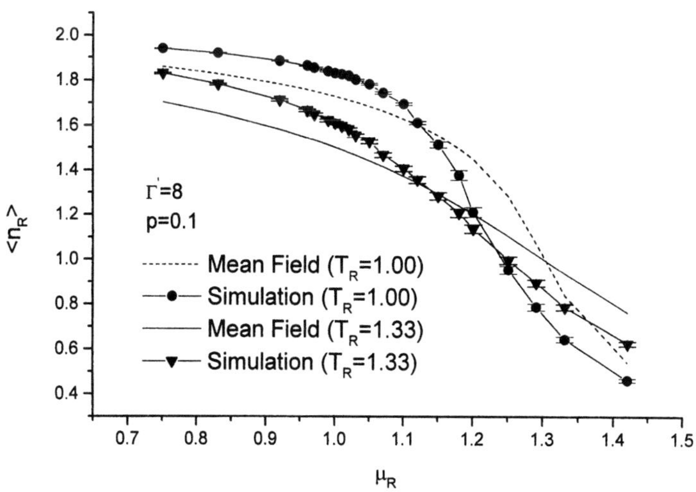

Figure 11.11 Comparison between mean-field predictions and GCMC simulation data fot the density isotherm of a fluid confined in a random quenched pore structure

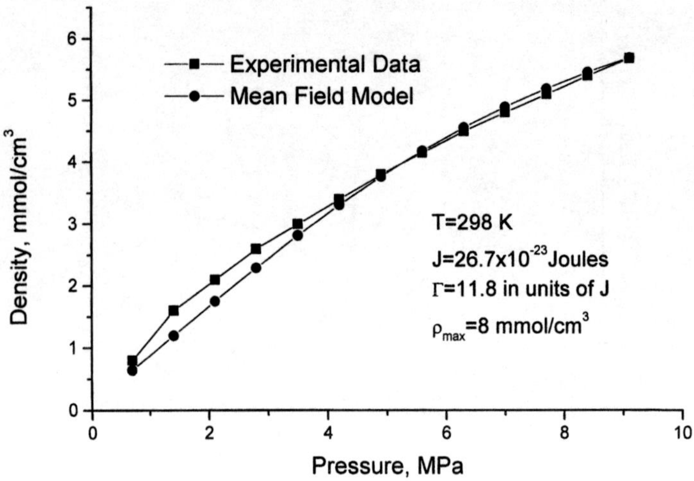

Figure 11.12 Mean-field equation-of-state representation of adsorption data for a fluid confined in a random quenched pore structure. Data for methane in silica aerogel [16]

11.7 Additional Exercises

Binomial Random Variables; Structure of Random Porous Media

1. If the outcome of a random event can be judged in one of two ways—say, a success or a failure—with the probability of the success being p, then the number x of successes k after n events is distributed as a *binomial* random variable, with a discrete probability density function defined as [12]

$$P\{x = k\} \equiv p(k) = \binom{n}{k} p^k (1-p)^{n-k} \quad (k = 0, \ldots, n)$$

Sketch $p(k)$ against k for values of the parameters $(n, p) = (10, 0.7)$. What do you notice about this function as n gets larger?

2. Prove that, for the binomial random variable,

$$\frac{P\{x = k\}}{P\{x = k - 1\}} = \frac{(n - k + 1)p}{k(1 - p)}$$

3. Prove, for a binomial random variable defined in additional exercise 1, that

$$\sum_{k=0}^{n} \binom{n}{k} p^k (1-p)^{n-k} = 1$$

4. Show, for a binomial random variable with a given value of n and p, that $P\{x = k\}$ is a maximum for $k \leq (n + 1)p$.
5. In a 2-d random pore structure, what is the probability of finding an open site completely surrounded by solid porous material? In the language of porous systems, this would be a blocked region.

6. What is the probability of finding *at least* one open site surrounding a confined molecule in a cubic 3-d random porous medium?
7. In a membrane structure represented by a 3-d random cubic lattice with matrix particle density p (i.e., with porosity $1-p$), what is the most probable configuration of matrix and open-cell structures surrounding a diffusing molecule at any given time for the values $p = 0.1$ and $p = 0.25$?
8. Consider a 3-d random cubic lattice pore structure with two distinctive types of porous sites—think of them as labeled red and green, respectively. If the average porosity for these sites is given by p_{r} and p_{g}, respectively, write the probability expression for the structural configuration whereby a molecule finds itself adjacent to one of each of these types of site.
9. In a randomly structured medium like that studied in this chapter, we assign a value $\varepsilon = 0$ with probability p to imply a matrix site and the value $\varepsilon = 1$ with a probability $1 - p$ to indicate an open site. If we define a correlation function for the medium as

$$C(\boldsymbol{r}) \equiv \langle \varepsilon(\boldsymbol{R})\varepsilon(\boldsymbol{R}+\boldsymbol{r})\rangle - \langle \varepsilon \rangle^2$$

show that, at a value of porosity equal to p, we have

$$\langle \varepsilon \rangle = 1 - p$$

and that the correlation function can be expressed as

$$C(\boldsymbol{r}) = p(1-p)\delta(\boldsymbol{r})$$

If we now define the structure factor $s(\boldsymbol{k})$ of the medium to be the Fourier transform of its correlation function, show that, in a random medium,

$$s(\boldsymbol{k}) = \frac{1}{(2\pi)^{3/2}}$$

10. In a cubic random pixelized medium of linear dimension L, show that the probability of a void–solid interface is given by

$$p = 2\frac{N}{N_0}\frac{N_0 - N}{N_0}$$

and that the total number of internal faces is given by

$$1 - \frac{N_i N_j}{N_i}$$

If the interfacial area per unit volume in a lattice with pixel size l is defined as S_{v}, then show that

$$S_{\mathrm{v}} l = 6\varepsilon(1-\varepsilon)$$

where ε is the porosity of the medium. It can also be shown [13] that

$$S_{\mathrm{v}} = \frac{4\varepsilon}{d_0}$$

where d_0 is the mean pore diameter. How would you attempt to find d_0 for a random pixelized medium? [Note: $N_i = L/\ell$, $N_0 = N^3$.]

Mean-Field Equation of State for Confined Fluids; Ising-Model Analogies

11. What is the relationship between the critical-point temperature conditions in the confined lattice gas (equation (11.51)) and its mean-field equivalent Ising equation of state? Do these models show critical behavior at identical absolute temperatures?
12. How does one establish the pressure change for a pore-confined fluid from computer simulations in the grand canonical ensemble?
13. The two-phase region of the confined fluid is a property of fundamental interest in dealing with porous materials. This can be found with the mean-field model developed in this chapter, equation (11.51). At a reduced temperature $T' = 6$, for $p = 0.03$, calculate the adsorption isotherm in the porous system. How do we know that this is a supercritical isotherm?
14. Calculate the confined fluid's critical temperature $T_c' \equiv kT/\Im$ for a 2-d cubic lattice with $p = 0.05$ as a function of the fluid–site interaction parameter $\Gamma' \equiv \Gamma/\Im$. Is T_c' symmetric about any particular value of Γ'? Use equation (11.51) as the basis for the calculations.
15. For $p = 0.1$, the temperature $T' = 4$ is subcritical for the confined-fluid mean-field model given in equation (11.51). Calculate the van der Waals loop in the system at this temperature and use it to find the value of the coexisting confined-fluid densities.
16. De et al. [14], calculated the critical temperatures in a random porous medium, for the Hamiltonian given in equation (11.37) with $\Gamma' = 2$, using computer simulation. Their data are as follows.

p:	0	0.1	0.2
T_c':	4.5114	3.8664	3.4957

Show that the mean-field model developed earlier (equation (11.51)) provides an excellent description of this system when $T_R(p) \equiv T_c'(p)/T_c'(p = 0)$ is represented against p. Would you expect this to be as good for other values of Γ'? If not, why not?
17. If we define a modified Ising model similarly to our lattice-gas model (equations (11.44)–(11.51)) studied in this chapter, show that the local mean-field Ising equation of state is given by the equation

$$\langle m_i \rangle = \tanh\left[\beta \sum_{\langle j \rangle_i} (\varepsilon_j J m_j + (1 - \varepsilon_j) K + B)\right]$$

where the term $\langle j \rangle_i$ denotes a sum over neighboring sites of i, B is a uniform external field and K the spin-surface interaction constant. Assume that, for every site, $\langle m_i \rangle \equiv m$; then use averaging over all possible neighborhood configurations of i to show that we obtain the mean-field equation

$$m = \sum_{n=0}^{z} \binom{z}{n} (1 - p)^{z-n} p^n \tanh\{\beta[J(z - n)m + nK + B]\}$$

where

$$\binom{z}{n} \equiv \frac{z!}{(z-n)!n!}$$

This equation bears a resemblance to the usual mean-field result for an (unconfined) Ising system, but contains additional variables like p and K. Show that, in the low-p limit, by taking the terms of first order in p in this equation, it becomes

$$m = (1 - zp)\tanh[\beta(Jzm + B)] + zp\tanh[\beta(J(z-1)m + K + B)]$$

or

$$m = (1 - zp)\tanh[\beta'(zm + B')] + zp\tanh[\beta'((z-1)m + K' + B')]$$

where we have defined the dimensionless quantities

$$\beta' = \beta J,\ K' = K/J,\ B' = B/J$$

For these results, show that, when $K = 0$, the critical temeperature is $T_c = z(1-p)$, which is the known result for the site-diluted Ising model (SDIM) in the mean-field regime. [Note that $K = 0$ physically implies an Ising lattice with certain sites blocked out (i.e., the spins are diluted) with no interaction between the blocked sites and remaining spins. The site-diluted Ising model is a well-known model in the physics literature].

18. For a cubic confined Ising lattice, in the limit $p \to 0$, show that $T_c' = 6$ for its mean-field model. Prove that, when $p > 0$, for any value of K (both positive and negative), the critical temperature for finite p is always predicted to be less than the zero-p case.
19. Using the results of this chapter, compare the critical exponents for the "bulk" lattice-gas model to its confined-fluid counterpart in the mean-field regime. Now do the same for the "bulk" Ising model and its site-diluted counterpart. What are your main conclusions? If the critical exponents for these exact models are compared (i.e., not just in the mean-field regimes), what would you conclude?
20. Following the analysis given in this chapter, develop a mean-field lattice-gas equation of state for a *binary confined fluid* in a random porous substrate of porosity $1 - p$. Find the critical temperature and density of this system at a given fluid composition and porosity. Discuss how these differ from the constituent pure component critical properties given by equations (11.71)–(11.73).

Bibliography

Reference [4] is a useful article on capillarity, on which the first part of this chapter is based, while in references [7] and [8] scaling in well-defined geometric structures is analyzed. Experimental data in a confined near-critical system is discussed in reference [5] and an analytic mean-field equation of state in porous random structures is presented in reference [11].

[1] J. V. Behren, E. H. Chimowitz, and P. M. Fauchet, "Critical behavior and the processing of nanoscale porous materials," *Adv. Mater.*, vol. 9, p. 921, 1997.

[2] R. F. Cracknell, P. Gordon, and K. E. Gubbins, "Influence of pore geometry on the design of microporous materials for methane storage," *J. Phys. Chem.*, vol. 97, p. 494, 1993.

[3] G. Afrane and E. H. Chimowitz, "Experimental investigation of a new supercritical fluid—inorganic membrane separation process," *J. Membr. Sci.*, vol. 116, p. 293, 1996.

[4] D. Nicolaides and R. Evans, "Monte Carlo study of phase transitions in a confined lattice gas," *Phys. Rev. B*, vol. 39, p. 9336, 1989.

[5] M. Thommes and G. H. Findenegg, "Critical depletion of a pure fluid in controlled pore-size glass. Experimental results and grand canonical ensemble Monte Carlo simulation," *Langmuir*, vol. 11, p. 2137, 1995.

[6] M. Fabrizio, C. Martin, and E. H. Chimowitz, "A statistical mechanical model for hydrogen absorption in porous substrates for use in solid-oxide fuel cells," *Proceedings of the Sixth European Solid Oxide Fuel Cell Forum, 28 June–2 July, Lucerne, Switzerland*, vol. 2, p. 746, 2004.

[7] M. E. Fisher and N. Nakanishi, "Scaling theory for the criticality of fluids between plates," *J. Chem. Phys.*, vol. 75, p. 5857, 1981.

[8] T. MacFarland, G. T. Barkema, and J. F. Marko, "Equilibrium phase transitions in a porous medium," *Phys. Rev. B*, vol. 53, p. 148, 1996.

[9] M. Plischke and B. Bergersen, *Equilibrium Statistical Physics*. Singapore: World Scientific, 1994.

[10] W. C. Barber and D. P. Belanger, "Monte Carlo simulations of the random field Ising model," *J. Magn. Magn. Mater.*, vol. 1, p. 1, 2001.

[11] S. De, Y. Shapir, E. H. Chimowitz, and V. Kumaran, "Critical behavior in quenched random structures: mean-field lattice-gas approach," *AIChE J.*, vol. 47, p. 463, 2001.

[12] S. Ross, *A First Course in Probability*. New York: Macmillan, 1988.

[13] V. N. Burganos, "Gas diffusion in random binary media," *J. Chem. Phys.*, vol. 109, p. 6772, 1998.

[14] S. De, Y. Shapir, and E. H. Chimowitz, "Diffusion in random structures in the critical region," *Mol. Simul.*, vol. 29, p. 167, 2002.

[15] H. G. Ballesteros, L. A. Fernandez, V. Martin-Mayor, and A. M. Sudupe, "Critical exponents of the three-dimensional diluted Ising model," *Phys. Rev. B*, vol. 58, p. 2740, 1998.

[16] S. Masukawa and R. Kobayashi, "Adsorption equilibrium of the system methane–ethane–silica gel at high pressures and ambient temperatures." *J. Chem. Eng. Data*, vol. 13, p. 197, 1968.

12 Transport in the Critical Region

The behavior of dynamic properties in the critical region is important in many engineering applications, and in this chapter we investigate this topic, focusing upon diffusion. In the literature, the term *critical slowing down* is used to describe the long relaxation times that occur when criticality is approached. Does this mean that diffusion processes *per se* come to a halt and, if not, how does slowing down manifest itself in fluids? We see that, in spite of the nonequilibrium nature of this topic, equilibrium concepts still play a key role in describing dynamics in the critical region.

12.1 Self-Diffusion in Fluids

To begin this discussion, we investigate the dynamic behavior of a tagged fluid molecule as it experiences random fluctuations in its position in the fluid. These fluctuations would be induced by random thermally induced collisions between the tagged species and other fluid molecules. This type of dynamics is referred to as *Brownian* motion or a *random walk.* A schematic showing two 10-step trajectories depicting a random walk in two dimensions is shown in figure 12.1, and its analysis leads naturally to the definition of the *self-diffusion coefficient.*

Brownian Dynamics

Brownian dynamics is often described stochastically using a probability density function $P(\boldsymbol{r}, t)$ [1]. This function describes the probability that a tagged molecule is in a neighborhood of $\boldsymbol{r}$ at time t, given that it was initially at the origin at time zero. This probability is given by the product $P(\boldsymbol{r}, t)d\boldsymbol{r}$, where $d\boldsymbol{r}$ is the (infinitesimal) volume of the neighborhood. Following the development given by McQuarrie [1], we further assume that the time evolution of $P(\boldsymbol{r}, t)$ is given by a *Fickian-type* partial differential equation of the form

$$\frac{\partial P}{\partial t} = D_{\mathrm{s}} \nabla^2 P \tag{12.1}$$

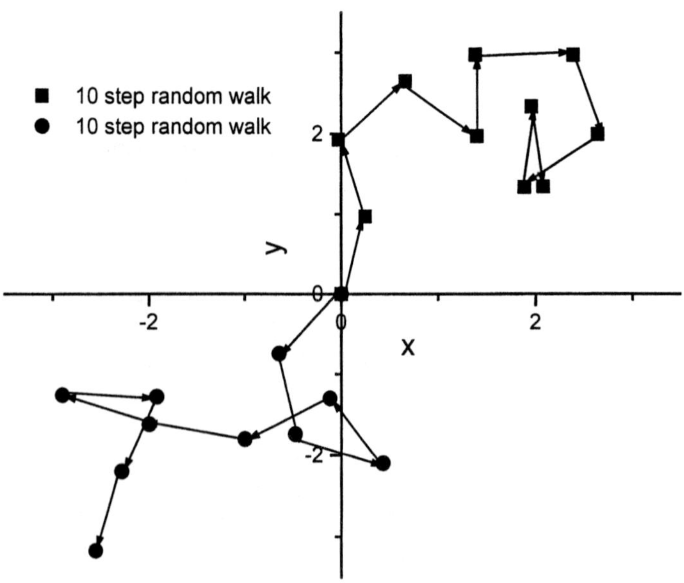

Figure 12.1 Schematic of a 2-d random walk

with the initial boundary condition

$$P(\boldsymbol{r}, t) = \delta(\boldsymbol{r}) \quad at \quad t = 0 \tag{12.2}$$

Equation (12.1) is used to define a diffusion coefficient D_s which we call the *self-diffusion coefficient*. If we define the Fourier transform of $P(\boldsymbol{r}, t)$ as

$$\bar{P}(\boldsymbol{k}, t) \equiv \int_{-\infty}^{\infty} e^{i\boldsymbol{k}\cdot\boldsymbol{r}} P(\boldsymbol{r}, t) d\boldsymbol{r} \tag{12.3}$$

then Fourier-transforming (12.1) and (12.2), in the isotropic case, leads to the equation

$$\frac{\partial \bar{P}(\boldsymbol{k}, t)}{\partial t} = -k^2 D_s \bar{P}(\boldsymbol{k}, t) \tag{12.4}$$

with boundary condition

$$\bar{P}(\boldsymbol{k}, 0) = 1 \tag{12.5}$$

The solution of equations (12.4) and (12.5) is given by

$$\bar{P}(\boldsymbol{k}, t) = e^{-k^2 D_s t} \tag{12.6}$$

By using the Fourier transform identity

$$F_r[\exp(-a^2 r^2)] = \frac{1}{\sqrt{2}a} \exp \frac{-k^2}{4a^2} \tag{12.7}$$

we can immediately invert (12.6) in the isotropic case to produce the solution

$$P(r, t) = \sqrt{\frac{1}{2D_s t}} \exp\left(-\frac{r^2}{4D_s t}\right) \tag{12.8}$$

$P(r, t)$ is a normal random-variable density function, to within a Fourier factor $(2\pi)^{-1/2}$, which would have resulted had we used this convention in the Fourier transform of equation (12.3). However, we don't want to stop at this point, since there is a more interesting aspect to this situation. We write the wave vector $\boldsymbol{k}$ in terms of a magnitude k and a *unit vector* $\boldsymbol{w} : \boldsymbol{k} = k\boldsymbol{w}$. In this case, equation (12.3) becomes

$$\bar{P}(k, t) = \int_{-\infty}^{\infty} e^{ik\boldsymbol{w}\cdot\boldsymbol{r}} P(\boldsymbol{r}, t) d\boldsymbol{r} \tag{12.9}$$

Taking the derivative of $\bar{P}(k, t)$ with respect to k, we get

$$\frac{\partial \bar{P}(k, t)}{\partial k} = \int (i\boldsymbol{w} \cdot \boldsymbol{r}) e^{ik\boldsymbol{w}\cdot\boldsymbol{r}} P(\boldsymbol{r}, t) d\boldsymbol{r} \tag{12.10}$$

$$\frac{\partial^2 \bar{P}(k, t)}{\partial k^2} = \int (i\boldsymbol{w} \cdot \boldsymbol{r})^2 e^{ik\boldsymbol{w}\cdot\boldsymbol{r}} P(\boldsymbol{r}, t) d\mathbf{r} \tag{12.11}$$

In the limit $k \to 0$, for an isotropic system, we have that

$$\left(\frac{\partial^2 \bar{P}(k,t)}{\partial k^2}\right)_{k=0} = -\int_0^\infty 4\pi r^2 P(r,t) w^2 r^2 dr \tag{12.12}$$

where $3w^2 = 1$. Therefore

$$\left(\frac{\partial^2 \bar{P}(k,t)}{\partial k^2}\right)_{k=0} = -\tfrac{1}{3}\int_0^\infty r^2 P(r,t)\pi r^2 dr \tag{12.13}$$

The integral in (12.13) can be interpreted as the mean of the particle's squared displacement from the origin, written as

$$\left(\frac{\partial^2 \bar{P}(k,t)}{\partial k^2}\right)_{k=0} = -\tfrac{1}{3}\langle r^2\rangle \tag{12.14}$$

where we implicitly assume the normalization condition on $P(\boldsymbol{r}, t)$:

$$\int_0^\infty P(\boldsymbol{r},t)d\boldsymbol{r} = 1 \tag{12.15}$$

From equation (12.6), we can show that

$$\left(\frac{\partial^2 \bar{P}(k,t)}{\partial k^2}\right)_{k=0} = -2D_{\mathrm{s}}t \tag{12.16}$$

Therefore, from equations (12.14) and (12.16),

$$\frac{\langle r^2\rangle}{6t} = D_{\mathrm{s}} \tag{12.17}$$

which is famously known as the *Einstein equation*. It predicts that the mean distance traveled by a tagged molecule experiencing Brownian dynamics, given by the quantity $\sqrt{\langle r^2\rangle}$, is proportional to $t^{1/2}$. This is different to the predictions of classical mechanics, where distance traveled scales linearly with time. The self-diffusion coefficient can be found from a plot of $\langle r^2\rangle$ versus t using (12.17), an example of which we now describe.

Self-Diffusion from a Random-Walk Calculation in Gases at Low Densities

We now analyze this mechanistic view of Brownian dynamics using a tagged gaseous species within a larger ensemble of such molecules at low density. We assume that collisions occur between the tagged molecule and other molecules in the ensemble. These occur after the tagged species has moved a distance equal to its *kinetic mean free path*. Furthermore, after each collision the tagged molecule's new direction is found from a *uniform random sampling of directions*. In this way the tagged species is said to experience dynamics corresponding to a random walk.

We look at this problem in two dimensions, since the essential elements are not changed by dimensionality. Let the change in position of the molecule at the ith step be given by the equations

$$x_i = l\cos\theta_i, \qquad y_i = l\sin\theta_i \tag{12.18}$$

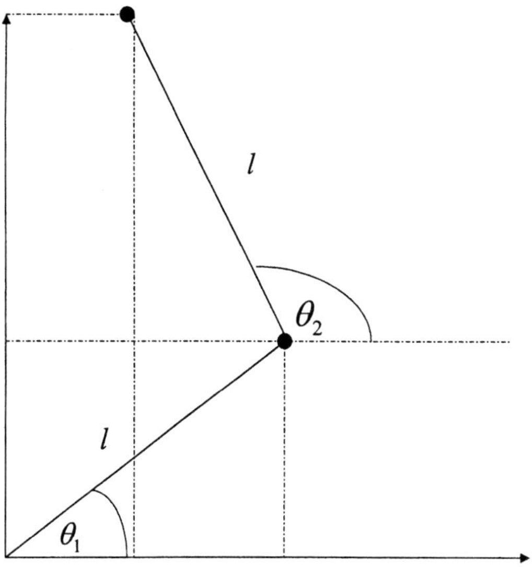

Figure 12.2 Co-ordinates used to describe a 2-d random walk

where figure 12.2 describes these angles. By assumption, the angles θ_i $(i = 1, \ldots, n)$ are independent, uniformly distributed, random variables in the interval $(0, 2\pi)$, with l the distance of the *mean free path*. After n random steps, the position of the tagged species is given by the coordinates $(\sum_{i=1}^{n} x_i, \sum_{i=1}^{n} y_i)$. The square of the particle's distance from the beginning point, designated by R_n^2, is given by the equations

$$R_n^2 = \left(\sum_{i=1}^{n} x_i\right)^2 + \left(\sum_{i=1}^{n} y_i\right)^2 \tag{12.19}$$

$$= \sum_{i=1}^{n}(x_i^2 + y_i^2) + \sum_{i \neq j}\sum_{j}(x_i x_j + y_i y_j) \tag{12.20}$$

$$= nl^2 + l^2 \sum_{i \neq j}\sum_{j}(\cos\theta_i \cos\theta_j + \sin\theta_i \sin\theta_j) \tag{12.21}$$

The average (expectation) of $\cos\theta_i$ in the interval $(0, 2\pi)$ is zero, and asserting the statistical independence of the angles θ_i and θ_j, we get the following result for the mean of R_n^2 from (12.21):

$$\langle R_n^2 \rangle = nl^2 \tag{12.22}$$

If we assume that the total time taken by the random walk is T, with the time taken per step t, then

$$n = T/t \tag{12.23}$$

$$t = l/\bar{v} \tag{12.24}$$

where $\bar{v}$ is the average molecular velocity. These definitions lead to the result

$$\frac{\langle R_n^2 \rangle}{6T} = \frac{1}{6}\bar{v}l = D_s \tag{12.25}$$

which is the result given by kinetic theory, applied to a 2-dimensional dilute gas, with D_s related to the physical quantities $\bar{v}$ and l. Typical values for these parameters in low-pressure gases are $l \approx 10^{-8}$ m, $\bar{v} \approx 500\,\mathrm{ms^{-1}}$, which implies a collision frequency of $\approx 10^9$–$10^{10}\,\mathrm{s^{-1}}$. Diffusion coefficients at these conditions are thus $\approx 10^{-5}$–$10^{-6}\mathrm{m^2\,s^{-1}}$.

EXERCISE 12.1

Prove that, if θ_i is a uniform random variable over the interval $[0, 2\pi]$, then the expectation (mean) value of $\cos\theta_i$ is 0.

The Difference between a Single Molecule and the Ensemble's Diffusion Coefficients

In working out the mean of the displacement squared for a *single molecule* $\langle R_1^2 \rangle_n$ after n steps, we use the formula

$$\langle R_1^2 \rangle_n = \langle (\Delta x_1)_n^2 + (\Delta y_1)_n^2 + (\Delta z_1)_n^2 \rangle \tag{12.26}$$

where

$$(\Delta x_1)_n = \sum_{i=1}^{n} \Delta x_{1i}, \qquad (\Delta y_1)_n = \sum_{i=1}^{n} \Delta y_{1i}, \quad \ldots \tag{12.27}$$

with the self-diffusion coefficient for this single molecule given by

$$\frac{\langle R_1^2 \rangle_n}{6nt} = D_s \tag{12.28}$$

with t the time taken per step. However, if for some reason we were interested in the mean displacement squared of the *entire ensemble of molecules* over the same number of steps n, denoted by $\langle R_{\mathrm{total}}^2 \rangle_n$, it would be given by the equation

$$\langle R_{\mathrm{total}}^2 \rangle_n = \langle (\Delta x_{\mathrm{total}})^2 \rangle_n + \langle (\Delta y_{\mathrm{total}})^2 \rangle_n + \langle (\Delta z_{\mathrm{total}})^2 \rangle_n \tag{12.29}$$

where

$$\Delta x_{\mathrm{total}} = \sum_{k=1}^{N}\sum_{i=1}^{n} \Delta x_{ki}, \quad \Delta y_{\mathrm{total}} = \sum_{k=1}^{N}\sum_{i=1}^{n} \Delta y_{ki}, \quad \cdots \tag{12.30}$$

where we have numbered the molecules $1, \ldots, N$. If we designate the x position of molecule 1 after n steps by the definition

$$(\Delta x_1)_n \equiv \sum_{i=1}^{n} \Delta x_{1i}, \ldots \tag{12.31}$$

then

$$(\Delta x_{\text{total}})^2 = [(\Delta x_1)_n^2 + \cdots + (\Delta x_N)_n^2] + 2(\Delta x_1 \Delta x_2)_n + \text{ other cross terms} \quad (12.32)$$

and

$$\langle(\Delta x_{\text{total}})^2\rangle = \langle[(\Delta x_1)_n^2 + \cdots + (\Delta x_N)_n^2] + 2(\Delta x_1 \Delta x_2)_n + \text{ other cross terms}\rangle \quad (12.33)$$

with similar terms for the y and z components, respectively. If we further define a self-diffusion coefficient for the entire ensemble after n steps (referred to as the *mass* or *bulk self-diffusion constant*) by the equation

$$D_{\text{s,total}} \equiv \frac{\langle R_{\text{total}}^2\rangle_n}{6nt} \quad (12.34)$$

we see that, in general,

$$\frac{D_{\text{s,total}}}{N} \neq D_{\text{s}} \quad (12.35)$$

Thus, in general, the average of the bulk self-diffusion coefficient is not equal to the self-diffusion coefficient of a single tagged species.

EXERCISE 12.2

At which conditions do we expect the following equation to hold?

$$\frac{D_{\text{s,total}}}{N} = D_{\text{s}}$$

When is this likely to be the case?

Self-Diffusion and the Langevin Equation

McQuarrie [1] presents another interesting route to the Einstein result using the *Langevin equation* which we now discuss. This approach represents the random movement of a particle in a fluid medium by the dynamic equation

$$\boldsymbol{u}' + \xi\boldsymbol{u} = \boldsymbol{A}(t) \quad (12.36)$$

where $\boldsymbol{u}$ is the particle's velocity, $\xi\boldsymbol{u}$ the *Stokes force* on the particle, and $\boldsymbol{A}(t)$ a random time-dependent contribution to the force. Equation (12.36) is a *stochastic differential equation* along the lines of Newton's second law, with the standard integrating-factor solution given by the equation:

$$\boldsymbol{u} = e^{-\xi t}\left[\int e^{\xi t}\boldsymbol{A}(t)dt + c\right] \quad (12.37)$$

and boundary condition $\boldsymbol{u} = \boldsymbol{u}_0$ at $t = 0$. The solution of (12.37) with this boundary condition becomes equal to

$$\boldsymbol{u} = e^{-\xi t}\boldsymbol{u}_0 + e^{-\xi t}\left[\int_0^t e^{\xi x}\boldsymbol{A}(x)dx\right] \quad (12.38)$$

If we look at the position of the particle after time t defined by the equation

$$\boldsymbol{r} - \boldsymbol{r}_0 = \int_0^t \boldsymbol{u}(t')dt' \tag{12.39}$$

then, using (12.38) for the velocity in (12.39), we get the following equation for the net particle displacement:

$$\boldsymbol{r} - \boldsymbol{r}_0 = \int_0^t e^{-\xi t'} \boldsymbol{u}_0 \, dt' + \int_0^t e^{-\xi t'} \left[\int_0^{t'} e^{\xi x} \boldsymbol{A}(x) dx \right] dt' \tag{12.40}$$

which becomes, upon doing the first of these integrals,

$$\boldsymbol{r} - \boldsymbol{r}_0 = \xi^{-1} \boldsymbol{u}_0 (1 - e^{-\xi t}) + \int_0^t dt' e^{-\xi t'} \int_0^{t'} e^{\xi x} \boldsymbol{A}(x) dx \tag{12.41}$$

Further integrating equation (12.41) by parts leads to the equation [1]

$$\boldsymbol{r} - \boldsymbol{r}_0 = \xi^{-1} \boldsymbol{u}_0 (1 - e^{-\xi t}) + \int_0^t \xi^{-1} \left[1 - e^{-\xi (t' - t)} \right] \boldsymbol{A}(t') dt' \tag{12.42}$$

EXERCISE 12.3

Prove equation (12.42). [Hint: Consider its predecessor integral in (12.41) to be of the form $\int_0^t e^{-\xi t'} f(t')dt'$, where $f(t') = \int_0^{t'} e^{\xi x} \boldsymbol{A}(x)dx$.] Now do this integral by parts, leading to (12.42).

From this equation and with the assumption $\langle \boldsymbol{A}(t) \rangle = 0$, it follows immediately that

$$\langle \boldsymbol{r} - \boldsymbol{r}_0 \rangle = \xi^{-1} \boldsymbol{u}_0 (1 - e^{-\xi t}) \tag{12.43}$$

We now use this equation to look at the quantity $\langle |\boldsymbol{r} - \boldsymbol{r}_0|^2 \rangle$: the mean of the squared displacement of the particle at time t. This is found by taking the ensemble average of $|\boldsymbol{r} - \boldsymbol{r}_0|^2$, with the aid of (12.42), leading to the equation

$$\begin{aligned} \langle |\boldsymbol{r} - \boldsymbol{r}_0|^2 \rangle &= \xi^{-2} u_0^2 (1 - e^{-\xi t})^2 \\ &+ \xi^{-2} \int_0^t \int_0^t [1 - e^{-\xi (t' - t)}][1 - e^{-\xi (t'' - t)}] \langle \boldsymbol{A}(t') \cdot \boldsymbol{A}(t'') \rangle dt' dt'' \end{aligned} \tag{12.44}$$

It is safe to assume that the covariance term $\langle \boldsymbol{A}(t') \cdot \boldsymbol{A}(t'') \rangle$ is a peaked function of the difference in absolute times $|t' - t''|$, since $\boldsymbol{A}(t)$ represents a random force on the tagged species. So, for large $|t' - t''|$, the force should be uncorrelated. We write this [1] as $\langle \boldsymbol{A}(t') \cdot \boldsymbol{A}(t'') \rangle = \psi(|t' - t''|)$, with $\psi(|t' - t''|) \propto \delta(|t' - t''|)$.

Equation (12.44) is evidently a complicated equation whose integration, however, can be simplified by using variable transformations along the following lines. Consider figure 12.3(a), where we have a square domain $0 \leq p, \ q \leq t$. If we define the variables

$$x = p - q \tag{12.45}$$

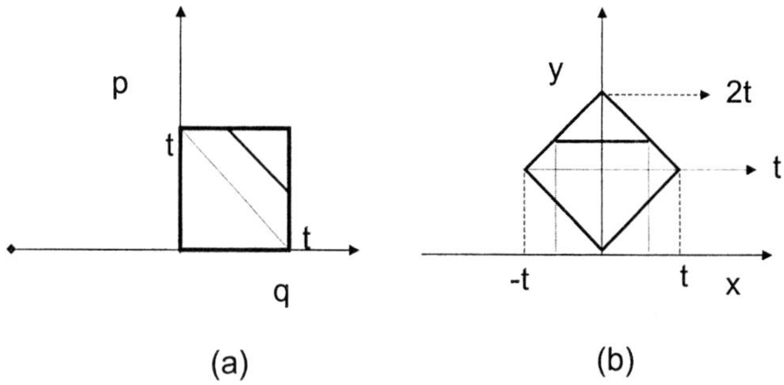

Figure 12.3 Schematic showing limits of a 2-d integral

and

$$y = p + q \tag{12.46}$$

then the (p, q) domain is transformed into the "tipped-square" domain shown in figure 12.3(b) A double integral of the form

$$\int_0^t \int_0^t f(p+q)g(p-q)dp\,dq \tag{12.47}$$

can be rewritten in the transformed system as

$$\iint_D f(y)g(x)dx\,dy \tag{12.48}$$

with the domain of integration equivalent to that shown in figure 12.3(b). Applying these ideas to equation (12.44) leads to the following equation [1] (see the additional exercises for a related problem):

$$\langle|\boldsymbol{r} - \boldsymbol{r}_0|^2\rangle = \frac{u_0^2}{\xi^2}(1 - e^{-\xi t})^2 + \frac{3kT}{m\xi^2}(2\xi t - 3 + 4e^{-\xi t} - e^{-2\xi t}) \tag{12.49}$$

This is a key result with important limits. In the small-time limit (i.e., $t \to 0$), we have that

$$\langle|\boldsymbol{r} - \boldsymbol{r}_0|^2\rangle = |u_0^2|t^2 \tag{12.50}$$

which can be found from expanding the first term in (12.49) to leading order in ξt. At large t,

$$\langle|\boldsymbol{r} - \boldsymbol{r}_0|^2\rangle \to \frac{6kTt}{m\xi} \tag{12.51}$$

If we take the term $\langle|\boldsymbol{r} - \boldsymbol{r}_0|^2\rangle/6t$ as the (usual) basis for the definition of D_s, consistent with our earlier choices, we see that, in this case,

$$D_s \equiv \frac{kT}{m\xi} \tag{12.52}$$

Note that, while the generic form of the self-diffusion constant is preserved in equation (12.52), its relationship to physical quantities in (12.52) is much different to that for the dilute-gas case discussed earlier. One may think of the definition for the self-diffusion coefficient given in (12.52) as being more appropriate for a dense fluid (why?).

D_s from the Velocity–Velocity Autocorrelation Function

Another very useful way to find the property D_s is from the ensemble average of a quantity called the *velocity autocorrelation function* defined as $\boldsymbol{u}(t) \cdot \boldsymbol{u}(0)$. The velocity is given by (12.38), and we consider the quantities $\boldsymbol{u}_0$ and $\boldsymbol{A}(x)$ to be uncorrelated. This immediately leads to the result

$$\langle \boldsymbol{u}(t) \cdot \boldsymbol{u}(0) \rangle = \langle u_0^2 \rangle e^{-\xi t} \tag{12.53}$$

Since, from kinetic theory, $\langle u_0^2 \rangle = 3kT/m$ and $D_s \equiv kT/m\xi$, we find that

$$D_s = \frac{\langle u_0^2 \rangle}{3\xi} \tag{12.54}$$

which implies that

$$\frac{1}{3}\langle \boldsymbol{u}(t) \cdot \boldsymbol{u}(0) \rangle = -\frac{d(D_s e^{-\xi t})}{dt} \tag{12.55}$$

or

$$D_s = \frac{1}{3}\int_0^{\infty} \langle \boldsymbol{u}(t) \cdot \boldsymbol{u}_0 \rangle dt \tag{12.56}$$

Equation (12.56) is useful for finding D_s from computer simulation methods for *molecular dynamics*, which calculate $\boldsymbol{u}(t)$ based upon integrating Newton's laws of motion.

In this section we have presented three ways of calculating the self-diffusion coefficient in systems: using (1) random-walk ideas, (2) a "force" equation leading to the Langevin equation, and (3) integration of the velocity–velocity autocorrelation function. In all of these cases, we see that D_s is intimately related to the root mean square of the distance traveled by a tagged particle. However, the analysis of transport phenomena in fluid systems usually employs the *Fickian* description of molar fluxes, an issue we now look at, before focusing upon the main business of this chapter which is diffusion processes in the critical region.

The Difference between Self-Diffusion and Fickian Diffusion Coefficients

Having focused upon self-diffusion, we now consider the Fickian diffusion coefficient and explore the confusion that often surrounds these two quantities. *Fick's law* of diffusion states that the flux $J(\boldsymbol{r}, t)$ of a species is given by the differential equation

$$J(\boldsymbol{r}, t) = -D_F \nabla \rho(\boldsymbol{r}, t) \tag{12.57}$$

which defines the *Fickian diffusion constant* D_F. The particle density at position $\boldsymbol{r}$ is given by $\rho(\boldsymbol{r}, t)$ at time t. If we use the equation of continuity

$$\frac{\partial \rho(\boldsymbol{r}, t)}{\partial t} = -\nabla J(\boldsymbol{r}, t) \tag{12.58}$$

in (12.57), we get the time-dependent Fickian transport equation

$$\frac{\partial \rho(\boldsymbol{r}, t)}{\partial t} = D_F \nabla^2 \rho(\boldsymbol{r}, t) \tag{12.59}$$

We now relate the conditional probability function $P(\boldsymbol{r}, t; \boldsymbol{r}_0, t_0)$ for particles experiencing *Brownian* motion to their density as follows:

$$\rho(\boldsymbol{r}, t) = \int P(\boldsymbol{r}, t; \boldsymbol{r}_0, t_0)\rho(\boldsymbol{r}_0, t_0) d\boldsymbol{r}_0 \tag{12.60}$$

where $P(\boldsymbol{r}, t; \boldsymbol{r}_0, t_0)$ is the probability density that a particle is at position $\boldsymbol{r}$ at t given that at time t_0 it was at $\boldsymbol{r}_0$. It is important to note that (12.60) applies only to the random (uncorrelated) motion of particles. In this situation we can use (12.60) in (12.59) to derive the equation

$$\frac{\partial P(\boldsymbol{r}, t)}{\partial t} = D_F \nabla^2 P(\boldsymbol{r}, t) \tag{12.61}$$

with boundary condition

$$P(\boldsymbol{r}, t = 0) = \delta(\boldsymbol{r}) \tag{12.62}$$

which is identical to equations (12.1) and (12.2) derived earlier for self-diffusion of a tagged species. Thus we see that only in the situation of uncorrelated motion is there an equivalence between the self-diffusion and Fickian diffusion coefficients, namely

$$D_s = D_F = \frac{\langle r^2 \rangle}{6t} \tag{12.63}$$

Since these conditions are only true for uncorrelated motion, they strictly apply only to ideal gases. When particle correlations exist, these two quantities are not equivalent, and one must always be careful to distinguish between both cases. Cooperative (correlated) diffusion is described by the Fickian diffusion equation, which is why D_F is sometimes referred to as the *cooperative diffusion coefficient.*

12.2 Diffusion in the Critical Region

We now focus on the main point of this chapter, which concerns the effects of proximity to the critical point on diffusion in fluid systems. We use the Onsager regression hypothesis for this analysis, an approach sometimes called *linear response theory.*

Fick's Law and the Onsager Relationships

Purely diffusive systems

The diffusive flux J_i for species i with respect to its density ρ_i can be written in both Fickian and Onsager terms as follows:

$$J_i = -D_{ii} \nabla \rho_i \tag{12.64}$$

$$J_i = -L_{ii} \nabla \mu_i \tag{12.65}$$

where D_{ii} is the Fickian (transport) diffusion constant and L_{ii} the Onsager regression constant for species i in a purely diffusive system, which would be represented by a lattice-gas fluid for example. Equation (12.65) represents the Onsager hypothesis, with L_{ii} called the Onsager coefficient. (We use double subscripts in these equations to allow the subsequent consideration of multicomponent mixtures.) For the time being, however, consider a pure lattice-gas species.

If we use (12.65) and consider a gradient in the chemical potential μ_i for species i over the spatial coordinate x, we find that

$$J_i = -L_{ii} \left(\frac{\partial \mu_i}{\partial x} \right)_T = -L_{ii} \left(\frac{\partial \mu_i}{\partial \rho_i} \right)_T \left(\frac{\partial \rho_i}{\partial x} \right)_T \tag{12.66}$$

in which case it follows that

$$D_{ii} = L_{ii} \left(\frac{\partial \mu_i}{\partial \rho_i} \right)_T \tag{12.67}$$

The Onsager coefficient can be found from microscopic parameters using the Einstein equation [2]

$$L_{ii} = \frac{\beta \rho_i}{6N_i} \lim_{t \to \infty} \frac{1}{t} \left\langle \left| \sum_{k=1}^{N_i} (\boldsymbol{r}_{ki}(t) - \boldsymbol{r}_{ki}(0)) \right|^2 \right\rangle \tag{12.68}$$

where $\boldsymbol{r}_{ki}(t)$ is the position of the kth molecule of species i at time t. We now extend this analysis into the realm of binary mixtures. Once again think in terms of a binary lattice fluid with the third "component" being empty sites.

Binary diffusion in mixtures

In a binary mixture, the fluxes can be written in matrix form:

$$\begin{bmatrix} J_i \\ J_j \end{bmatrix} = -\begin{bmatrix} D_{ii} & D_{ij} \\ D_{ji} & D_{jj} \end{bmatrix} \begin{bmatrix} \nabla \rho_i \\ \nabla \rho_j \end{bmatrix} = -\begin{bmatrix} L_{ii} & L_{ij} \\ L_{ji} & L_{jj} \end{bmatrix} \begin{bmatrix} \nabla \mu_i \\ \nabla \mu_j \end{bmatrix} \tag{12.69}$$

from which it follows that

$$\begin{bmatrix} D_{ii} \\ D_{ij} \end{bmatrix} = \begin{bmatrix} \left(\dfrac{\partial \mu_i}{\partial \rho_i} \right)_{T,\rho_j} & \left(\dfrac{\partial \mu_i}{\partial \rho_j} \right)_{T,\rho_i} \\ \left(\dfrac{\partial \mu_j}{\partial \rho_i} \right)_{T,\rho_j} & \left(\dfrac{\partial \mu_j}{\partial \rho_j} \right)_{T,\rho_i} \end{bmatrix} \begin{bmatrix} L_{ii} \\ L_{ij} \end{bmatrix} \tag{12.70}$$

The matrix in equation (12.70),

$$\begin{bmatrix} \left(\dfrac{\partial \mu_i}{\partial \rho_i}\right)_{T,\rho_j} & \left(\dfrac{\partial \mu_i}{\partial \rho_j}\right)_{T,\rho_i} \\ \left(\dfrac{\partial \mu_j}{\partial \rho_i}\right)_{T,\rho_j} & \left(\dfrac{\partial \mu_j}{\partial \rho_j}\right)_{T,\rho_i} \end{bmatrix} \tag{12.71}$$

is a thermodynamic stability matrix whose determinant vanishes at a stability limit, as can be seen from the results derived in chapter 3. Therefore, at the critical point, equation (12.70) is singular. To obtain further useful information relating Fickian and Onsager coefficients at the critical point requires that more be known about the behavior of the Onsager coefficients there. For this, we need to consider more sophisticated *dynamic scaling* theories that apply to critical region dynamics [3].

Onsager Coefficients in the Critical Region: The Pure Lattice Gas

Conventional theory

In the conventional theory of dynamic critical phenomena, L_{ii} is taken to be *finite at the critical point.* This result is considered to apply to the kinetic (diffusive) behavior of spins on the uniaxial, Ising ferromagnet which corresponds to simple fluids like the lattice gas, belonging to the same critical universality class; dynamics for these systems is often referred to as *Kawasaki dynamics* following the seminal papers [4] in this area by that author.

Combining this assumption with equation (12.67), we see that D_{ii} in this case is predicted to approach criticality according to the scaling law:

$$D_{ii} \propto \frac{L_{ii}}{K_T} \tag{12.72}$$

where

$$\left(\frac{\partial \mu_i}{\partial \rho_i}\right)_T = \frac{1}{\rho_i^2 K_T} \tag{12.73}$$

with K_T the familiar isothermal compressibility (susceptibility) defined as

$$K_T \equiv \frac{1}{\rho_i}\left(\frac{\partial \rho_i}{\partial P}\right)_T \tag{12.74}$$

These results imply that, along the coexistence direction in the fluid, the Fickian diffusion constant is predicted to scale as follows:

$$D_{ii} \propto t^{\gamma} \tag{12.75}$$

with $t \equiv (T - T_c)/T_c$ and $\gamma (\approx 1.25)$ the usual scaling exponent for K_T in this direction. Using the correlation-length scaling relationship $\xi \propto t^{-\nu}$ in equation (12.75), we find the following scaling relationship for the Fickian D_{ii} in terms of the correlation length ξ, as the critical point is approached:

$$D_{ii} \propto \xi^{-2+\eta} \tag{12.76}$$

where the scaling-exponent equation

$$\gamma = \nu(2 - \eta) \tag{12.77}$$

has been used to get (12.76). Equation (12.76) predicts that the Fickian diffusion coefficient will *vanish* at the critical point and is the provenance of the term *critical slowing down* used to describe near-critical diffusional dynamics.

Dynamic scaling theory

The conventional theory, however, is now thought to be incorrect for describing dynamics near the vapor–liquid critical point. Critical dynamics in these systems is now considered to be described by the *Model H universality class* discussed in detail in the classic review of dynamic critical phenomena by Hohenberg and Halperin [3]. This theory predicts the divergence of the Onsager coefficient L_{ii} at the critical point, according to the scaling relationship

$$L_{ii} \propto \xi^{y} \tag{12.78}$$

with the scaling exponent y given by a *renormalization-group* calculation as follows:

$$y \approx \frac{18}{19}\varepsilon(1 - 0.033\varepsilon + \cdots) \tag{12.79}$$

where ε is related to the system dimension d by

$$d = 4 - \varepsilon \tag{12.80}$$

From these equations, we find that $y = 0.916$, and this theory also predicts the vanishing of the Fickian D_{ii} at the critical point in a pure 3-d fluid. However, it predicts a weaker anomaly than does conventional theory with the asymptotic scaling of D_{ii} now given by

$$D_{ii} \propto \xi^{-2+\eta}\xi^{0.916} = \xi^{-1.044} \propto t^{0.68} \tag{12.81}$$

with $\eta = 0.04$. Table 12.1 summarizes these Fickian diffusion results for the purely diffusive fluid case.

Table 12.1 Predictions for the Fickian diffusion coefficient in a pure fluid as the critical temperature is approached along the critical isochore

	Conventional Theory	*Dynamic Scaling*
$D_{ii}(T \to T_c)$	0	0
$\lim_{T \to T_c} \left(\frac{\partial D_{ii}}{\partial T}\right)$	0	∞

Scaling in mixtures

In homogeneous mixtures the limiting behavior of the D_{ij} depends upon the relative scaling exponents of the L_{ij} and the quantity $(\partial\rho_j/\partial\mu_i)_{T,\rho_i}$. Since the latter type of derivative is predicted to diverge weakly (see chapter 3) this leads to a situation potentially different to that for the pure-fluid case. However, to the present time, there are no precise results for the dynamic scaling exponents of either of these types of properties in a real fluid mixture, so this remains in the realm of conjecture.

We now consider computer simulation methods for calculating these transport coefficients in near-critical fluids.

12.3 The Calculation of Diffusion Coefficients by Computer Simulation

In the realm of transport phenomena, the use of computer simulation has assumed a dominant role for studying system behavior, especially for complex fluids like those confined in porous media. In this section, we explore simulation approaches, beginning with the simple self-diffusion case before progressing to the more complicated case involving Fickian diffusion in confined porous structures.

Self-Diffusion Coefficients in Pure Lattice Fluids: Kinetic Monte Carlo

The self-diffusion coefficient for a single species can be found from either molecular dynamics or *kinetic Monte Carlo* simulations. Molecular dynamics is described in many computer simulation texts [5] and uses equations like (12.56) to calculate D_s. However, given our focus on lattice models in this text, we dwell upon diffusion calculations using kinetic Monte Carlo methods applied to lattice gases.

We first illustrate what it means to do kinetic Monte Carlo simulations in lattice gases. The idea of kinetic Monte Carlo at first might sound like an oxymoron given the usual association of Monte Carlo methods for simulations in equilibrium systems. We now describe how we might carry out a kinetic Monte Carlo calculation in a lattice fluid. For a given temperature and particle density in the lattice, a particle and adjacent site are randomly chosen. If the adjacent site is vacant, an attempt is made to move the particle there. The *acceptance probability* for this Monte Carlo move from a configuration with energy E_i to that with energy E_f is given, for example, by the expression:

$$P_{i\to f} = \frac{1}{1 + e^{\beta(E_f - E_i)}} \tag{12.82}$$

which obeys the principle of detailed balance, namely that

$$\frac{P_{1\to 2}}{P_{2\to 1}} = e^{\beta(E_1 - E_2)} \tag{12.83}$$

Time in the kinetic Monte Carlo sense is now considered to be proportional to the number of such successful moves. One complete set of n attempted moves in an $(n \times n)$ lattice is often designated one *Monte Carlo time step* (MCS). The Hamiltonian used to find the energy in (12.82) could be given by the usual lattice-fluid equation

$$E = -4J \sum_{\langle ij \rangle}{}' n_i n_j \tag{12.84}$$

where n_i can take on either the value 1 (presence of a particle) or 0 (absence of a particle) at site i. The symbol $\sum'_{\langle ij \rangle}$ denotes the usual summation over nearest neighbors throughout the lattice.

Using this approach, all the particles' positions could be tracked, thereby providing data essential for calculating self-diffusion coefficients. For example, we could calculate the following quantities for a given particle: $[\Delta x(m) - \Delta x(0)]^2$, $[\Delta x(2m) - \Delta x(m)]^2, \ldots,$ and similarly for the y coordinate, and so on, over literally millions of lattice sweeps. From these data, we could find the quantity $\langle R_1^2 \rangle_m$, at a given value of m (see (12.26)) which would yield a single data point on the $\langle R_1^2 \rangle_m$ against m graph. This process could be repeated for many values of m, and the data for $(R_1^2)_m$ averaged. It is predicted that a graph of $\langle R_1^2 \rangle_m$ against m at large m should be linear, with the average value of the slope taken to be equal to $6D_{\mathrm{S}}$.

EXERCISE 12.4

How would you adapt the kinetic Monte Carlo algorithm described above for the situation where we are attempting to calculate $D_{\mathrm{s,total}}$ for the lattice?

The more difficult quantity to calculate in simulations is D_{F}, an issue we now address using recently proposed relaxation-dynamic ideas.

12.4 Relaxation Dynamics in the Critical Region

The calculation of Fickian coefficients is significantly more complicated, since these properties are not related in an obvious way to the particle configurations, as is the situation with D_{S}. However, an attractive way to do this calculation was recently proposed using *relaxation-dynamic* ideas [6]. It requires following the dynamics of a system as it relaxes from an initial nonequilibrium state. A two-chamber ensemble schematically shown in figure 12.4 is used for these purposes an entity that has been called a *Fickian ensemble*, given its phenomenological construction, chosen to mimic density-driven diffusion processes. The square-wave profile in the ensemble represents preequilibrated high-density and low-density regimes separated, initially, by an impermeable partition. The two profiles meet at a discontinuous-density interface. Once the partition is removed, diffusion between both chambers will occur as the system proceeds to relax toward its equilibrium state.

The dynamics in such an ensemble can be analyzed by studying the time-dependent Fickian diffusion equation. For illustrative purposes, we look at one-dimensional isotropic diffusion in a pure fluid which can be represented as

$$\frac{\partial \rho(x,t)}{\partial t} = D_{\mathrm{F}} \frac{\partial^2 \rho(x,t)}{\partial x^2} \tag{12.85}$$

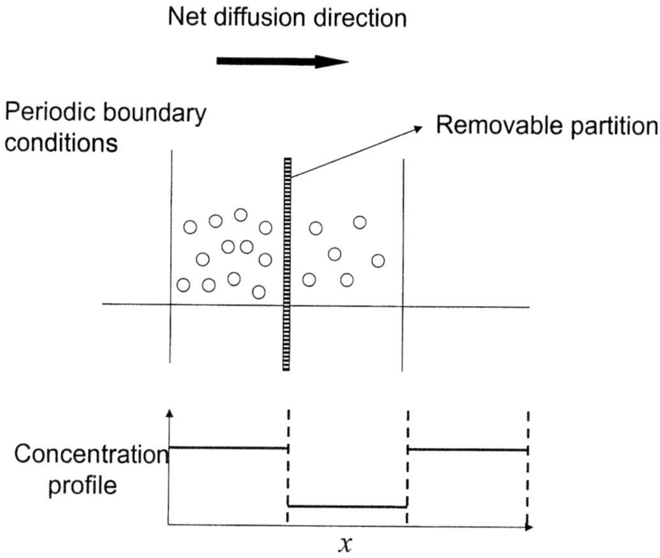

Figure 12.4 Schematic describing the Fickian ensemble for performing relaxation-dynamic simulations of transport properties

where D_F is the Fickian diffusion constant, $\rho(x, t)$ the fluid density and x the coordinate in the diffusion direction. This equation can be solved for a variety of initial conditions including the one shown in figure 12.4.

For a periodic initial density profile, the temporal and spatial dependence of the density profile in the system can be found from the general equation

$$\rho(x, t) = a_0 + \sum_{n=1}^{\infty} a_n(\cos nkx)e^{-D_F n^2 k^2 t} + \sum_{n=1}^{\infty} b_n(\sin nkx)e^{-D_F n^2 k^2 t} \tag{12.86}$$

where $k \equiv 2\pi/L$ is the wave number, and a_0, a_n, and b_n are Fourier coefficients. For the square-wave initial density profile shown in figure 12.4, we find that

$$\rho(x, 0) = \frac{4\varepsilon}{\pi} \sum_{\text{odd } n} \frac{1}{n} \sin nkx + \bar{\rho} \tag{12.87}$$

Here $\bar{\rho}$ is the average density of the fluid in the entire ensemble, and ε is equal to the magnitude of the initial density perturbation in each chamber from $\bar{\rho}$. In this case, the solution to (12.85) with this initial condition takes the particular form

$$\rho(x, t) = \frac{4\varepsilon}{\pi} \sum_{\text{odd } n} \frac{1}{n}(\sin nkx)e^{-D_F n^2 k^2 t} + \bar{\rho} \tag{12.88}$$

We now define a quantity Δm in each chamber of (length $L/2$) as follows:

$$\Delta m \equiv \langle m(t) \rangle - \langle m(\infty) \rangle \tag{12.89}$$

where $m(t)$ represents the number of particles in a given chamber at time t, and $m(\infty)$, the equilibrium condition. We can find Δm from the following integral:

$$\Delta m = L^2 \int_0^{L/2} (\rho - \bar{\rho})dx \tag{12.90}$$

which becomes, upon the use of equation (12.88),

$$\Delta m = \frac{4\varepsilon}{\pi} L^2 e^{-Dn^2k^2t} \int_0^{L/2} \frac{1}{n}(\sin nkx)dx \tag{12.91}$$

leading to the result

$$\Delta m = \frac{4\varepsilon L^3}{\pi^2} \sum\nolimits_{\text{odd } n} \frac{1}{n^2} e^{-D_{\text{F}}n^2k^2t} \tag{12.92}$$

For $t = 0$, the value of the term within the summation is 1.233; while for $t = 0.5/D_{\text{F}}k^2$, its value is 0.608 with the first term, namely $e^{-D_{\text{F}}k^2t}$, equal to 0.607. In other words, when Δm equals about half of its initial value, all the Fourier components, except for the first one, have already decayed considerably. Therefore, at a time $t \geq t_0$, with t_0 precisely chosen in advance, we have to a good approximation that

$$\Delta m = \frac{4\varepsilon L^3}{\pi^2} e^{-D_{\text{F}}k^2t} \tag{12.93}$$

So, for two values of $t > t_0$, namely t_1 and t_2, we get that

$$\frac{\Delta m_2}{\Delta m_1} = e^{-D_{\text{F}}k^2(t_2 - t_1)} \tag{12.94}$$

from which it follows that

$$D_{\text{F}} = \frac{\ln(\Delta m_1/\Delta m_2)}{k^2(t_2 - t_1)} \tag{12.95}$$

This simple equation is the basis for the relaxation-dynamic simulation method. It can be further refined to provide useful error estimates for the simulation results.

Error Bounds: A Controlled Approximation for D_{F}

In order to estimate the error associated with ignoring the higher wave numbers in reaching the result shown in equation (12.95), we can approximate this equation for small differences in time by the differential form

$$D_{\text{F}} = -\frac{1}{k^2 \Delta m} \frac{d(\Delta m)}{dt} \tag{12.96}$$

Now using (12.92) in (12.96) gives

$$D'_{\text{F}} = D_{\text{F}} \frac{\sum_{\text{odd n}} e^{-D_{\text{F}}n^2k^2t}}{\sum_{\text{odd n}} \frac{1}{n^2} e^{-D_{\text{F}}n^2k^2t}} \tag{12.97}$$

If the higher-order terms are small compared to the leading term, then we may write

$$D'_{\mathrm{F}} = D_{\mathrm{F}} + \delta D_{\mathrm{F}} \tag{12.98}$$

where

$$\frac{\delta D_{\mathrm{F}}}{D_{\mathrm{F}}} = \sum_{\text{odd } n \neq 1} \left(1 - \frac{1}{n^2}\right) e^{-D_{\mathrm{F}}(n^2-1)k^2 t} \tag{12.99}$$

EXERCISE 12.5

Prove equation (12.99) and explain why it is an exact bound on the simulation data found for D_{F} using the relaxation-dynamics method.

Clearly we see that $\delta D_{\mathrm{F}}/D_{\mathrm{F}} \to 0$ as $t \to \infty$. If we begin data acquisition in the simulations when Δm equals about half of its initial value with $t \geq t_0 = 0.5/D_{\mathrm{F}}k^2$ and $e^{-D_{\mathrm{F}}k^2 t_0} \sim 0.607$, then we have from (12.99) that

$$\frac{\delta D_F}{D_F} \propto 1.6\% \tag{12.100}$$

In this case, we have a very precisely controlled approximation, since the exact error implicit to neglecting terms for which $k > 1$ is given by (12.100).

A detailed prescription for the relaxation-dynamic algorithm implementing these ideas is now given.

Relaxation-Dynamic Simulation Algorithm

We define the number of Monte Carlo cycles by N_{eq} with the number of cycles in the preequilibration stage by N_{peq}. The total number of particles in the system is designated N with an index n that runs from 1 to N. The simulation algorithm is as follows:

1. Start loop n_{eq} which runs from 1 to N_{eq}.
2. Start loop n which runs from 1 to N.
3. Randomly choose both a particle labeled by n and one of its six neighboring sites.
   - *If the site and the particle are in the two different chambers and $n_{\mathrm{eq}} \leq N_{\mathrm{peq}}$, then abandon this attempted move, increment n by one, and return to step 3.
   - Otherwise, if that neighboring site is filled with another particle, then abandon this attempted move, increment n by one and return to step 3.
   - Otherwise, if that site is empty, then calculate the present value of the energy E_i and the value of the energy E_f that corresponds to the state of the system in the event that the particle moves to the chosen empty neighboring site.
4. Calculate the probability

$$p \equiv P_{i \to f} = \frac{1}{1 + e^{\beta(E_f - E_i)}} \tag{12.101}$$

as follows:

- Generate a random number r uniformly distributed between 0 and 1.
- If $r < p$, then accept the move so that particle n moves to the chosen empty neighboring site.
- If $r > p$, then reject the move.

5. If $n_{eq} > N_{peq}$, then calculate the number of particles throughout the system, and their positions, and return to step 3.
6. End loop n.
7. End loop n_{eq}.

The first condition in step 3, denoted by *, ensures that there is no particle exchange between the two chambers during the preequilibration stage. At the end of this stage, both chambers are independently equilibrated at their respective densities, at the given temperature. During the next ("kinetic") part of the algorithm, Monte Carlo cycles are run in the same manner as described above, however, without the restriction in the step denoted by * above; thus, particle exchange between the two chambers now becomes possible. During this stage of the simulation, the whole system approaches equilibrium, with the two chambers tending to approach the same final density $\bar{\rho}$. We now analyze this method for calculating transport coefficients in the fluid's critical region.

Relaxation Dynamics in the Critical Region of a Pure Lattice-Gas Fluid

Critical slowing down

In figure 12.5, we show relaxation-dynamic simulation results for transport fluxes in a 3-d lattice-gas fluid as its critical point is approached along its critical isochore. The phenomenon of critical slowing down is clearly seen in these results, and the critical temperature, found from extrapolating to the zero flux limit, has been shown to be in excellent agreement with that obtained from independent equilibrium simulations [6]. Critical scaling can also be studied with this approach, and finding critical exponents this way provides a stringent test of the computational procedure.

For small t and k and large system size L, finite-size-scaling theory discussed in a previous chapter postulates the following scaling law for the Fickian diffusion coefficient:

$$D_F(t, k, L) \propto b^{-(2-\eta)} \Delta(tb^{1/\nu}, L/b, kb) \tag{12.102}$$

where b is a rescaling factor and Δ a universal scaling function. At the critical temperature, the correct scaling in equation (12.102) is found by replacing b by L, which leads to the result

$$D_F(t = 0, k, L) \propto L^{-(2-\eta)} \Delta(0, 1, kL) \tag{12.103}$$

Now, $k = 2\pi/L$ for each L; thus the scaling of the diffusion coefficient becomes

$$D_F(t = 0, k, L) \propto L^{-(2-\eta)} \Delta(0, 1, 2\pi) \tag{12.104}$$

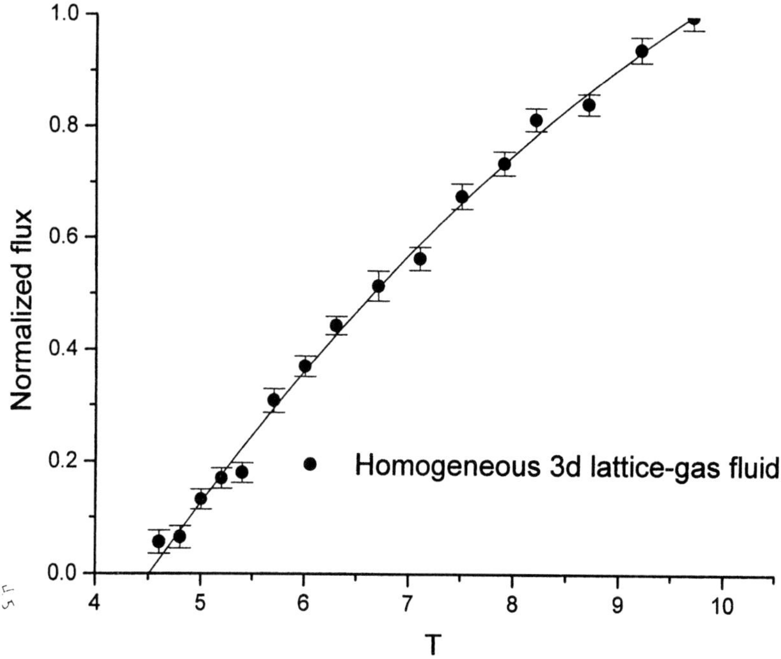

Figure 12.5 Simulation data for fluxes in the critical region of a 3-d homogeneous lattice-gas fluid

where $\Delta(0, 1, 2\pi)$ is a constant so that

$$D_{\mathrm{F}}(t = 0, k = 2\pi/L, L) \propto L^{-(2-\eta)} \tag{12.105}$$

Equation (12.105) can be used to analyze relaxation-dynamic simulations at various system sizes L, at the system's critical temperature, if it is known *a priori*. In figure 12.6, we show a graph of some simulation data in the form of $\ln D_{\mathrm{F}}$ versus $\ln L$ at the critical temperature of a pure 3-d lattice fluid system for which T_{c} is known. The slope of the line in figure 12.6 yields a value for the quantity $y \equiv -(2-\eta)$ equal to -1.971 ± 0.09, where η is the correlation function's critical exponent. This result is in excellent agreement with the value for $y = -1.972$ found independently using Monte Carlo simulation in an equilibrium ensemble [7] and illustrates the numerical power and accuracy of this relaxation-dynamics approach.

The Importance of Long-Wavelength Fourier Modes near the Critical Point

By combining equation (12.66) with the *continuity equation*

$$\frac{\partial \rho}{\partial t} = -\nabla \cdot J \tag{12.106}$$

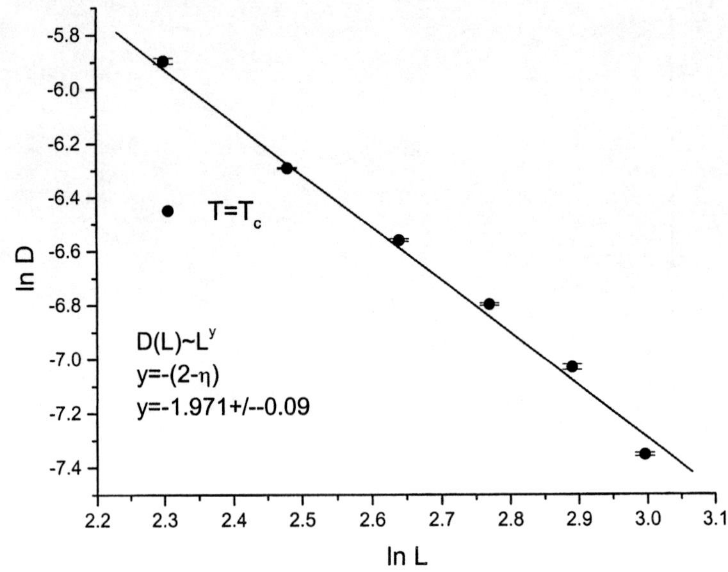

Figure 12.6 Simulation data for the scaling of the diffusion coefficient at the critical point of a 3-d homogeneous lattice-gas fluid

we get the following time-dependent partial differential equation, in 1-d, for example, describing density relaxation dynamics in the fluid:

$$\frac{\partial \rho}{\partial t} = \lambda \cdot \frac{\partial^2 \rho}{\partial x^2} \tag{12.107}$$

where $\lambda \equiv L/\chi_T$ is the Onsager-coefficient–susceptibility ratio corresponding to that given in equation (12.67) if we assert the equivalence between L and L_{ii}, χ_T, and $(\partial \rho_i / \partial \mu_i)$. The function $\rho(x, 0)$ is the initial density profile which can be arbitrarily specified. This equation can be Fourier-transformed into the ordinary differential equation

$$\Omega'(q, t) + \lambda q^2 \Omega(q, t) = 0 \tag{12.108}$$

where

$$\Omega(q, t) \equiv \frac{1}{\sqrt{2\pi}} \int_{-\infty}^{\infty} \rho(x, t) e^{-iqx} dx \tag{12.109}$$

The solution to equation (12.108) with specified boundary condition is given by [8]

$$\Omega(q, t) = \frac{1}{\sqrt{2\pi}} \int_{-\infty}^{\infty} \rho(u, o) \exp(-iqu - q^2 \lambda t) du \tag{12.110}$$

with u a dummy variable. It follows that

$$\rho(x, t) = \frac{1}{\sqrt{2\pi}} \left(\int_{-\infty}^{\infty} \exp(iqx) \Omega(q, t) dq \right) \tag{12.111}$$

Note that ordinarily, at large times, one would expect the Fourier coefficient contribution to $\rho(x, t)$ to vanish rapidly for all x. However, near the critical point, $\lambda \to 0$, which largely mitigates this effect, especially for small wave numbers q (i.e., long-wavelength modes). The complex integral in (12.111) can be done using special techniques borrowed from complex-variable theory [8] and yields the solution

$$\rho(x, t) = \frac{1}{\sqrt{4\pi\lambda t}} \int_{-\infty}^{\infty} \rho(u, 0) \exp\left[-\frac{(x-u)^2}{4\lambda t}\right] du \qquad (12.112)$$

Computer Simulation of Fickian Diffusion Coefficients in Porous Structures in the Critical Region

The relaxation-dynamic method described in the previous section for simulation of Fickian diffusion coefficients in homogeneous systems can just as easily be used for porous media. One simply imbeds the fluid molecules in a porous matrix that makes up the Fickian ensemble and proceeds to calculate relaxation rates from a prepared initial nonequilibrium state in a manner identical to the homogeneous case.

To illustrate this approach, we present results with the confined-lattice-gas Hamiltonian described in the previous chapter with the value $\Gamma = 2\Im$ for the fluid–pore-surface interaction parameter. In this case the relaxation-dynamic results can be *exactly* benchmarked against those of the randomly site-diluted Ising model done in an equilibrium ensemble (see exercise 12.6).

EXERCISE 12.6

Prove that the value $\Gamma = 2\Im$ for the fluid–pore-surface interaction parameter in the confined lattice-gas fluid corresponds exactly to a Hamiltonian isomorphic to that of the *randomly site-diluted Ising model.*

In figure 12.7, we show comparative results between the relaxation-dynamic simulation data developed in this study and an equilibrium-based indirect method using the Onsager approach that led to (12.67). The agreement between both sets of results is good. The advent of critical slowing down as the critical point is approached implies that, once again, the critical temperature can be estimated from extrapolating the data to the zero diffusion limit. These results for various values of p are seen in table 12.2, compared with results found independently for the randomly site-diluted Ising system by equilibrium simulations. The reduced temperature in the systems is defined as $T^* \equiv k_B T/\Im$, $L \to \infty$.

Agreement between both sets of simulation results is excellent and serves to underscore the validity of the relaxation-dynamic approach even for more complicated fluid structures. It is conjectured that critical dynamics for homogeneous and confined fluids belong to different critical universality classes. Notwithstanding this view, it is tempting to look for a *characteristic relaxation time* τ, at a given p, in terms of the temperature "distance" from the critical point, namely $|T^* - T_c^*(p)|$, for these respective systems. If such a characteristic time were available, then we should find that an expression of the form

$$\Delta m_2 = \Delta m_1 e^{-k^2(t_2-t_1)/\tau} \qquad (12.113)$$

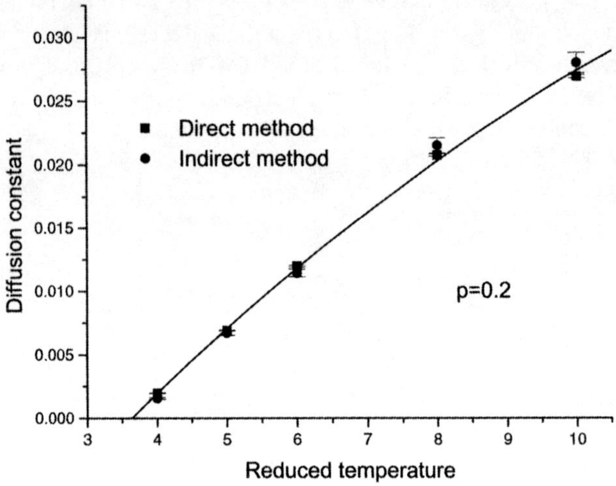

Figure 12.7 Simulation data for the scaling of the diffusion coefficient of a fluid, confined in a random quenched pore structure, at its critical point

would represent all of the dynamic data in these various systems, where we define

$$\frac{1}{\tau} \equiv \phi(p) f(T^* - T_c^*(p)) \tag{12.114}$$

In (12.114), $\phi(p)$ is an amplitude factor while f is a universal function of $T^* - T_c^*(p)$. From (12.113) and (12.114), we find that

$$\phi(p) \equiv \frac{\ln \Delta m_1/\Delta m_2}{k^2 f(T^* - T_c^*(p))(t_2 - t_1)} \tag{12.115}$$

To complete this definition for $\phi(p)$, we need only specify the temperature used in (12.115), a natural one being the high temperature limit, where we require that

$$\lim_{T^* - T_c^*(p) \to \infty} f \to 1 \tag{12.116}$$

in which case,

$$\phi(p) = D_h \equiv \lim_{T^* - T_c^*(p) \to \infty} D_F \tag{12.117}$$

Table 12.2 Comparison of critical temperatures in quenched random porous structures at various values of porosity *p*

	T_c (*Relaxation Dynamics*)	T_c (*Independently Found*) [10]
0	4.52	4.53
0.1	4.04	4.02
0.2	3.6	3.5
0.35	2.8	2.7

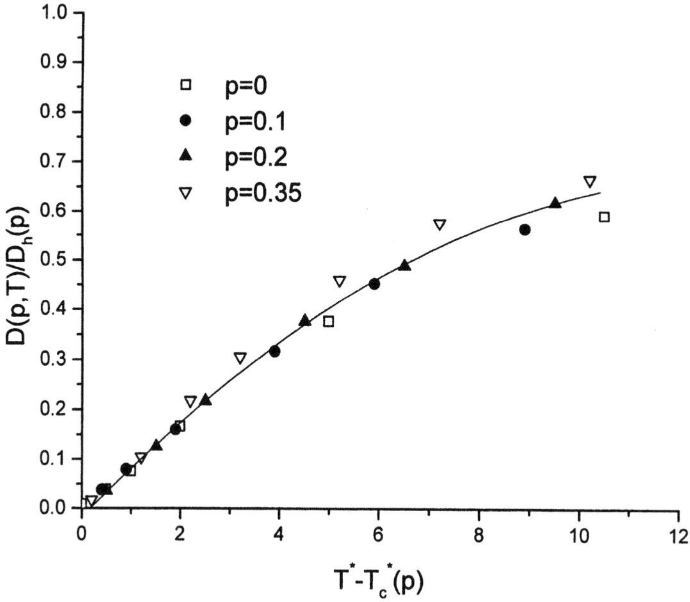

Figure 12.8 Universal curve for describing simulation data of the diffusion coefficients of pore-confined fluids in their critical regions

In this case, the characteristic relaxation time is given by

$$\frac{1}{\tau} \equiv D_{\text{h}} f(T^* - T_{\text{c}}^*(p)) \tag{12.118}$$

The efficacy of this approach depends upon the applicability of the product $D_{\text{h}} \cdot f(T^* - T_{\text{c}}^*(p))$ in (12.118) for describing the relaxation dynamics of many different systems. If this approach holds, then we should find from (12.95) and (12.115)–(12.118) that

$$\frac{D_{\text{F}}}{D_{\text{h}}} = f \tag{12.119}$$

should be a universal function of $|T^* - T_{\text{c}}^*(p)|$ for simulation data from disparate systems. This result appears to be borne out, to a very considerable degree, by the results presented in figure 12.8, irrespective of the value of p used. This shows that there appears to be a strong similarity between near-critical dynamics in both homogeneous and confined systems like those used here.

The relaxation-dynamic approach is not restricted to isotropic randomly structured media, but is applicable to any confined system. It has, for example, also been used in an *anisotropic system* consisting of nanowire bundles. These are linear strands of matrix material, one lattice unit thick, separated by 4 lattice units (see figure 12.9), oriented perpendicular to the x–y plane. The results for these simulations are shown in figure 12.10 and illustrate the fact that the anisotropic diffusion coefficients D_{xy} and D_{yz} are different and, as we might intuitively expect, $D_{xy} > D_{yz}$. However, as we approach the critical temperature in this system, we expect both D_{xy} and D_{yz} to vanish because of the divergence of the susceptibility (isothermal compressibility) there. The values of

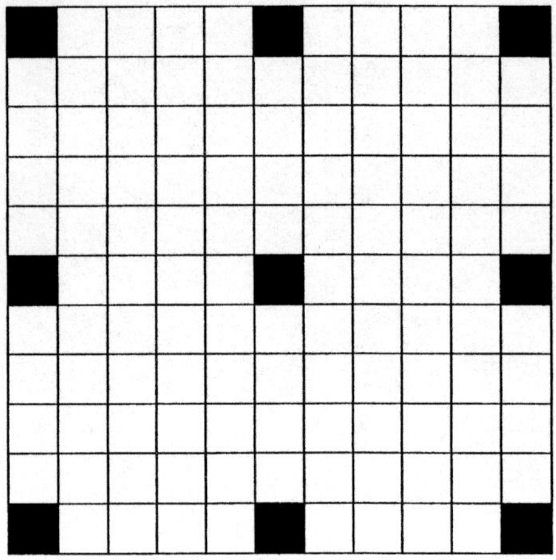

Figure 12.9 Schematic of nanowire arrays

both diffusion constants D_{xy} and D_{yz} appear to approach one another in this respect, with the prospect of vanishing in the asymptotic limit as would be theoretically expected.

12.5 Binary Mixtures

Diffusion and Scaling Ideas

The general equations governing the diffusive fluxes in a two-component binary mixture are given by

$$j_\rho = -D_{\rho\rho}\nabla\rho - D_{\rho\sigma}\nabla\sigma \tag{12.120}$$

$$j_\sigma = -D_{\sigma\rho}\nabla\rho - D_{\sigma\sigma}\nabla\sigma \tag{12.121}$$

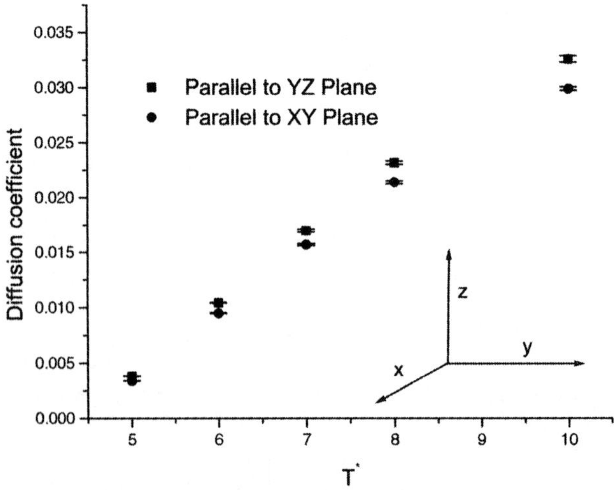

Figure 12.10 Simulation data for the diffusion coefficients of a near-critical fluid confined in the anisotropic nanowire system

where we denote the solvent and the solute densities by ρ and σ, respectively. In these equations, $D_{\sigma\rho}$ and $D_{\sigma\sigma}$ for the solute are effective scalar diffusivities which can be expressed as follows:

$$D_{\sigma\rho} = \left(\frac{\partial \mu_\rho}{\partial \sigma}\right)_{\rho,T} L_{\sigma\sigma} + \left(\frac{\partial \mu_\rho}{\partial \rho}\right)_{\sigma,T} L_{\sigma\rho} \tag{12.122}$$

and

$$D_{\sigma\sigma} = \left(\frac{\partial \mu_\sigma}{\partial \sigma}\right)_{\rho,T} L_{\sigma\sigma} + \left(\frac{\partial \mu_\sigma}{\partial \rho}\right)_{\sigma,T} L_{\sigma\rho} \tag{12.123}$$

where the various subscripted L quantities are the Onsager coefficients with μ_ρ and μ_σ the chemical potentials of the solvent and solute species, respectively. The diffusivities of the solvent can be expressed in a similar manner; however, here we are interested in the behavior of the dilute solute at the critical point of the confined solvent. For both homogeneous lattice-gas and continuum-fluid binary mixtures, the dynamic scaling of the various terms in equation (12.123) remains an open issue, with only conjectures to go by. In porous media, the situation is similar, although the lack of momentum conservation in the confined fluids allows, currently at least, for a marginal improvement in the confidence of the conjectures as now explained.

Confined Binary Mixture

The Onsager coefficients are finite for a fluid mixture embedded in a quenched porous structure, since there are no conserved momentum modes for the fluid components, and such a system is thought to belong to the "model B" class (kinetic Ising dynamic universality class) [3]. Hence, the limiting critical properties of these diffusion constants will depend on how the derivatives of the chemical potentials with respect to the densities, as shown in equations (12.122) and (12.123), behave as the critical point is approached.

We see from (12.122) and (12.123) that, in each case, a derivative of one chemical potential with respect to one of the densities is taken, with the other density being kept constant. At present we are not aware of any particular results for the limiting behavior of these quantities in confined fluids, the closest available analogy being the results for a binary near-critical homogeneous fluid [11]. In the case of a binary fluid at finite solute composition along the critical line, the inverses of these derivatives are supposed to diverge weakly. Keeping in mind that the weak exponent α is negative for a random field system, and the analogy often employed between the confined fluid and the random-field Ising models, we conjecture that, for a binary fluid confined in such random structures, the derivatives mentioned above will stay finite at the critical line (finite solute composition), in which case $D_{\sigma\sigma} > 0$. The situation near the pure solvent's critical point is once again special as now discussed.

Near the Solvent's Critical Point, $x \sim 0$

Binary mixture scaling near the pure solvent's critical point is a special case [11], and once again employing the homogeneous fluid analogy would lead us to conjecture that the dependence of the derivatives in equations (12.122) and (12.123) on the small solute concentration x can be written as follows [11]:

$$\left(\frac{\partial \mu_\sigma}{\partial \sigma}\right)_{\rho,T} \propto \left(\frac{\partial \mu_\sigma}{\partial x}\right)_{P,T} \propto x^{-2} \tag{12.124}$$

$$\left(\frac{\partial \mu_\rho}{\partial \sigma}\right)_{\rho,T} \propto \left(\frac{\partial \mu_\rho}{\partial x}\right)_{P,T} \propto x^{-1} \tag{12.125}$$

$$\left(\frac{\partial \mu_\sigma}{\partial \rho}\right)_{\sigma,T} = \left(\frac{\partial \mu_\rho}{\partial \sigma}\right)_{\rho,T} \propto x^{-1} \tag{12.126}$$

$$\left(\frac{\partial \mu_\rho}{\partial \rho}\right)_{\sigma,T} = 0 \tag{12.127}$$

This leads to a simple relationship between the two diffusion constants $D_{\sigma\sigma}$ and $D_{\sigma\rho}$ controlling the diffusion of the solute throughout the porous structure. Putting the above relationships into equations (12.122) and (12.123), we immediately obtain that

$$\frac{D_{\sigma\rho}}{D_{\sigma\sigma}} \propto x \tag{12.128}$$

in the dilute confined binary mixture near the solvent's critical point. It follows then that

$$D_{\sigma\rho} = cx D_{\sigma\sigma} \tag{12.129}$$

where c is a constant of proportionality that should depend on the identity of the solvent and solute species as well as the embedding porous structure; it should, however, be independent of x. Equation (12.129) turns out to be a potentially important relationship, since it allows us to express the diffusive flux of the solute in terms of only one of the two diffusion constants that govern it, namely $D_{\sigma\sigma} > 0$. In fact, it follows that solute diffusion in the structure can be represented by the equation

$$j_\sigma \propto -D_{\sigma\sigma}\nabla\sigma \tag{12.130}$$

where $D_{\sigma\sigma}$ can be found from computer simulation [9].

12.6 Diffusion in Porous Media

So far, in this chapter, we have focused upon analyzing diffusion in the critical region of a fluid. This usually occurs at high pressures, yet, in many process engineering applications, transport occurs within the porous medium at lower pressure conditions. We now briefly look at this issue using concepts that have proven useful in analyzing this situation. Fortunately, many of the computer simulation methods and ideas described previously apply fairly directly to an analysis of this problem.

Low Pressures: The Self-Diffusion Constant

At low pressures, we assume the equivalence between Fickian and self-diffusion constants and introduce the concept of *Knudsen diffusion*, which is used to describe diffusion at low pressure in porous structures. In the Knudsen regime, the diffusing species are considered to collide more often with the pore walls than with one another. The resistance to diffusion is now due to the fact that, after a wall collision, the molecule is just as likely to reverse its direction as it is to continue in the direction of its previous flight [12]. In cylindrical pores with mean diameter d_0, the *Knudsen diffusion coefficient* D_K is given by

$$D_K = \frac{1}{3} d_0 \bar{v} \tag{12.131}$$

where the mean velocity $\bar{v} = \sqrt{8k_B T/\pi m}$ can be obtained from elementary kinetic theory results with m the molecule's mass.

We see that D_K is independent of pressure and proportional to $T^{1/2}$ in the Knudsen regime. This should be compared to the result for gaseous bulk diffusion at low-pressure conditions, where the diffusion coefficient is found from an equation similar to (12.25) with l given by kinetic theory:

$$l = \frac{0.707RT}{\pi\sigma^2 P} \tag{12.132}$$

Here σ is the molecule's *collision diameter*, and we see that, with regular low-pressure gas diffusion, $D \propto T^{3/2}$ and $D \propto P^{-1}$.

However, diffusion in pore structures is usually characterized as being in one of three regimes: bulk, Knudsen, or an intermediate regime where both mechanisms contribute. These are usually characterized in terms of a dimensionless number called the *Knudsen number* K_n which is defined as

$$K_n \equiv \frac{l}{d_0} \tag{12.133}$$

A large value for K_n implies the dominance of the Knudsen contribution to D and vice versa. While d_0 applies to an idealized single pore, in a more complicated medium the Knudsen diffusion constant will often be approximated by (12.131) with d_0 now taken to be the "average" pore size in the medium defined in some reasonable way.

At low to modest pressures, simulation approaches for this problem have often used the Einstein equation. Given an assumed pore structure (e.g., the random pixelized medium described in this chapter) we define an effective self-diffusion constant D_e for a test molecule, representing a sufficiently large number of sample molecule trajectories by the equation

$$D_e \equiv \lim_{t\to\infty} \frac{\langle R^2 \rangle}{6t} \tag{12.134}$$

In the Knudsen regime of a porous medium, an effective Knudsen coefficient D_e^K can be defined as

$$D_e^K \equiv \frac{D_e}{D_K} = \lim_{\frac{s}{d_0}\to\infty} \frac{\langle R^2/d_0^2 \rangle}{2s/d_0} \tag{12.135}$$

where s is the total distance traveled by the molecule, with $D_K = d_0 v_0/3$ and $v_0 = s/t$.

Similarly, in the bulk regime, we can define an equivalent property D_e^b:

$$D_e^b \equiv \frac{D_e}{D_b} = \lim_{\frac{s}{\lambda_0}\to\infty} \frac{\langle R^2/\lambda_0^2\rangle}{2s/\lambda_0} \tag{12.136}$$

These quantities can be found by simulating tens of thousands of molecular trajectories in the media. The mean free path is found from an exponential distribution with mean λ_0, a statistical realization of which is found from the formula (see additional exercises)

$$\lambda = -(\ln U)\lambda_0 \tag{12.137}$$

with U a uniform random variable over the interval (0, 1). After molecule–molecule collisions, *direction cosines* are assigned to the molecule for its next flight, with the cosine law for the *polar angle* ϕ and the *azimuthal angle* θ (see figure 12.12) being sampled from a uniform distribution over the interval $[0, 2\pi]$. If a solid face is encountered, the molecule is assigned to undergo a diffuse reflection, where the new direction cosines are sampled according to specific distributions. This process is repeated, all the while testing for the possibility of molecule–molecule and molecule–wall collisions en route. Calculations of this nature were done [12] for up to 10,000 test molecules in 3-d pixelized media in 50 × 50 × 50 cubic structures to generate statistically reliable results for the averages.

Using this approach, both D_e^K and its bulk counterpart can be found by doing simulations at the required Knudsen-number conditions. In the intermediate regime, we can find the effective overall diffusion coefficient D_e^o in the porous structure making use of a Bosanqueth-type of equation:

$$\frac{1}{D_e^o} = \frac{1}{D_e^b} + \frac{1}{D_e^K} \tag{12.138}$$

EXERCISE 12.7

Rationalize the basis for equation (12.138).

The Cosine Law for Molecule–Molecule Collisions

Consider a molecule–molecule collision centered at the origin of the sphere shown in figure 12.11. If the number of reflected molecules passing through a given area along the surface of the sphere is to be equal regardless of the reflected angle (i.e., there is no directional bias to the reflections over many steps), then we must have, from geometrical considerations, that the surface area element $d\Omega$ is given by the equation

$$d\Omega = R d\phi \cdot r\, d\theta \tag{12.139}$$

Since, by construction, $d\Omega$ must be the same *irrespective of the polar angle*, and given that $r = R\sin\phi$, it follows that

$$d\Omega = -2\pi R^2 d\cos\phi \tag{12.140}$$

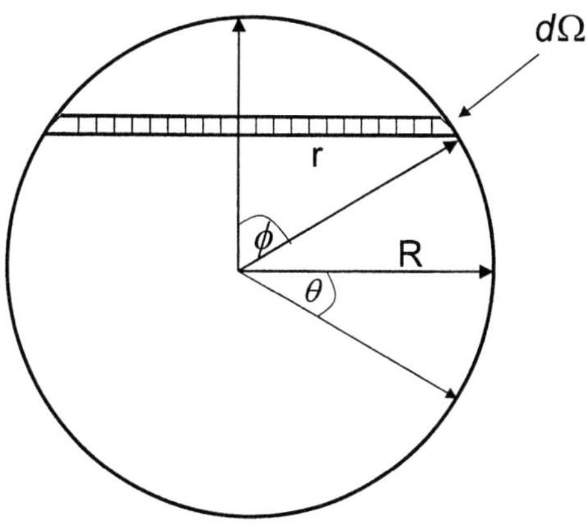

Figure 12.11 Schematic showing the cosine reflection law applied to molecule-molecule collisions

where we have integrated over the angle θ. From this equation, we get that $\cos\phi$ must be uniformly distributed over the interval $[0, \pi]$, which implies that ϕ must be found from the equation

$$\phi = \cos^{-1} U \tag{12.141}$$

with U a uniform random variable over the interval (0, 1). The azimuthal angle θ is found uniformly in the interval $[0, 2\pi]$. In the molecule-wall case, show that ϕ is found from the equation

$$\phi = \cos^{-1} \sqrt{U} \tag{12.142}$$

The approach outlined here can be used at any thermodynamic conditions of interest for any type of porous material structure. It is thus a powerful way of studying transport in

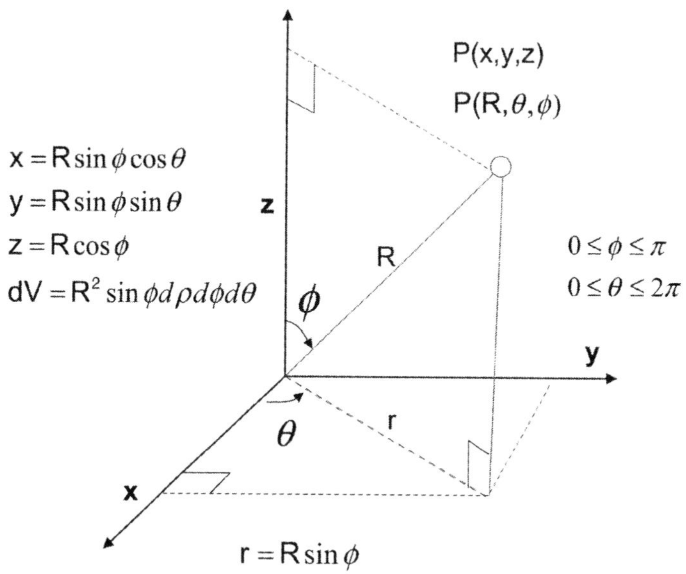

Figure 12.12 Schematic of coordinates used in the spherical coordinate system

complex structures at low pressures where self-diffusion and Fickian diffusion constants are taken to be equal.

12.7 Chapter Review

Transport in the critical region is characterized by a large increase in time for fluctuations to decay to the equilibrium state—a phenomenon known as critical slowing down. This implies that mass diffusion fluxes markedly decrease in this region—an issue relevant to engineering applications where the efficiency of a process is often defined by the magnitude of mass transfer rates.

We discussed these issues and introduced the two most important phenomenological constants used to characterize mass transport in fluid systems, the self-diffusion and Fickian diffusion coefficients, respectively. The self-diffusion constant characterizes the random movement of a single tagged molecular species undergoing Brownian motion and can be found from the Einstein equation which states that its mean squared displacement increases linearly with time at given thermodynamic conditions. The Fickian coefficient must, on the other hand, be found from flux measurements and/or calculations across a concentration gradient, and these two constants are not equivalent except in the ideal-gas limit, where molecular motions are not correlated.

In the critical region, the self-diffusion constant stays finite, as observed in many sets of experimental data. The Fickian diffusion constant, however, is predicted to asymptotically vanish as the critical point is approached. This vanishing is governed by scaling laws, and we looked at how these scaling predictions differed between conventional theory (which assumes the Onsager coefficient to be finite at the critical point) and dynamic scaling, which predicts the divergence of the Onsager coefficient there. Dynamic scaling predicts a weaker scaling for the diffusion constant than does conventional theory, a result only recently verified at the gas–liquid critical point in a real continuum fluid using molecular dynamics simulations [13].

Finally, we described various computer simulation methods for calculating both types of diffusion coefficients in homogeneous and confined-fluid systems: a random-walk method for D_S and relaxation dynamics for D_F. The power and accuracy of the relaxation-dynamic method was illustrated with finite-size scaling results at the critical point of a 3-d lattice fluid, which yielded excellent results for the dynamic scaling exponent for D_F. In addition, the method is capable of accurately simulating transport properties for fluids confined in complex porous media, an important engineering problem.

12.8 Additional Exercises

More Fourier-Transform Problems

1. The *complex form* of the Fourier series approximation to a function $f(x)$ is given by the series

$$f(x) = \sum_{n=-\infty}^{n=\infty} c_n e^{inx}$$

with

$$c_n = \frac{1}{2\pi}\int_{-\pi}^{\pi} f(x)e^{-inx}dx \quad (n = 0, \pm 1, \pm 2, \cdots)$$

Show that the Fourier approximation to an *odd function* is purely imaginary. An *odd* function is one for which $f(x) = -f(-x)$. How about the result for an *even* function defined as a function where $f(x) = f(-x)$?

2. Show that the Fourier transforms of the quantities $\partial c(x,t)/\partial x$ and $\partial^2 c(x,t)/\partial x^2$ are given by the functions $-ik\bar{c}(k,t)$ and $-k^2\bar{c}(k,t)$ where the Fourier transform of $c(x,t)$ is defined as

$$F_x[c(x,t)] = \bar{c}(k,t) \equiv \int_{-\infty}^{\infty} e^{ikx}c(x,t)dx$$

Do these results depend upon the boundary conditions on $c(x,t)$ and $\partial c(x,t)/\partial x$ at the limit $x \to \pm\infty$?

3. Show that

$$F_x\left[\frac{\partial^2 c(x,t)}{\partial t^2}\right] = \frac{\partial^2 \bar{c}(k,t)}{\partial t^2}$$

where $F_x[\bullet]$ denotes the Fourier transform as in exercise 2.
How does this result depend upon the boundary conditions on $c(x,t)$ and $\partial c(x,t)/\partial x$?

4. Prove the following very useful identities of the Fourier transform defined in exercise 2:

$$F_x[xf(x)] = -i\frac{d\bar{f}(k)}{dk}$$

$$F_x\left[\frac{\partial^n}{\partial x^n}f(x)\right] = (-ik)^n\bar{f}(k)$$

$$F_x[\exp(-a^2x^2)] = \frac{\sqrt{\pi}}{a}\exp\frac{-k^2}{4a^2}$$

Using one of these identities prove equation (12.108).

5. A spherical coordinate system is shown in figure 12.12. Given a Fourier transform of a function $Y(\boldsymbol{r})$ defined as

$$\bar{Y}(\boldsymbol{k}) = \int_{-\infty}^{\infty}\int_{-\infty}^{\infty}\int_{-\infty}^{\infty} Y(\boldsymbol{r})e^{-i\boldsymbol{k}\cdot\boldsymbol{r}}d\boldsymbol{r}$$

use spherical coordinates to show that for $Y(r)$ isotropic

$$\bar{Y}(k) = \frac{2}{k}\int_0^{\infty}\frac{Y(r)\sin kr}{k}dr$$

[Hint: Use $\boldsymbol{k}$ as a vector pointing in the usual z direction]. What happens if $\boldsymbol{k}$ points in the x direction and why would you not pursue this approach?

6. The Fickian diffusion equation is sometimes augmented as follows by a "field-driven" term:

$$\frac{\partial c(x,t)}{\partial t} = D\frac{\partial^2 c(x,t)}{\partial x^2} + \alpha c(x,t)$$

where the field couples through the term $\alpha c(x,t)$. Solve this equation, using a Fourier transform, with the boundary conditions

$$\lim_{t\to 0} c(x,t) = \delta(0,t)$$

$$\lim_{x\to\infty} c(x,t) = 0$$

$$\lim_{x\to\infty} \frac{\partial c(x,t)}{\partial x} = 0$$

Is it possible to use the *Laplace transform* to solve this problem? If so, find the Laplace transform of this equation with the given boundary conditions.

7. Solve the previous problem, using Fourier transforms, where the first boundary condition is changed to *a pulse function* centered at the origin of width and height one unit.

Brownian Dynamics and Sampling of Random Variables

8. Find the self-diffusion constant of a tagged species in a low-pressure 3-d gas, given that the mean free path is l and the average molecular velocity is $\bar{v}$.
9. Show that, in Brownian motion, the covariance of the particle position at times s $(\neq 0)$ and $(s+t)$ given by $r(s)$ and $r(s+t)$ is nonzero. We define this covariance as

$$\text{covar}\{r(t+s), r(s)\} = E[\{r(s+t) - E[r(s+t)]\}\{r(s) - E[r(s)]\}]$$

10. If we want to sample a random variable with a density function given by $p(x)$, then show that it is given by ξ, the solution to the integral equation

$$\int_{-\infty}^{\xi} p(x)dx = \gamma$$

where γ is *uniformly distributed* over the interval [0, 1].

11. Use the result of exercise 10 to show that a random variable x with an exponential density function given by the equation

$$p(x) = \lambda e^{-\lambda x}$$

with λ a constant, can be sampled using the equation

$$x = -\frac{1}{\lambda}\ln(1-U)$$

where U is uniformly distributed over the interval [0, 1]. What is the mean value of x found this way?

12. Write a computer program to test the Einstein equation using a random-walk algorithm in 2 dimensions with a path length of unity. On what basis do you conclude that your simulation is working accurately?

13. The function NORMSINV in EXCEL generates normal random variables with the properties $\mu = 0, \sigma = 1$; that is, with the pdf

$$P_N(X) = 1/\sqrt{2\pi}e^{-X^2/2}.$$

Use this function to generate 500 such data points and use the HISTOGRAM tool in EXCEL to "bin" these data. How would you ascertain that these data satisfy the given distribution?

14. Random-walk concepts are found to be very useful throughout the physics literature. Consider a 1-d random walk in which the probability of jumping left or right is given by the probability p. Show that after t total jumps the probability that the random walker has taken y steps to the right is given by the binomial distribution

$$P(y, t) = \binom{t}{y} p^y (1-p)^{t-y}$$

What is the expression for the probability that $y \in [0, t/2]$ after t steps? Show that, for long times (i.e., $t \to \infty$), for $p = 0.5$, the right displacement $x \equiv 2y - t$ of the random walker is given by a normal distribution. What are the mean and variance of this distribution? How would scaling up to 3-d space affect this problem?

15. Using transformations along the lines suggested in figure 12.3 with equation (12.37), prove that

$$\langle u^2 \rangle = \frac{a}{2\xi}(1 - e^{-2\xi t})$$

where

$$a \equiv \int_{-\infty}^{\infty} \psi(x)dx$$

16. *The Maxwell–Boltzmann* probability density $\phi(\mathbf{p}_i^N)$ for a system of $i = 1, \ldots, N$ identical particles is a famous result in the kinetic theory of gases. It provides that the momentum $\mathbf{p}_i$ (velocity) distribution of the particles (of mass m) at temperature T is given by

$$\varphi_{MB}(\mathbf{p}_i^N) \equiv \frac{\prod_i \exp[-\beta \mathbf{p}_i^2/2m]}{\int \exp[-\beta \mathbf{p}_i^2/2m] d\mathbf{P}^N}$$

where $\mathbf{p}_i^2 = p_{x_i}^2 + p_{y_i}^2 + p_{z_i}^2$. Show that

(a) The average kinetic energy per particle is equal to $3k_BT/2$.
(b) The average momentum of the system is zero.
(c) The most probable speed of a particle is given by $\sqrt{2k_BT/m}$.

Fickian Transport through Membranes and Other Assorted Geometries

17. If diffusion is occurring in a membrane structure represented by a 3-d random cubic lattice with matrix particle density p (i.e., porosity $1 - p$), what is the most probable configuration of matrix and open-cell structures surrounding a diffusing molecule at any given time?

18. Adsorption occurs through a thin membrane in contact with two gaseous reservoirs at constant concentrations c_1^0 and c_2^0, respectively. The adsorbed species' concentration on the membrane surface is given by a linear relationship of the form

$$c = Kc_i^0$$

Show that, if Fick's law applies, then the *steady-state flux* j through the membrane is given by

$$j = \frac{DK}{l}(c_1^0 - c_2^0)$$

where D is the diffusion coefficient in the membrane, and l its thickness. The product DK is often referred to as *the membrane permeability*.

19. Consider the results of exercise 18 in light of the apparatus depicted in figure 12.13. Assume that the initial concentration of the species in each chamber is given by c_1^0 and c_2^0, respectively. If diffusion is allowed to occur between the chambers, with the flux, at any given time, represented by a linear equation like that developed in exercise 18, show that the time-dependent concentration in chamber 2 is given by the expression

$$c_2(t) = c^\infty + (c_2^0 - c^\infty)e^{-\alpha t}$$

where

$$\alpha = \frac{DA}{l}\left[\frac{V_0}{V_1 V_2}\right]$$

$$V_0 = V_1 + V_2$$

$$c^\infty = c_2^0 + \frac{V_1}{V_0}(c_1^0 - c_2^0)$$

This equation leads to a direct way of estimating the diffusion coefficient from such time-dependent concentration data.

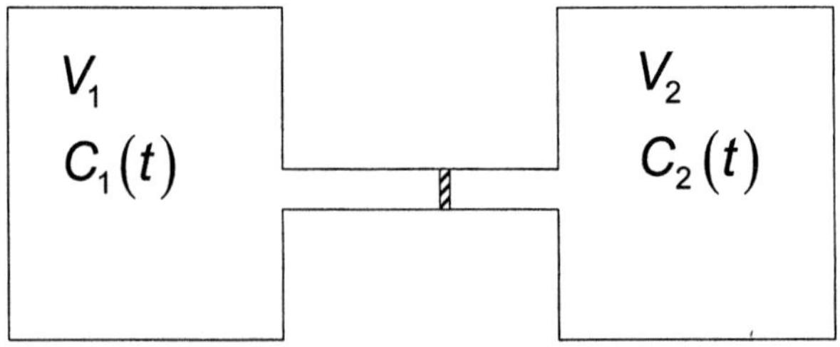

Figure 12.13 Schematic of an experimental diffusion chamber

20. Solute diffusion with adsorption is occurring through a host fluid in a semi-infinite membrane. Solute adsorption is modeled by the equation $q = Kc$, with K the adsorption constant (the units of q and c are moles per volume). Show that the time-dependent Fickian diffusion equation governing this process is given by

$$\frac{\partial c}{\partial t} = D' \frac{\partial^2 c}{\partial z^2}$$

with boundary conditions

$$c = 0 \quad \text{at } t = 0 \text{ for all } z$$
$$c = c_0 \quad \text{at } z = 0 \text{ for all } t > 0$$
$$c = 0 \quad \text{at } z = \infty$$

where $D' \equiv D/\varepsilon(1 + \alpha K)$, $\alpha \equiv (1 - \varepsilon)/\varepsilon$, ε is the porosity of the structure, and D the Fickian diffusion coefficient in the medium. Show that the flux at any given z is given by the equation

$$j_c = \sqrt{\frac{D'}{\pi t}} e^{-z^2/4D't} c_0$$

If the confined fluid system is near its critical point, how will this affect its flux through the porous structure? [Hint: Use the transformed variable $x \equiv z/\sqrt{4Dt}$ in the partial differential equation to convert it to an ordinary differential equation.]

21. Cussler [14] discusses a common separation problem involving diffusion–adsorption through a membrane of thickness l with the following boundary conditions:

$$c = 0 \quad \text{at } t = 0 \text{ for all } z$$
$$c = c_0 \quad \text{at } z = 0 \text{ for all } t > 0$$
$$c = 0 \quad \text{at } z = l$$

Show that

$$\frac{c}{c_0} = 1 - \frac{z}{l} - \frac{2}{\pi} \sum_{n=1}^{\infty} \frac{\sin(n\pi z/l)}{n} e^{-Dn^2\pi^2 t/l^2}$$

Does the technique used for exercise 20 work for these boundary conditions?

22. In this exercise, we analyze the experimental technique of Duffield and Harris [15] who used the weak beta emitter $^{14}CO_2$ to measure the self-diffusion of CO_2 in a tracer diffusion experiment using a capillary diffusion chamber as shown in figure 12.13. These authors reported the existence of a *divergent critical anomaly* in the self-diffusion constant of a pure fluid—something that has uniformly been disputed in the subsequent literature. The data in the paper and the evident care with which Duffield and Harris did their experiments do, however, raise the issue of what the origin of the reported anomaly might have been. Upon closer examination of their data analysis, we observe that Duffield and Harris used the Fickian description for capillary diffusion to get their result for the self-diffusion constant. We showed earlier, though, that this equivalence is not valid except at the low density limit where $\rho_i \rightarrow 0$. Using the Einstein equation, show how the self-diffusion coefficient can be found from data obtained by beta-emitter tracer diffusion in a capillary chamber. Prove that this equation predicts a linear relationship between time and the counter rate in the second chamber, which is entirely consistent with the data of Duffield and Harris.

23. *Biased diffusion models* [16] have the form

$$J = -(D + \gamma c)\nabla c$$

where J is the flux, c the concentration of species, and γ the bias constant. Show that the time-dependent diffusion equation in this system is given by

$$J = -(D + \gamma c)\nabla^2 c + \gamma(\nabla c)^2$$

What is the effect of the sign of γ on the solution of this model compared to the $\gamma = 0$ case?

24. For the relaxation-dynamic simulation algorithm described in this chapter, prove equations (12.97)–(12.99).

25. Equimolar counterdiffusion can be used to describe the flux of two interdiffusing species when the species' fluxes are equal and opposite. Why would this pertain in the A–B diffusion circumstances represented schematically in figure 12.14? The usual flux equation of species A is written as

$$N_{\mathrm{A}} = -cD_{\mathrm{AB}}\frac{dx_{\mathrm{A}}}{dz} + x_{\mathrm{A}}(N_{\mathrm{A}} + N_{\mathrm{B}})$$

and, in this situation, $N_{\mathrm{A}} = -N_{\mathrm{B}}$. At steady-state diffusion, therefore, show that

$$D_{\mathrm{AB}} = \frac{VL}{2At}\left[\frac{2x_{\mathrm{A}}(t) - (x_{\mathrm{A}10} + x_{\mathrm{A}20})}{(x_{\mathrm{A}10} - x_{\mathrm{A}20})}\right]$$

The quantities V, A, c, D_{AB}, and x_A denote the volume of each bulb, cross-sectional area of the connecting tube, total molar concentration, binary diffusion coefficient, and mole fraction of species A, respectively. Demonstrate how a knowledge of t and $x_{\mathrm{A}}(t)$ is sufficient to yield an estimate of D_{AB}. What is the disadvantage of using this approach to develop a computer simulation algorithm for calculating D_{AB} in a binary diffusing system? For example, concentration changes could in principle be found by monitoring the change in mole fraction of the species in two chambers along the lines suggested by the relaxation-dynamic method discussed in this chapter; the counterdiffusion dynamics could be generated either by a Monte Carlo or a molecular-dynamic simulation method. Why is the relaxation-dynamic method intrinsically superior to this idea?

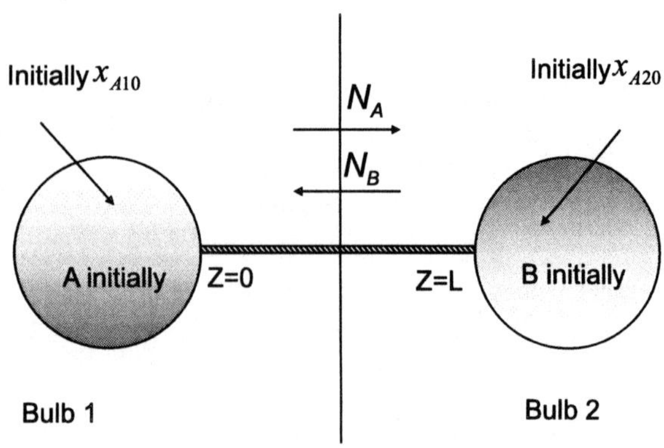

Figure 12.14 Schematic of a counter diffusion process

26. Prove equation (12.95), the central equation for the relaxation-dynamic method.
27. Clark and Rowley [17] measured mutual diffusion coefficients near the liquid–liquid consolute point in the methanol–n-hexane system. They reported the mutual diffusion coefficient data along the coexistence direction as shown in table 12.3.

Table 12.3

$D(10^{-10}\ \mathrm{m^2\ sec^{-1}})$	$t \equiv (T - T_c)/T_c$
1.053	2.8
1.31	0.632
1.057	0.546
0.937	0.456
0.368	0.274
0.105	0.111

What does it mean to be along the coexistence direction in this system? Is the scaling exponent for the mutual diffusion coefficient available from these data consistent with the predictions of dynamic scaling theories? Does it make sense in the first place to make this comparison? If not, why not?
28. Compare the simulation critical temperature data in table 12.2 with the mean-field model predictions of chapter 7 for the quantity $T_R \equiv T_c(p)/T_c(p=0)$. Your results should show that agreement between both approaches is excellent.
29. Show that if a particle experiences random walk diffusion, superimposed with a constant drift velocity, then

$$\frac{\langle r^2 \rangle}{t} = 2D + v^2 t$$

where D is the particle's self-diffusion coefficient, v is the drift velocity, and r is its displacement. Is this equation valid at long times?
30. Ion transport through reverse-micelle aggregates consisting of aqueous cores encapsulated by surfactant solvent molecules in a supercritical fluid solvent like carbon dioxide is potentially an attractive method for remediating heterogeneous solid waste contaminated by metallic ions [18]. One mechanistic view of transport in this process suggests that the micellar aggregates may, in some conditions, form percolation clusters that provide diffusion pathways for the ionic species in the presence of an externally imposed electric field. The actual ion transport is then modeled by the Einstein self-diffusion equation through the micellar cluster [19, 20]. In addition, near the solvent critical point, the cluster structures are thought to be quite stable with long rearrangement times (compared to the ion diffusion time scale). Analyzing this problem involves an interesting conflation of critical phenomena and percolation theory, some first steps for which we now consider. Percolation theory provides that the probability of a micelle being in the spanning cluster, through which diffusion occurs, denoted by P_∞, is given by [21]

$$P_\infty \approx (P - P_c)^{\beta_P}$$

where P_c is the aggregate micellar structure's percolation threshold with associated critical exponent β_p. In addition, the percolation correlation length is $\xi \approx (P - P_c)^{-\upsilon_P}$ with scaling exponent υ_p. Try and show that, with these assumptions, the mean square distance traveled by an ion through the cluster, R^2_{ion}, in the presence of an external field, scales as follows:

$$R^2_{\text{ion}} \approx (p - p_c)^{-(2\upsilon_p - \beta_p)}$$

[Hint: Use the result that the probability of a micelle being in the spanning cluster is also given by the relationship

$$P_\infty \approx \frac{\xi_p^{d_f}}{\xi_p^d}$$

where d_f is the fractal dimension of the cluster structure and d is the system dimension.]

How would you use computer simulation to study the validity of this scaling analysis if the clustering structure is represented by nearest neighbor up-spins (say) in a 2-d Ising lattice?

Bibliography

The subject of transport phenomena is covered in many excellent texts and publications in the area of statistical mechanics. Some of the ones used in this chapter are listed below.

References [2, 6, 7, 9] are articles specifically dealing with the intricacies and subtleties surrounding computer simulation in the critical region. References [3, 4, 22, 23] are classical articles in critical dynamics. References [5, 11, 13, 19] are useful textbooks on simulation algorithms, equilibrium scaling results, and basic ideas in diffusion.

[1] D. A. McQuarrie, *Statistical Mechanics*. New York: Harper & Row, 1976.

[2] S. De, Y. Shapir, and E. H. Chimowitz, "Scaling of self and Fickian diffusion coefficients in the critical region," *Chem. Eng. Sci.*, vol. 56, p. 1, 2001.

[3] P. C. Hohenberg and B. I. Halperin, "Theory of dynamic critical phenomena," *Rev. Mod. Phys.*, vol. 49, p. 435, 1977.

[4] K. Kawasaki, "Diffusion constants near the critical point for time-dependent Ising models. III Self-diffusion constant," *Phys. Rev.*, vol. 150, p. 285, 1966.

[5] M. P. Allen and D. J. Tildesley, *Computer Simulation of Liquids*, Oxford Science Publications. Oxford: Clarendon Press, 1992.

[6] S. De, S. Teitel, Y. Shapir, and E. H. Chimowitz, "Monte Carlo simulation of Fickian diffusion in the critical region," *J. Chem. Phys.*, vol. 116, p. 3012, 2002.

[7] A. M. Ferrenberg and D. P. Landau, "Critical behavior of the three-dimensional Ising model: A high-resolution Monte Carlo study," *Phys. Rev. B*, vol. 44, p. 5081, 1991.

[8] A. Jeffrey, *Advanced Engineering Mathematics*. London: Academic Press, 2002.

[9] S. De, Y. Shapir, and E. H. Chimowitz, "Diffusion in dilute binary fluids confined in porous structures near the solvent critical point," *J. Chem. Phys.*, vol. 119, p. 1035, 2003.

[10] H. G. Ballesteros, L. A. Fernandez, V. Martin-Mayor, and A. M. Sudupe, "Critical exponents of the three-dimensional diluted Ising model," *Phys. Rev. B*, vol. 58, p. 2740, 1998.

[11] J. M. H. Levelt-Sengers, "Chapter 1: Thermodynamics of solutions near the solvent's critical point," in *Supercritical Fluid Technology: Reviews in Modern Theory and Applications,* T. J. Bruno and J. F. Ely (eds.). Boca Raton, FL: CRC Press, 1991.

[12] V. N. Burganos, "Gas diffusion in random binary media," *J. Chem. Phys.,* vol. 109, p. 6772, 1998.

[13] A. Chen, E. H. Chimowitz, S. De, and Y. Shapir, "Universal dynamic exponent at the liquid–gas transition from molecular dynamics." *Phys. Rev. Lett.*, 2004, in review.

[14] E. L. Cussler, *Diffusion Mass Transfer in Fluid Systems*, second ed. New York: Cambridge University Press, 1997.

[15] J. S. Duffield and M. Harris, "Self-diffusion of gaseous carbon dioxide in the critical region," *Ber. Bunsen-Ges. Phys. Chem.*, vol. 80, p. 157, 1976.

[16] J. Jorné and U. N. Safriel, "Linear and non-linear diffusion models applied to the behavior of a population of an intertidal snail," *J. Theor. Biol.*, vol. 79, p. 367, 1979.

[17] W. M. Clark and R. L. Rowley, "The mutual diffusion coefficient of methanol–*n*-hexane near the consolute point," *AIChE J.,* vol. 32, p. 1125, 1986.

[18] M. L. Campbell, D. L. Apodaca, M. Z. Yates, T. M. McCleskey, and E. R. Birnbaum, "Metal extraction from heterogeneous surfaces using carbon dixoide microemulsions," *Langmuir*, vol. 17, p. 5458, 2001.

[19] G. S. Grest, I. Webman, S. A. Safran, and A. L. R. Bug, "Dynamic percolation in microemulsions," *Phys. Rev. A,* vol. 33, p. 2842, 1986.

[20] C. Blattner, J. Bittner, G. Schmeer, and W. Kunz, "Electrical conductivity of reverse micelles in supercritical carbon dioxide," *Phys. Chem. Chem. Phys.,* vol. 4, p. 1921. 2002.

[21] D. ben-Avrahman and S. Havlin, *Diffusion and Reactions in Fractals and Disordered Systems*. Cambridge: Cambridge University Press, 2000.

[22] J. V. Sengers, "Transport properties of fluids near critical points," *Int. J. Thermophys.*, vol. 6, p. 203, 1985.

[23] K. Jagannathan and A. Yethiraj, "Molecular dynamics simulations of a fluid near its critical point," *Phys. Rev. Lett.*, vol. 93, p. 15701, 2004.

Index